VOLCANOES
AS LANDSCAPE FORMS

PRINTED IN NEW ZEALAND

VOLCANOES

AS LANDSCAPE FORMS

by

C. A. COTTON

D.Sc., F.R.S.N.Z.

Victoria University College

Wellington

WHITCOMBE & TOMBS LIMITED

CHRISTCHURCH AUCKLAND WELLINGTON DUNEDIN
AND INVERCARGILL, N.Z.
LONDON MELBOURNE PERTH SYDNEY

1944

To

FRANK A. PERRET

and

THOMAS A. JAGGAR

volcanologists of this century

Preface

An approach to volcanism may be made through either petrology or geomorphology, a volcano and its mechanism being regarded either as a geological crucible or as a builder of landforms. In this book, after a few brief introductory chapters, the latter approach is adopted, and the book is frankly concerned with landforms of volcanic origin and with their geomorphic development rather than with theories of volcanism and petrogenesis. Any full discussion of the physical phenomena that accompany eruptions or of the chemical problems arising from the interpretation of fumarolic and solfataric activity is also outside its scope.

For a book devoted to the geomorphology of volcanoes there is no precedent: no model work is available which may be followed in the setting out of the subject matter. The arrangement is experimental and tentative, therefore, and this whole book may (like *Moby Dick*) turn out to be "but a draught—nay, but the draught of a draught". I trust that it will find some interested readers, nevertheless, not only among geologists and geographers but also among those who, without a specialist's knowledge of geology and perhaps but little versed in systematic geomorphology, have yet a lively interest in landscape forms as natural phenomena.

Acknowledgments

I must acknowledge my indebtedness to my neighbour Dr P. Marshall, of Lower Hutt, and to my colleague Professor J. A. Bartrum, of Auckland, for allowing me to use their volcanological libraries, for guidance in the field, and for much helpful discussion of problems of both local and general interest. My thanks are due also to the American Geographical Society of New York, the Hawaiian Volcano Observatory, the New Zealand Geological Survey, and other authorities who have permitted the reproduction of illustrations, and to those correspondents and colleagues who have assisted me by supplying photographs for the same purpose.

C. A. Cotton

Wellington
June 1944

Contents

Contents

Part I: Introduction
The Mechanism of Volcanism

Part II: Volcanic Landscapes

CONTENTS

CONTENTS

VOLCANOES AS LANDSCAPE FORMS

PART I: INTRODUCTORY

THE MECHANISM OF VOLCANISM

Volcanism everywhere has unity; gas is the prime mover.
T. A. JAGGAR[1]

Gas is the active agent and the magma is its vehicle.
F. A. PERRET[2]

CHAPTER I

Types of Eruption

IT MAY BE STATED IN GENERAL THAT THE PHENOMENA OF volcanism are merely superficial manifestations of the deeper-seated processes of igneous injection or intrusion. Thus magma of deep-crustal or sub-crustal origin becomes, or gives rise to, superficial lava. Such highly debatable topics as the origin of magma, or fused earth-substance, and the causes which generate the forces that impel magma towards the surface of the earth may be touched upon only very lightly in a discussion which aims mainly at explaining superficial volcanic phenomena as they affect the construction and modification of landscape forms.

Some discussion of the properties of liquid and of newly solidified lava will be necessary, however, as well as of the products of the fragmentation of magma and the gases which escape from it, and it is clear that the nature of these depends on the properties of the magma from which they are derived. Various types and phases of volcanic activity, which give rise

[1] *The Volcano Letter,* 382, p. 3, 1932.
[2] *The Vesuvius Eruption of 1906,* p. 59, Washington, 1924.

to volcanic landscapes in considerable variety, and of kinds indeed which contrast strongly with one another, depend on the physical properties—and these in turn to some considerable extent on the chemical constitution—of lavas and ultimately of the magmas from which the lavas are derived.

Considerable use is made of terms such as "Hawaiian," "Plinian," "Vulcanian," and "Pelean" to characterise definite types of eruptive activity as differentiated by their superficial manifestations. These terms have descriptive value; but it must be stressed at the outset that volcanoes not uncommonly change their habit so as to exhibit different kinds of activity either successively or alternately. Nor is eruption of a particular type quite definitely associated with a special kind of lava. Though examples may be cited to prove this assertion, these may, on the other hand, be relegated to the category of exceptions to the general rule when at least the extreme types of activity are under consideration.

Hawaiian (non-explosive, lava-effusive) activity on the one hand and on the other Plinian and Pelean (both characterised by their highly explosive phenomena) are generally conceded to depend on differences in the chemical character of the magma associated with the activity. Some apparent exceptions to the rule that the chemical nature of the magma approaching the surface at the time of the eruption controls the type of activity exhibited may disappear when more is known of the effects which small variations in the proportions of magmatic constituents have upon the physical properties of the magma at temperatures in the vicinity of the melting point. It has been noted, for example, that quite a small departure from the basic, olivine-basalt composition of Hawaiian lavas introduces an appreciable modification of normal Hawaiian activity. In the case of explosive eruptions a wide range of magma types may be responsible for the phenomena, though these all have certain chemical characteristics in common.

It is, of course, only in so far as volcanic phenomena are of magmatic origin—i.e. are not affected by complications resulting from contact of hot rocks with water from an extraneous source, resulting in "phreatic" phenomena—that they may be expected to exhibit a close relation to the composition of the magma. It is obvious that explosions of the "steam-boiler" type may result from contact of extraneous water with molten material or even hot solid rock whatever the nature of the magma from which this is derived.

Types of Eruptive Activity

Making use of and defining terms most of which have already been employed by others, Lacroix[3] has distinguished four types of eruption: Hawaiian, Strombolian,[4] Vulcanian, and Pelean, though he is careful to point out that activity quite commonly changes from one type to another even during the course of a single eruption, and that it may combine the characteristics of more than one type ("mixed" phenomena).

The types recognised by Lacroix are as follows: (1) Hawaiian (Friedländer and Aguilar). Much fluid, effusive (as contrasted with explosive) basaltic lava is poured out. Explosive phenomena are rare, but scoria is piled into mounds around lava vents. (2) Strombolian (Mercalli). Active lava appears in the crater. It is basaltic, but rather more viscous than that emitted in Hawaiian eruptions. Liquid lava is thrown up by explosions or gas fountains, and some of this accumulates round about as spatter, scoria, and bombs. There is no dark smoke-like cloud emitted. (3) Vulcanian[5] (Mercalli). The lava is very viscous and, unlike that in (1) and (2), it does not long remain liquid in contact with the atmosphere but consolidates in the crater. The phenomena of eruption are explosive. Much fine ash is emitted, and ash-

[3] A. Lacroix, *La Montagne Pelée après ses éruptions,* pp. 74-93, Ac. Sci. Paris, 1908.

[4] The close relation Hawaiian and Strombolian is recognised by G. Mercalli (*I vulcani attivi della Terra,* p. 119, 1907), who classes the two under one head. He introduces also a "mixed" type, which specifically combines characteristics of Strombolian and Vulcanian eruptions as defined by Lacroix; he does not separate the Pelean as a type distinct from the Vulcanian.

[5] The characteristics described are those of Vulcano (Lipari Islands)—a rhyolite volcano—but the term is used generally so as to include a considerable variety of phenomena, including the explosive eruption of lava less acid and less viscous. The explosive eruptions of Vesuvius, indeed, which are generated in a lava that is very fluid when emitted, are regarded by some as the best examples of Vulcanian activity.

laden gases ascend so as to form voluminous "cauliflower" clouds; the ash is distributed widely by winds. "Breadcrust" bombs and angular blocks, both derived from new lava in the crater, are thrown out together with fragments of old lavas and the debris of prevolcanic rocks. (4) Pelean (Lacroix). The lava is extremely viscous and the chief characteristic of this type is the formation of *nuées ardentes.* These "glowing clouds" consist of a mixture of fine ash and coarser rock fragments (all incandescent) permeated by hot gases so as to make of the whole what Lacroix compares to an emulsion—an extremely mobile fluid, which is dense enough, however, to rush with great velocity down slopes, maintaining contact with the ground.

There are at least two distinct variants of the Pelean type. In one emission takes place vertically from an open crater—e.g. such an eruption took place in 1902 from the Soufrière crater on the island of St Vincent.[6] Though resembling in some respects a Vulcanian emission the *nuée ardente* in this case falls back immediately upon the mountain—or it simply froths over the rim of the crater—and streams down the slopes. The voluminous eruption of incandescent material from Mont Pelée, Martinique, that overwhelmed the town of St Pierre at the outbreak of the eruptions of 1902 was of this kind. Related to this type also were the eruptions which filled the Valley of Ten Thousand Smokes, near the volcano Katmai (Alaska), with a deposit which has been called a "sandflow." In this case the *nuée ardente* material was emitted apparently from many fissures. Lacroix[7] has included such eruptions with the Pelean type as a "Katmaian" variety of *nuée ardente.*

In the second main variant of the Pelean type, exemplified by very numerous successive eruptions of Mont Pelée, the volcanic orifice is plugged by stiff lava, which exudes as a tholoid or cumulo-dome like a stopper in, or a gigantic nail-head over, the volcanic pipe, and the flanks of this dome

[6] Separated as the "St Vincent" type by B. G. Escher (On a Classification of Central Eruptions according to Gas Pressure of the Magma and Viscosity of the Lava, *Leidsche Geol. Meded.,* 6 (1), pp. 45-49, 1933).

[7] A. Lacroix, *Liv. Jub. Cent. Géol. Soc. Fr.,* 1930. C. N. Fenner also uses the description "Katmaian" (*Trans. Am. Geophys. Union,* 18, pp. 236-239, 1937).

disintegrate explosively, driving out a series of *nuées ardentes* horizontally or obliquely.

The close relationship between types (1) and (2) on the one hand (*lava volcanoes* of Jaggar—Chapter II) and (3) and (4) on the other (the "explosive volcanoes" of Jaggar, or *pumice volcanoes,* as they may also be called) suggests that the fourfold classification of Lacroix must be recognised to be subsidiary to a more fundamental twofold division of volcanic phenomena. In most basaltic, i.e. effusive, or lava, volcanoes the production of pumice is rare and exceptional (Dana did not employ the term pumice for any of the products of Hawaiian volcanoes), whereas this solidified lava foam is the characteristic product of those volcanoes which are subject to magmatic explosion.

Other types have been proposed besides the four selected by Lacroix. Some authors class as of a distinct type the phenomena of those eruptions in which no material derived from new lava is ejected. This explosive type is variously called "semivolcanic," "ultravulcanian" and "indirect explosion" (see Chapter V), and has generally been attributed to contact of water of surface origin ("phreatic"—Suess) either with magma or (as in the case of the eruption of 1924 at Kilauea, Hawaii—Chapter IV) with a "hot hole" from which lava has just been withdrawn. The great explosion which blasted away half of the mountain Bandaisan (Japan) in 1888 is frequently cited as of similar origin. Other causes which may lead to such an explosion have been suggested, however (see Chapter V). Von Wolff[8] has proposed to explain it on the basis of experiments by Brun[9] as the result of heating by some means of a plug of viscid or semi-solid, gas-rich, glassy lava to a temperature at which it will explode.

Plinian Eruptions

The description "Plinian"[10] has been employed for Vulcanian (or ultravulcanian) eruptions of extreme violence like that described by Pliny, which took place in A.D. 79,

[8] F. von Wolff, *Der Vulkanismus,* pp. 94, 546, 1914.
[9] A. Brun, *Recherches sur l'exhalaison volcanique,* Geneva, 1911.
[10] F. von Wolff, *loc. cit.* (8), p. 541.

wrecked the ancient mountain centred approximately on the site of the present-day Vesuvius (Fig. 1), and buried the towns of Herculaneum and Pompeii under debris.[11] The characteristic Plinian phenomena have been only rather vaguely defined. They are summarised by Tyrrell[12] as follows: "A type characterised by steam explosions of colossal power taking

Fig. 1. Monte Somma at the left of the modern Vesuvius in this view is a relic of the great volcano destroyed by the Plinian eruption of A.D. 79.

place after long periods of absolute quiescence and without emission of lava. . . . The ejected materials consist entirely[13] of fragments of old lavas and the country rocks." The further statement that "the greatest volcanic events of the historical period belong to this type" is in conflict with that part of the definition quoted above which excludes material derived from new lava from a place among the ejected fragmentary debris.

[11] Recently re-named the "Perret" type by B. G. Escher, *loc. cit.* (6), who regards as comparable to this the 1906 eruption of Vesuvius, which was observed and described by F. A. Perret.

[12] G. W. Tyrrell, *Volcanoes,* pp. 88-89, 1931.

[13] The definition given by von Wolff, *loc. cit.* (8), is not quite so restrictive as this.

It is the definition, however, which must be amended, and these latter-day great eruptions, including that of Katmai (Alaska) may indeed be regarded as outbursts similar in a general way to the eruption described by Pliny—and, therefore, Plinian. Just as in the case of practically all the more modern catastrophic eruptions the material scattered by the Plinian eruption of Vesuvius actually consisted very largely of pumice and ash derived from either new magma or remelted lava (or from some mixture of the two); it contained only a rather small proportion of debris derived from the solid rocks of the ancient Vesuvius—the old volcanic mountain of which Somma (Fig. 1) is a remnant. Lacroix, who has examined the covering of Pompeii in detail, describes the lowest and by far the thickest layer of it as consisting mainly of roundish lumps of white pumice mixed with a much smaller proportion of other rock fragments. He notes that this material covering Pompeii differs from the Vesuvian debris which buried Ottajano in 1906 in few respects other than that the former consists of white pumice of the composition of leucite phonolite, while the latter is a similarly derived dark-coloured scoria of the composition of leucite tephrite.[14]

[14] A. Lacroix, *loc. cit.* [3], pp. 120-121.

CHAPTER II

Lava Volcanoes

OF THE LAVA ACTUALLY POURED OUT IN THE LIQUID STATE ON the surface of the earth the greatest bulk has been of olivine-basaltic composition both in recent times and in past ages.[1] By far the largest volcanic mountains have been built of successive sheets and tongues of this basalt with little admixture either of lavas of other types or of layers of material which has originated as fragments explosively ejected.[2]

On the active basaltic volcanoes of the island of Hawaii, especially at the Kilauea vent, protracted studies have been made not only of the lava itself in the liquid state and in the process of eruption but also of its behaviour underground as manifested by the earth tremors which result from its subterranean movements; and on the basis of these studies hypotheses have been formulated which not only serve to explain the behaviour of lava volcanoes but afford a definite contribution towards the explanation of volcanic phenomena in general.[3]

[1] H. S. Washington, Deccan Traps and other Plateau Basalts, *Bull. Geol. Soc. Am.*, 33, pp. 765-804, 1922; G. W. Tyrrell, Flood Basalts and Fissure Eruptions, *Bull. Volcanol.*, 1 pp. 89-111, 1937.

[2] This must not be interpreted as a statement that lava outflow is yielding a greater quantity of material at the present day than is the explosive emission of pumice and pumiceous ash. As the result of a worldwide survey K. Sapper (*Zeits. Vulk.*, 9 (3), 1928) has estimated that since A.D. 1500 volcanoes have emitted as fragmentary pumiceous material a bulk equal to five times their total output of effusive lava. The total figures are: 331·1 cubic kilometres of fragmentary material and 62·7 cubic kilometres of lava. Generally speaking, the pumice-producing volcanoes are in different regions from those which pour out lava abundantly.

[3] See especially R. A. Daly, The Nature of Volcanic Activity, *Proc. Am. Ac. Arts and Sci.*, 47 (3), pp. 47-122, 1911; A. L. Day and E. S. Shepherd, Water and Volcanic Activity, *Bull. Geol. Soc. Am.*, 24, pp. 573-606, 1913; and T. A. Jaggar, Seismometric Investigation of the Hawaiian Lava Column, *Bull. Seism. Soc. Am.*, 10, pp. 155-275, 1920, and The Mechanism of Volcanoes, in "Volcanology," *U.S. Nat. Res. Council Bull.*, 77, pp. 44-71, 1936. The classic investigation of Day and Shepherd into the nature of the gases emitted by Hawaiian lavas (which has been carried farther by Shepherd) was inspired by a desire to check the statements and theories of A. Brun after the publication by that author of the remarkable work *Recherches sur l'exhalaison volcanique* in 1911. These amounted to a categorical denial of the existence of water or of hydrogen in appreciable quantity among the volcanic emanations.

The essential characteristics of the lavas of Hawaiian and other similar volcanoes are closely related apparently to their chemical composition. Chief among these is the extreme liquidity of the lava as it is emitted, to which is related its habit of flowing out quietly—that is to say, with a minimum of explosive phenomena. By this means volcanic mountains are built up which consist almost entirely of numerous thin, discontinuous, interfingering lava sheets and tongues and incorporate very little material that has been ejected either as pasty or solid fragments or as dust. Many statements regarding the flow of basalt lava convey, however, an erroneous impression in that they tend to exaggerate the mobility of the fluent lava. Thus Becker[4] has calculated the viscosity of basalt which attained a velocity of 22 feet per minute when flowing down a 2 per cent gradient to be only 60 times as great as that of water; and in another case of rapidly flowing basalt Palmer[5] has estimated that the viscosity was only 15 times as great as that of water. It appears that results such as these are not even of the correct order of magnitude, and they give an entirely misleading idea of the liquidity of lava. According to Nichols[6] the reason is that such estimates are based on an erroneous assumption that the flow of lava is turbulent, like that of water, whereas it is laminar even at high velocities. It is the development of turbulence that reduces the velocity of flow of mobile liquids so that it is comparable to that of the laminar flow of lava. On the assumption of laminar flow certain observed rapid velocities in lava streams[6a] and, more generally, the velocity of flow required to convey lava from its source to the lower ground on which it accumulates, comes to rest, and solidifies can be accounted for without assuming a value for the viscosity lower than those which have been determined experimentally by Russian

[4] G. F. Becker, Some Inquiries into Rock Differentiation, *Am. Jour. Sci.*, 7, p. 29, 1897.

[5] H. S. Palmer, A study of the Viscosity of Lava, *Bull. Haw. Volc. Obs.*, 15, pp. 1-4, 1927. The velocity in this case was 11 miles per hour.

[6] R. L. Nichols, Viscosity of Lava, *Jour. Geol.*, 47, pp. 290-302, 1939.

[6a] Lava-river velocities up to 25 miles per hour are reported by G. A. Macdonald, The 1942 Eruption of Mauna Loa, Hawaii, *Am. Jour. Sci.*, 241, pp. 241-251, 1943.

and Japanese investigators, whose results are quoted by Nichols. Even at 1150° to 1200° C. the viscosity of basalt is found to be a million times as great as that of water at ordinary temperatures, and the viscosity of freely-flowing Hawaiian basalt has been compared with that of sugar at 109° C.—"a somewhat pasty liquid which nevertheless flows out of an inclined test-tube."

In the light of this information it is easy to explain the behaviour of lava in the crater, or lava lake, of Kilauea which, though it entered, failed to fill iron pipes immersed in it for four minutes in experiments conducted by Jaggar.[7] In these experiments, however, lava at 1200° C. was found to enter the pipes more freely than that at 1100° C. Laboratory tests have shown also that at temperatures not far above the melting point of crystalline basalt a lowering of temperature of 100° causes a tenfold increase of viscosity.

Hypomagma and Pyromagma

Jaggar is convinced that the liquid, free-flowing phase of basalt lava exists as such only near the surface and that the lava becomes liquid (i.e. mobile) only shortly before it is emitted. This mobile lava he terms "pyromagma." It is sometimes described as a foam, but this is an exaggeration. It is spongy—i.e. it tends to froth—with expanding bubbles of gas, and is still effervescing or giving off gas that has been in solution in it. Investigators in Hawaii generally assume that half the bulk of this effervescent lava is gas, so that its specific gravity is reduced to half that of gas-free lava. Pyromagma if chilled externally solidifies in "dermolithic"[8] fashion—a skin being formed on its surface—so as to make lava flows of the kind termed in Hawaii "pahoehoe" (Chapter IX).

The "lava column" in the depths, from which pyromagma is derived is judged by Jaggar to consist of *relatively very immobile* "hypomagma," which, "with reagent gases in

[7] T. A. Jaggar, Experimental Work at Halemaumau, *Bull. Haw. Volc. Obs.*, 9, pp. 26-29, 1921.

[8] T. A. Jaggar, *loc. cit.* (3).

solution, . . . is the actuating medium for the phenomena of volcanism."[9]

The fact that lava in separate vents, whether far apart or very near together, commonly stands at different levels is satisfactorily explained when it is assumed that the vents, though they may be related to the same lava column, are not connected underground by continuous mobile pyromagma. The long-debated independence of the free surfaces of the "lava lakes" in the vents on the adjoining Mauna Loa and Kilauea volcanoes, in Hawaii, between which there is a difference of elevation of nearly 10,000 feet, may be thus accounted for without necessarily assuming complete or even temporary separation of the lava columns beneath them.[10] The local independence of vents—i.e. the apparent absence of underground intercommunicating channels between them—is particularly noticeable on the summit of the Mauna Loa volcano, where Stearns has noted that rising lava is apparently confined in independently functioning fissures the walls of which must have become sealed by congealed pyromagma. "The molten magma, once it starts upward in a fissure, does not change its course until it reaches the surface unless its advance is seriously interrupted by leakage into intersecting fissures. Moreover, the altitude of the place at which the fissure intersects the surface does not seem to matter."[11]

Even "within the small area of the Halemaumau pit [containing the lava lakes of Kilauea volcano] several different lava pools and vents may maintain themselves in equilibrium at different heights through a range of fifty feet or more."[12] Independence of vents suggests that liquid pyromagma is present only to a shallow depth, and that effervescence, which is generally attributed to release of pressure on the hypomagma and which is the cause of the vesiculation or frothing that brings about the essential change to pyromagma, can take

[9] T. A. Jaggar, *loc. cit.* (3), p. 273.

[10] Such independence is alternatively explained by describing the lava (pyromagma) as a foam instead of a continuous liquid capable of transmitting hydrostatic pressure (G. W. Tyrrell, *Volcanoes,* p. 121, 1931).

[11] H. T. Stearns, Geology and Water Resources of the Kau District, Hawaii, *U.S. Geol. Surv. W-S. Paper,* 616, p. 49, 1930.

[12] T. A. Jaggar, *loc. cit.* (3), p. 160.

place only close to the surface.[13] Below the shallow depth at which pyromagma is generated the lava column is assumed by Jaggar to consist of hypomagma only. Such a conception of the lava column has met with rather general acceptance, as it is more easily reconcilable with the properties of magma as revealed by the laboratory experiments of geochemists than is the alternative hypothesis of a pipe containing mobile liquid. Thus Shepherd[14] refers to "the continuous solid—we have too long regarded it as liquid—[lava] column."

Besides effects that may be introduced by the development of either regional or locally generated geological compression, such as accompanies or is accompanied by movements of folding and faulting, there is a little-understood worldwide tendency of the hypomagma to push towards the surface—"magmatic pressure."[15]

Apart from this magmatic pressure, which may be due to some external cause, the observed phenomena of volcanism are the results of processes in, and properties of, the hypomagma. They are produced in the main by "the thermochemical processes in the magma itself, internal pressure-

[13] This contention is in contrast with the conclusion of R. A. Daly—bound up with the application of his "two-phase convection" theory of deep stirring of the lava column by vertical currents, one of which consists of lava specifically light because vesiculated with gas—that "bubbles of gas must form in the lava at depths of a few kilometres" (*Igneous Rocks and the Depths of the Earth,* p. 367, 1933).

[14] E. S. Shepherd, Notes on the Chemical Significance of Engulfment at Kilauea, *Jour. Wash. Ac. Sci.,* 15, pp. 418-420, 1925.

[15] T. A. Jaggar, Magmatic Gases, *Am. Jour. Sci.,* 38, pp. 313-353, 1940.

Some more or less tentative suggestions have been made in explanation of the development of magmatic pressure. The following is applicable to Hawaii: "If . . . we have a flexible crust bearing with full weight on the substratum, itself eruptible because vitreous, we need not look far to find explanation of much of the 'ascensive force' actuating Pacific volcanoes. . . . The dead weight of the crust is also competent to flood the Pacific floor with still more extensive plateau basalts." (R. A. Daly, The Sub-Pacific Crust, *Proc. V Pacific Sci. Cong.,* pp. 2503-2510, 1934).

B. Willis's theory of orogeny relies on the development of hot spots at depths exceeding 25 miles owing to local accumulation of heat from radioactivity and on the "buoyancy" of magma generated by fusion at such spots (B. Willis and R. Willis, Eruptivity and Mountain Building, *Bull. Geol. Soc. Am.,* 52, pp. 1643-1684, see pp. 1657, 1676, 1941). "It is suggested that this change may be the most significant in the development of a stress equal to the demands of orogenic deformation. Heat thus transformed into a simple gravitational differential may through growth over a long period of time become the force adequate to the task of promoting the displacement of mountain ranges," (p. 1657). Such is a mechanism proposed to account both for orogeny and for igneous injection in orogenic belts.

temperature-saturation factors as well as chemical composition governing the result."[16]

This hypomagma is contained in and rises through crowded fissures in "rift zones" (JAGGAR), which afford passage through the enclosing rocks (or "perilith"). Though the causes contributing to the invasion of the rift zones by hypomagma can only be guessed at, the phenomena of volcanism and of igneous injection are obviously closely related. A laccolithic intrusion of porphyrite is "a short-lived, temporary, violent vesiculate irruption of a volcanic dome in the midst of soft shales that would have been a Santorin eruption if it had been on the surface. . . . The injection of a granite body is a similar irruption in the midst of diastrophic continental material, also a sudden event, and vesiculate with gases."[17]

"Probably slow intrusion never ceases except during eruption" (JAGGAR). Intrusion thus gives place to volcanic outbreak only when yielding of the superincumbent edifice takes place; and accumulation of magma as sills or laccoliths within great domes such as Mauna Loa is suspected as the cause of observed upswellings of this and other volcanoes. Commensurate sinking may result from a drawing off of such magma.[18]

Phenomena of this kind are not confined to Hawaiian domes. Upheaval with tilting on a large scale at Usu (Japan) in 1910 is regarded as the effect of permanent injection. At Sakurajima, on the other hand, upward swelling, which occurred in 1914, seems to have been relieved by an eruption during which 2·2 cubic kilometres of material were ejected,

[16] T. A. Jaggar, *loc. cit.* (3), p. 161.

[17] T. A. Jaggar, *loc. cit.* (15).

[18] "During the rising-lava period of 1913-1922 the mountain-top surrounding Kilauea crater was lifted more than two feet. . . . In other words, the turtle-back of a volcano swells when the lava rises and flows out and shrinks when the lava sinks back" (T. A. Jaggar, *The Volcano Letter,* 360, p. 2, 1931). Jaggar has tentatively accepted wide variations in the estimated height of Mauna Loa made from 1794 onward as an indication of real fluctuation of level. Measurements made from 1885 onward show little or no change of level, however. "There may have been some vast outflows . . . in the last half of the eighteenth century . . . which accounted for the low level of the summit measured by Menzies," who in 1794 estimated the elevation as 13,564 feet as compared with an estimate of 13,750 feet made in 1841 and the present measured altitude of 13,680 feet.

and the mountain sank again.[19] Movements of a similar order have been measured after the Komagatake eruption, also in Japan[20], and Rittmann records that Stromboli Island rose temporarily one metre in the eruption of 1930.

Instrumental observations of tilts, swellings, and tremors indicate that the hypomagma at comparatively shallow depths within the great basalt-built domes of Hawaii is sensitive to "pressure effects. There is an incessant deep flow . . . towards the region of minimum load."[21] This does not imply that the hypomagma is by any means a mobile liquid, however. It is envisaged as possessing even a certain "rigidity,"[22] though it has that measure of mobility implied in the guarded description "more or less mobile vitreous silicate magma" (JAGGAR).

When by some means external pressure is relieved, resulting in the change of equilibrium conditions requisite for an outbreak or recrudescence of volcanic activity, and "the substance of the hypomagma loses gas [i.e. gas previously in solution in it makes its appearance as bubbles] . . . it warms itself with the bubbles, where heat reactions are in progress between the distinct gases that have been released from solution. This softened, frothing melt is pyromagma."[23]

The proportion of volatile constituents contained in basaltic and in granitic magmas has been found to be not very different.[24] Thus results obtained experimentally on granitic material are tentatively applicable to Hawaiian basaltic lava in so far as they indicate the relation of the solubility of water gas in magma to changing pressure and temperature. At 900° C. the solubility of water in the granite glass has been found to diminish from 9·35 per cent (by weight) at 4000 bars (atmospheres) pressure to 3·75 per cent at 500 bars. Increase of solubility with reduction of temperature, which has also been

[19] F. Omori, *Bull. Imp. Earthquake Investigation Committee* (Tokyo), 5 (1), 1911; 5 (3), 1913; 8 (2), 1916.

[20] S. Kôzu, The Great Activity of Komagatake in 1929, *Tchermaks M. u. P. Mitt.*, 45, pp. 133-174, 1934.

[21] T. A. Jaggar, *loc. cit.* (3), p. 274.

[22] T. A. Jaggar, *loc. cit.* (3), p. 168.

[23] T. A. Jaggar, *loc. cit.* (3), p. 164.

[24] E. S. Shepherd, Gases in Rocks and Volcano Gases, *Year Book Carnegie Inst. Wash.*, pp. 77-82, 1932.

measured, is insufficient to interfere to any serious extent with the effervescence of gas-rich magma which will follow a substantial release of pressure.[25]

In contrast with the conception of the distinctness of hypomagma and pyromagma as outlined above, Daly,[26] it may be noted, does not recognise the existence of a hypomagma differing in its properties from pyromagma. That is to say, he regards the pyromagma phase as primary—highly fluid magma rising from great depths already charged with swarms of gas bubbles, practically a froth.

Rift Complex and Perilith

Describing the hypomagma as "the actuating medium of volcanic action or eruption within a volcanic system," Jaggar explains that "a volcanic system occupies a rift complex."[27] At another place: "Cones and domes are superficial, . . . vents are profound. Fissures below are necessary [i.e. a deeply-extending rift complex] . . . unless we suppose mobile plastic matter immediately under our volcanoes."[28]

Discussing the "transference of vents from one place to another," which implies the opening of new underground conduits, Jaggar remarks: "There are undoubtedly fissures, but there are also layers, surfaces, slopes, cavities, and a machinery of pericentric accumulation followed by [laccolithic] intrusion. This mechanism is incessantly solving problems of least resistance"[29] The lava column, as noted earlier, is thus contained within walls of other rocks termed collectively the "perilith." These are not all prevolcanic, but comprise also the solidified lava flows and other materials of volcanic origin present especially in the basal parts of large volcanic mountains, the origin of which may be of ancient date as compared with the modern lava column traversing them.

[25] R. W. Goranson, The Solubility of Water in Granite Magmas, *Am. Jour. Sci.*, 22, pp. 481-502, 1931.

[26] R. A. Daly, *loc. cit.* (13).

[27] T. A. Jaggar, *loc. cit.* (3), pp. 169-170.

[28] *Loc. cit.* (3), p. 200.

[29] *Loc. cit.* (3), p. 200.

CHAPTER III

Volcanic Heat from Gas Reactions

MUCH ATTENTION HAS BEEN GIVEN TO THE QUESTION OF HOW A high temperature is maintained in lava which remains liquid with a free surface exposed to the atmosphere—a "lava lake." The problem has been raised especially by the flow and ebb—with only rare overflow—of liquid lava in the lakes which remain in existence for long periods in Hawaiian craters (Fig. 2). Little room is made by overflow for new hot magma

Fig. 2. Lava lake in the fire-pit Halemaumau, Kilauea Volcano, Hawaii, as it existed from time to time prior to the explosive eruption of 1924.

rising from the depths.[a] Some heat is expended in these lakes in the liquefaction of pyromagma and in the remelting of rock slides of perilith from the walls and of engulfed crusts; but a far greater amount of heat is expended in radiation from the free surface. Why does not the lava soon freeze?

[a] Such a question does not arise concerning lava vents from which long-continued outpourings of basalt take place—e.g. that from Matavanu (Samoa) in 1905-11 and that from the flank of Nyamlagira (E. Africa) in 1939-40.

One source that has been invoked to replenish the supply of heat in lake lava as it suffers this continuous loss is upward transference of heat by convection currents of some kind in the lava column so that a store of heat contained in magma at a hypothetical high temperature in the depths of the column may be drawn upon. Vigorous streaming of superficial lava in the lakes has frequently been observed, but it is not known whether the convectional circulation indicated by lava currents extends at all deeply into the column beneath. The most favoured hypothesis of the mechanism of convection in lava is the "two-phase" theory of Daly,[1] who has pointed out that a rising current of vesiculate, and therefore very light, lava will replace a downward or return current of lava which is much denser because gas bubbles have been released from it. The convection which stirs the pyromagma and causes visible currents in lava lakes is certainly of this nature. If effervescence is only superficial, however, two-phase convection can extend to no great depth.

Magmatic Gases

If the supply of heat necessary to keep the pyromagma liquid in a lava lake is not derived from the depths, heat must be generated in the lava itself; and there is an actual source of heat within the vesicles of pyromagma, where reactions are taking place between gases newly liberated from solution.[2] Opinions differ as to the quantitative importance of these reactions, and geochemists are not all convinced that they are of paramount importance in the development of volcanic phenomena or furnish even a great part of the heat necessary to maintain long-continued activity in lava lakes. They are at least of local importance close to the surface, however, and wide and rapid fluctuations of the lava-lake temperatures at Kilauea have been confidently correlated with varying rate of escape of the reacting gases from the pyromagma.[3]

[1] R. A. Daly, The Nature of Volcanic Action, *Proc. Am. Ac. Arts and Sci.*, 47 (3), pp. 47-122 (p. 76), 1911.

[2] R. A. Daly, *loc. cit.* (1), p. 88. Daly here originated the expression "volcanic furnace" (p. 85).

[3] A. L. Day and E. S. Shepherd, Water and Volcanic Activity, *Bull. Geol. Soc. Am.*, 24, pp. 573-606, 1913.

Jaggar claims that the "reagent gases in solution" in hypomagma are "dominantly hydrogen, carbon compounds, and sulphur,"[4] and asserts confidently that "hydrogen first, and carbon gases second, are the dominant chemical agents of volcanism;"[5] but such a claim is based on the assumption that much water, which is found to be the most abundant gas in many samples of volcanic emanations, has been produced in the pyromagma by oxidation of hydrogen, and the possible sources of oxygen for such combustion must be examined. Locally, if not more generally also, oxygen is supplied by the ferric oxide of superficially oxidised or "rusty" lava crusts as well as by slides of the older, partly-weathered rocks forming the perilith, which are from time to time engulfed and may be remelted and digested by the pyromagma. It has been doubted, however, whether a supply of oxygen sufficiently abundant and generally available has been found in this reaction with the ferric oxide of weathered material to support the theory that combustion of hydrogen is the chief and not merely an auxiliary source of volcanic heat; and the quantitative importance of other heat-producing gas reactions among juvenile gases is also questioned, though such reactions certainly take place.[6] If it is not for the most part a product of the oxidation of hydrogen in the pyromagma a considerable proportion of the water in gases escaping from volcanoes must have been present rather abundantly as such in solution in hypomagma, as many petrologists assume to be the case.[7]

The well-known equation between the ferrous compounds in magma (generally stated as ferrous oxide) and water on the one hand and triferric tetroxide and hydrogen on the

[4] T. A. Jaggar, Seismometric Investigation of the Hawaiian Lava Column, *Bull. Seism. Soc. Am.*, 10, pp. 155-275 (p. 273), 1920. See, however, *infra* for Jaggar's revised opinion regarding sulphur.

[5] *Loc. cit.* (4), p. 167.

[6] E. S. Shepherd, *Bull. Nat. Res. Council,* 61, pp. 261-262, 1927.

[7] A. L. Day has pointed out that elements which appear eventually in volatiles cannot be thought of as present in combination as such in the magma any more than can those elements which finally appear in combination in crystallised minerals. "The combinations that escaping elements form among themselves are largely conditioned by the fact that these specific compounds are able to escape." (Ann. Rep. Director Geophys. Lab., *Carnegie Inst. Wash. Year Book*, 33, p. 64, 1934.) C. N. Fenner writes: "We have little knowledge of the form of combination of the volatiles with each other and with the non-volatiles." (The Katmai Magmatic Province, *Jour. Geol.,* 34, pp. 673-772 (p. 739), 1926.)

other ($3\ FeO + H_2O = F_3O_4 + H_2$) is a reversible one, and the concentrations and equilibrium conditions (if attained) depend largely on the temperature. If, therefore "we assume either water or hydrogen to have been originally present in the ferriferous . . . [magma], *the other must be original in the same sense.*"[8]

Halemaumau Gases

The principal gases emanating from the pyromagma in the lava lake Halemaumau at Kilauea (Hawaii) as found by Day and Shepherd[9] in their classic investigation of gases collected in 1912 were nitrogen, water, carbon dioxide, carbon monoxide, sulphur dioxide, hydrogen, and sulphur. As these authors remark:

This particular group of gases cannot possibly be in equilibrium [at 1000° C. or higher.] The interreactions between the gases . . . set free [in the pyromagma] in constantly increasing quantities as the surface is approached are accompanied by evolution of heat. . . . The heat generated by these reactions in the region near the surface, where the amount of gas is large, may well be more than sufficient to counteract the cooling effect of the expansion [of the liberated gases] within the rising lava column, which may thus become hotter and not cooler as it approaches the surface. . . . We have happened here on an enormous store of volcanic energy which reaches its maximum temperature at the surface itself.

A number of supplementary collections of gases from the same source have since been made as opportunities—i.e. suitable conditions in the lava lake at Kilauea—have presented themselves, especially in 1919,[10] and the proportions of the gases in different samples have been found to be very variable. In many cases, however, hydrogen, carbon, and sulphur have been almost completely oxidised, and the average content of water has been high—70 per cent.[11] Shepherd remarks:

The general inference is . . . that . . . [the gas] reaches the surface

[8] E. T. Allen, Chemical Aspects of Volcanism, *Jour. Franklin Inst.*, 193, pp. 29-80 (p. 43), 1922.

[9] *Loc. cit.* (3).

[10] E. S. Shepherd, Kilauea Gases, 1919, *Bull. Haw. Obs.*, 9, pp. 83-88, 1921.

[11] E. S. Shepherd, *loc. cit.* (10), p. 88; see also E. S. Shepherd, Note on the Chemical Significance of Engulfment at Kilauea, *Jour. Wash. Ac. Sci.*, 15, pp. 418-420, 1925.

almost completely burnt or else is actively burning in the surface layer of the lake. . . . The water present may be partly due to oxidation of evolved hydrogen, but such oxidation must occur in the body of the lava lake—presumably near the surface.

Some observers doubt, therefore, whether the Kilauea samples of gases, collected as they are from a basaltic pyromagma lake, are representative magmatic gases. Dilution and reactions with atmospheric air result from the lava-lake condition "in which the lava is peculiarly exposed to the air at all times when gases can be collected" and from the "downward transport of air by floating crust and infalling talus." When, on the other hand, "gases arise from lava which stands at a level well below the surface of the ground, and issue in large volumes from narrow funnels, oxidation like that at Kilauea ought to be practically obviated." Such gases are obtainable, however, only from fumaroles and are there diluted by much steam derived from water of surface origin. They are also contaminated by and generally oxidised by reactions with atmospheric air. "The hypothesis is that air is drawn through pores and crevices of the conduits of fumaroles by the suction of the hot rising gases."[12]

When gauging the significance of the oxidised condition of Kilauean emanations and the predominance of water among their constituents one must bear in mind that such gases have already passed through the upper levels of pyromagma in which the combustion of magmatic gases may be expected to take place if the reaction with ferric oxide is really of quantitative importance. Also, the collection of gases from molten pyromagma is a difficult and uncertain undertaking; it is well understood that some of the samples obtained must fail to be representative because of the unavoidable (and at the same time unsuspected) inclusion of superficially oxidised gases. These result from combustion of the hot juvenile gases as they make contact with atmospheric oxygen in some way beyond the control of the collector. Some, though not necessarily all, of the variability in composition of analysed gas samples is to be attributed to this cause.

[12] Quotations from E. T. Allen *loc. cit.* [8] pp. 30, 45.

In a late contribution to the discussion Jaggar[13] has applied a rigorous method of selection to the numerous published analyses of Kilauean gases and by eliminating results believed unreliable because the samples have been contaminated he has found support for the contention that uncombined hydrogen has been a very important constituent of the mixture of gases present in solution in the hypomagma. It is conceivable, however, that some, perhaps much, of the hydrogen may have come up from the depths in combination as water. Much water is produced, moreover, in the non-explosive high-temperature reaction between hydrogen and carbon dioxide,[14] which latter is undoubtedly another abundant gas of magmatic origin.

Jaggar reaches the conclusion indeed that hydrogen, carbon dioxide, and nitrogen are the most abundant gases escaping from hypomagma. Another important reaction in the magma, which on the other hand frees hydrogen from water, and is itself exothermic, is that already referred to which takes place between water and ferrous oxide yielding hydrogen and triferric tetroxide (magnetite).[15] The reaction is reversible, however.

This review of the Kilauean gas analyses has led Jaggar to make also the important suggestion that the rather large proportions of sulphur and compounds of sulphur which the gases contain give a misleading idea of the quantitative importance of sulphur as a constituent of the gas mixture and, it follows, as fuel in the volcanic furnace. He suspects that concentration of sulphur takes place in the gases of the superficial pyromagma, sulphur being the least volatile constituent of the volcanic gases.

As has already been stated, it has been seriously questioned whether the available evidence supports the theory that combustion of hydrogen within the pyromagma is the main source of water in the emanations and incidentally makes a considerable contribution to the output of heat from com-

[13] T. A. Jaggar, Magmatic Gases, *Am. Jour. Sci.*, 38, pp. 313-353, 1940.

[14] $H_2 + CO_2 = H_2O + CO$.

[15] R. T. Chamberlin, *The Gases in Rocks*, p. 66, Carnegie Inst., Washington, 1908.

bustion of gases which gives to pyromagma its mobility. The faith of the veteran Jaggar, however, based on many years of close observation at Kilauea, remains pinned to the theory of the volcanic furnace. It was thus expressed in 1932:

> Hydrogen and olivine basalt are fundamental constituents; water is a secondary oxidation product. . . . The gases rising are mostly hydrogen, carbon monoxide, and sulphur. They make flames. Water, carbon dioxide, and sulphur dioxide are in greatest volume, because the burning goes deep.[16]

The Engulfment-Combustion Theory

At one time Jaggar inclined to the belief that solid and semi-solid hot-lava crusts thrust up temporarily, or left stranded by ebb of pyromagma, above a lava lake and then rapidly rusted by exposure to atmospheric oxygen served as the chief vehicle for introduction of oxygen to the vesiculate pyromagma.[17] More recently, however, he has developed the "engulfment-combustion" theory, which suggested itself after vast quantities of rock from the walls (perilith) had collapsed as rock slides into the Halemaumau pit at Kilauea in 1924. This event followed the ebb, or withdrawal, of the lava column to a great depth and the total disappearance of the lava which had occupied the pit for many years. The pit was greatly enlarged and its depths were filled progressively by rock-sliding talus from the walls. According to this theory in its general application large-scale engulfment of such perilithic rock, consisting in part of oxidised material, must be a volcanic process of frequent occurrence.[18] It is not confined to basaltic craters, but has occurred in most large volcanoes of the "explosive" kind also and on an even larger scale, as is indicated by the forms of many large calderas (Chapter XVI). "Collapsing of crater regions with or without explosion has volumetrically been the biggest volcanic process of post-Tertiary time" (Jaggar).

[16] T. A. Jaggar, *The Volcano Letter,* 382, p. 3, 1932.

[17] T. A. Jaggar, Volcanologic Investigation at Kilauea, *Am. Jour. Sci.,* 44, pp. 161-219 (p. 165), 1917.

[18] T. A. Jaggar, Plus and Minus Volcanicity, *Jour. Wash. Ac. Sci.,* 15, pp. 416-417, 1925.

The estimated volume of the rock-slide talus precipitated into the Kilauea pit in 1924 is seven thousand million cubic feet. Even this is less than an eighth part of the estimated volume of material which, according to Finch,[19] has collapsed into the pit since 1823.

The engulfment-combustion theory has been stated by Jaggar[20] as follows:

> Every short cycle of slow lava-rising ends with rapid collapse and engulfment. Engulfment down vertical chasms [is] conceived as a normal process for introducing to the magma oxidised rock matter containing ferric oxide. This holds an excess of oxygen from rusting of old lava exposed to the air and acids. . . . Given a moderate amount of hydrogen in the new magma rising, and oft-repeated engulfment of rusty crater rock, enough heat would be supplied to keep lava volcanoes in activity without much fresh lava from great depths. . . . Doubtless other oxygen was entrapped as air. . . . If 3 per cent of the breccia [engulfed at Kilauea as a result of the rock-sliding which began in 1924] were ferric oxide reduced back [by reaction with magmatic hydrogen] to ferrous iron, enough heat would be supplied to raise seventy million cubic feet of the debris from air temperature, 20° C., to the temperature of liquid lava, 1200° C. . . . The upper portion of the breccia is renewed by every crateral subsidence.

It is obviously implicit in the theory of the volcanic furnace as outlined in this chapter that fresh supplies of unoxidised gases must be available. In the case of volcanoes that emit lava flows space thus becomes available for fresh supplies of hypomagma drawn from the deeper part of the lava column and fully charged with gases in solution. The maintenance of volcanic heat in the long intervals which sometimes intervene between overflows of lava has to be explained; but the pyromagma which is withdrawn below during the occasional ebbings or lowerings of lake level does not apparently return. No doubt the pyromagma which rises and fills the pit anew is newly derived from the deeper lava column.

According to an alternative hypothesis advocated by Daly,

[19] R. H. Finch, *The Volcano Letter,* 470, 1940.

[20] T. A. Jaggar, *The Volcano Letter,* 26, 1925; see also E. S. Shepherd, Note on the Chemical Significance of Engulfment at Kilauea, *Jour. Wash. Ac. Sci.,* 15, pp. 418-420, 1925.

which has already been referred to as the "two-phase convection" theory, gases are derived from magma at a much greater depth than seems possible according to Jaggar's conception of pyromagma and hypomagma, and the gases rise in mobile currents of froth which are of a deep-seated origin. Suess[21] also has pictured the juvenile (and extremely hot) gases, largely hydrogen, as derived from a source deeper even than that of the lava itself.

With such views may be compared the pronouncement of Allen,[22] however, that the solubility of a gas in a magma "ought to increase with pressure as it does for all similar cases." Bubbling cannot exist, therefore, in the depths, for "gas bubbles would be expected to disappear entirely at the enormous pressure in the great depths of the earth."

Gas-Fluxing

If the rising current of froth (pyromagma) which, according to Daly's theory, takes part in a deeply-extending two-phase convection is already very hot it will convey heat to the superficial pool of lava in a crater lake. The gases escaping from it may possibly be so hot that the heat they supply (even without, or perhaps with the help of, the additional heat from gas reactions) may be sufficient to melt superficial frozen crusts of lava, and the walls of craters and fissures may also be melted so as to enlarge these as pipes. This possible "blow-piping" activity has been termed by Daly[23] "gas-fluxing."

Gas-fluxing seems to play an important part in many volcanic processes, though which of its two sources of heat is the more important remains in doubt. Daly[24] has pointed out that any approach to accurate measurement of the "true furnace effect" produced by exothermic gas reactions in a

[21] E. Suess, *The Face of the Earth* (English edition), 4, pp. 548-555, 1909. Suess remarks: "From this point of view the difference in the height of the two volcanic [lava] lakes of Hawaii becomes more intelligible. A difference of two or three thousand metres in the relative level becomes as nothing in comparison with the depths from which the gases are derived; it only reveals a certain degree of independence of the two funnels."

[22] E. T. Allen, *loc. cit.* (8), p. 56.

[23] R. A. Daly, *loc. cit.* (1), p. 68.

[24] R. A. Daly, *Igneous Rocks and the Depths of the Earth,* p. 377, 1933.

volcano is a matter of extreme difficulty. As a result of re-heating frozen lava has been observed to melt and flow,[25] and Perret has fully recognised the gas-fluxing process as the principal agency in the mechanism that reopens a way upward for lava through a crater that has been blocked by rock-slide debris during a period of quiescence following explosive activity. The rising magma eats its way upward through the obstruction. Collapse of the crater-floor of Vesuvius in 1910 and 1911 prior to the first reappearance of lava in the crater after the great eruption of 1906 is attributed to a kind of overhead-stoping process in which gas-fluxing took an important part.[26] The persistence in position of the principal vent within the crater through which lava has subsequently been emitted is taken to indicate that "this part of the crater floor is the site of a gas-fluxed pipe that has persisted for many years" (H. S. Washington) (see Fig. 8).

The hypothesis that the pipes as a whole which feed volcanoes like Vesuvius have been opened to their present size by fusion of the walls has been rejected, however, by Perret,[27] despite his general support of the gas-fluxing hypothesis. "The excess of heat will not generally be sufficient," in his opinion, "to melt the containing walls to any great extent; otherwise the conduit would continuously widen and the volcanic edifice would cease to exist."

A remarkable instance of reheating of rock by a jet of exceedingly hot gas was observed in the crater of Ngauruhoe, New Zealand, by Dr P. Marshall, who was accompanied by Professor E. W. Skeats, during a period of unusually strong activity of that volcano in the latter part of January, 1913. In a letter to the author dated August, 1943, Dr Marshall has described this phenomenon as follows:

> We reached the edge of the crater on the western side, where the activity is always greatest. Steam was being emitted in considerable volume from an orifice in the red-hot rock at the bottom of

[25] F. A. Perret, *The Vesuvius Eruption of 1906*, p. 27, 1924.

[26] F. A. Perret, *loc. cit.* (25), pp. 113-114.

[27] F. A. Perret, The Ascent of Lava, *Am. Jour. Sci.*, 36, p. 605, 1913; *loc. cit.* (25), p. 113.

the crater. One very large jet from the southern side was issuing with a roaring noise. Its direction frequently changed, but for some time it impinged against a projection of rock on the north-east side of the crater. This speedily became red-hot, and in a few minutes was dazzling white even when seen in the bright sunlight. The steam obviously had a very high temperature. The surface of the rock of course fused, and small pieces were carried away as white-hot shards by the steam jet.

CHAPTER IV

Lava Volcanoes (continued)
Epimagma and Bench Magma

WHEN PYROMAGMA ESCAPES AS A LAVA FLOW AND ITS CHILLED surface becomes a "pahoehoe" skin (dermolithic solidification) the lava retains gas in solution as well as enclosing it in round vesicles (frozen bubbles). The process of cooling and solidification frequently takes a different course, however, even in lava flows, producing the clinker-like "aa" lava when the process is complete (Chapter X). A change similar to that from pyromagma to aa in a lava flow goes on also in the pyromagma which forms the lava lake sometimes present and exposed to the atmosphere at the top of the lava column (Fig. 2).

Depletion in gas-content, together with cooling which may be in part due to radiation at the free surface but must be largely the result of expansion of outbursting gases, results in the production of a gas-free, rather refractory magmatic residue which Jaggar has termed "epimagma."[1] This is semi-solid, pasty, cheese-like in consistency, and, being much heavier specifically, will sink through the frothy pyromagma—a process termed by Jaggar "clastolithic sedimentation."[2] As in the case of the clinkery scoria forming the surface of an aa lava flow, "its bubbles are dead vesicles, deformed, and subject to aeration"[3] if exposed to the atmosphere. Epimagma is not readily remelted by mere contact with pyromagma.

When visible in a vent open to the surface, epimagma is, "except for its cooled crust, evenly incandescent throughout,

[1] "While the gas-reaction of pyromagma, especially when atmospheric oxygen becomes available, is a heating process, the gas release from pyromagma to form epimagma is a cooling process" (T. A. Jaggar, Seismological Investigation of the Hawaiian Lava Column, *Bull. Seism. Soc. Am.*, 10, p. 165, 1920).

[2] *Loc. cit.* (1), p. 164.

[3] T. A. Jaggar, *loc. cit.* (1), p. 162.

sharply delimited from the containing crater walls. Its heavy blocks tilt up and form crags"[4] (Figs. 3 and 4). In such a lava lake the common succession of events includes "a crusting and foundering of crusts," which consist in part of epimagma and in part of congealed overflows of pyromagma. These together form "bench" lava or magma. When pyromagma floods the surface of the bench lava it deposits a layer of granules of epimagma over it by the process of clastolithic sedimentation.

Both sinking of crusts of bench lava and continued clastolithic sedimentation result in accumulation of epimagma as a thick floor beneath a superficial shallow lake or lakes of pyromagma (Fig. 3), so as to separate this from the hypomagma of the lava column below. "Wells are kept open by the rising pyromagma through the inert epimagma, and the walls of these tubes are probably glazed and gas-tight. . . . The epimagma . . . extends downward to the limit of the pyromagma tubes, and is the dense magma freed of gas by pyromagmatic effervescence."[5]

The theory of the coexistence of hypomagma, pyromagma, and epimagma in and under lava lakes like that in the Halemaumau pit at Kilauea was formulated prior to a cataclysmal sinking of the lava of Halemaumau to a great depth (followed by an explosive eruption) in 1924. After these events the open pit was 1335 feet deep, whereas from time to time previously Halemaumau had been full to the brim, though intermittently the lava level had fluctuated to the extent of several hundred feet. Tyrrell[6] argues that deep withdrawal of the mobile pyromagma should have left a sponge of epimagma with visible wreckage of crags and broken crusts if the proportion of pyromagma in the column had been as small as required by Jaggar's theory (Figs. 3 and 4). Was the epimagma, though very viscous, still capable of withdrawal through open

[4] T. A. Jaggar, *loc. cit.* (1), p. 163.
[5] T. A. Jaggar, *loc. cit.* (1), pp. 164-165.
[6] G. W. Tyrrell, *Volcanoes,* p. 116, 1931.

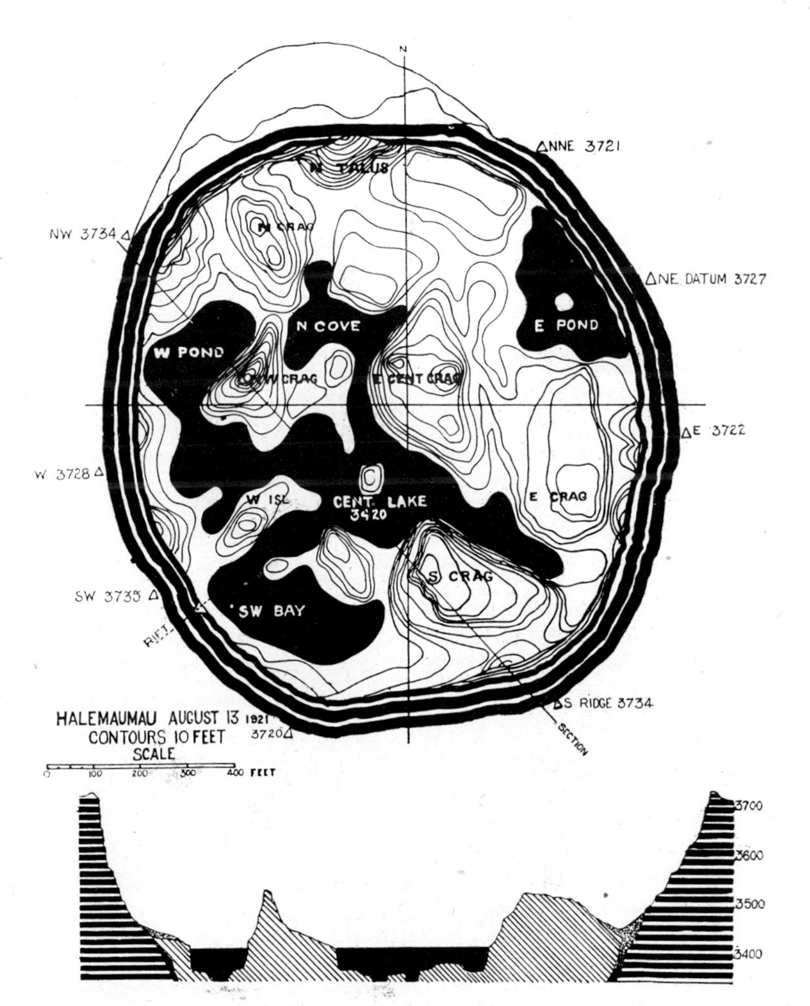

Fig. 3. Map and section of the Halemaumau pit, containing lava lakes, at Kilauea, Hawaii, on 13th August, 1921. Heavy black and white bands represent the walls or perilith. The crags are bench lava consisting largely of epimagma (shaded in the section). Pyromagma in lakes, wells, and wall cracks (under talus) is shown black. (From *Bull. Haw. Volc. Obs.*, 9 (8), 1921.)

Fig. 4. Northwest peak (crag) and west pond (lava lake) in Halemaumau pit (compare Fig. 3) on 20th September, 1921. (From *Bull. Haw. Volc. Obs.*, 9 (10), 1921.)

fissures;[7] was the withdrawal preceded by digestion or remelting of the sponge of epimagma; or was the epimagma less voluminous and continuous than had been supposed?

Actually it does not seem necessary to answer any of these questions in the affirmative; for the open pit 1335 feet deep was floored only with talus of perilith from the walls, which had collapsed as enormous rock slides. The actual lava that was withdrawn into the depths, whatever was its nature, was buried perhaps to a depth of 3000 feet by the rock-slide debris, if Jaggar's estimate of the depth of withdrawal is correct; though it seems possible that a drained sponge of epimagma may have occupied part of this 3000 feet. Pyromagma at least was withdrawn to a great depth so as to allow access of ground water from the perilith—available only near sea-level in permeable basalt structures[8]—for such an invasion by ground water is the only acceptable explanation that has been given of the cause of the short series of explosions in the Halemaumau pit which followed the collapse accompanying the withdrawal of the lava. It has been estimated that only $\frac{1}{253}$ of the available debris already in the pit was ejected by these steam explosions.[9]

The explosion of 1924 was the only one of its kind at Kilauea since that of 1789 or 1790, which was brought about in all probability in the same way.[10] It was in the interval between these two unusual eruptions that a lava lake was gradually formed and also its underlying sponge of epimagma. Several years after the explosion of 1924 the magma column under Kilauea had risen again to such a level that pyromagma flowed out into the Halemaumau pit from a fissure concealed under a cone of talus fallen from the walls; and in the

[7] The conception of a wide-open pipe extending to an indefinite depth beneath a Hawaiian lava lake or of the existence of a feeding channel of this nature anywhere under a basalt dome has been abandoned by common consent because of the observed basaltic habit of emission of pyromagma from chinks and crevices at high levels, which has already been mentioned. It is by such infilling from the side that the level of liquid lava in a pit is generally raised.

[8] In Oahu H. T. Stearns states that the ground-water level rises inland from the sea margin at a gradient of not more than 3 feet per mile (Geology and Ground-water Resources of the Island of Oahu, *T.H. Div. Hydrog. Bull*, 1, pp. 236-237, 1935).

[9] T. A. Jaggar, *Bull. Haw. Volc. Obs.*, Dec., 1924; *The Volcano Letter*, 26, 1925.

[10] J. D. Dana, Kilauea, *Am. Jour. Sci.*, 35, p. 31, 1888.

succeeding years the open part of the pit became 565 feet shallower as a result of the piling up of many sheets of lava over its wide floor.[11] Thus new lava now buries and interfingers with an immense quantity of old-lava talus (p. 23 and Fig. 5). It seems probable that, in the course of the cycle of

Fig. 5. A section exposed in a fresh fault scarp which shows much lava talus (rock-slide debris) under the (frozen) lava lake in an extinct pit crater, one of the Puna Craters, Hawaii. (Drawn from a photograph by R. A. Daly.)

activity, after about 100 years such flows and rock-slide talus undergo remelting; so that they become incorporated in a semi-mobile sponge of epimagma and pyromagma, which rises and sinks in the pit from time to time—a filling such as Jaggar pictured in the Halemaumau pit prior to 1924.

Hypomagma, Pyromagma, and Epimagma

The relation of hypomagma to both pyromagma and epimagma has been stated as follows. The words are Jaggar's:[12]

If we imagine the tunnels and wells [through the epimagma sponge] to change below to a spongy ramification of pyromagma within hypomagma so that the former becomes infinitely distributed through the mass within the zone of transition from gas-saturation to the bubble phase, we have in the sponge so formed a curious substance. The matrix material, hypomagma, is slightly cooled by gas expansion and so becoming increasingly viscous, an endothermic process so far as the gas is concerned; the fused filling or pyromagma, is heating by gas reaction and hence becoming increasingly fluent with bubbling melt. The cooling hypomagma is becoming denser and the warming pyromagma less dense. Hence the pyro-

[11] R. H. Finch, *The Volcano Letter,* 471, 1941.
[12] T. A. Jaggar, *loc. cit.* (1), p. 165.

magma rises and surficially loses its gas, and the residue sinks to merge with the epimagma. The reactions of the gaseous elements while in solution are problematical [according to the opinion of Day and Shepherd[13] no reaction will take place as long as the gases remain dissolved in the magma], but hypomagma is probably hotter than epimagma. But the epimagma is denser than the hypomagma below and sinks to displace it, releasing new gas-concentrate to the region of gas-release. The two convections may be said to be assisting each other at a critical level: pyromagmatic convection actuated by a rising temperature is dynamic; epimagmatic convection actuated by a fall of temperature is a static overturning; the first is swift, the second is slow; they feed each other. In the lower convection isothermal conditions probably prevail and pressure-saturation controls; in the upper, polythermal or exothermic mechanism is the controlling factor, with greater chemical instability.

The records of observed temperatures indicate that pyromagma is rarely much hotter than the temperature at which, under suitable conditions, it will begin to solidify as crystalline basalt (about 1150° C.); while bench lava (largely epimagma) is at a temperature about 900° C. "The pyromagma is hottest, the epimagma less hot, and presumably the hypomagma somewhat hotter than epimagma;" but it is "unnecessary to postulate any extremely high temperature for the deeper substance of volcanic magma."[14]

Hawaiian Cycles of Activity

A cyclical succession of volcanic phenomena has been observed continuously over a period of many years at Kilauea. The cycle, as described by Jaggar[15] begins with the development of tension in the interior of, or beneath, a basalt dome. Such tension results from gravitational adjustment of sinking blocks of wedge shape separated by rifts; it induces swelling of the hypomagma with emission of gas and production of pyromagma as effervescence proceeds, accompanied by the opening of fissures and the outflow of much lava into the large

[13] A. L. Day and E. S. Shepherd, Water and Volcanic Activity, *Bull. Geol. Soc. Am.*, 24, pp. 573-606 (p. 598), 1913.

[14] *Loc. cit.* (1), p. 169.

[15] Many references to the cycle theory are to be found in notes by T. A. Jaggar in the publications of the Hawaiian Volcano Observatory, especially in *The Volcano Letter;* for a summary account see G. W. Tyrrell, *Volcanoes,* pp. 121-124, 1931.

"sinks" (sometimes termed craters) on the summits of the domes.[16] The weight of the solidified lava which thus accumulates next causes subsidence, the first effect of which is to jam and obstruct the summit outflow channels, so that pyromagma thenceforth has to find a way of escape through fissures that open on the outer slopes of the dome. Such fissures have perhaps been wedged open by the sinking of the central lava-weighted core. As pyromagma escapes the lava column subsides, as is indicated by an ebb of pyromagma in the lava lake at the summit.

Every few years this cycle is repeated; but, if the lava column in an exceptional case subsides sufficiently to leave open voids into which ground water may flow so as to make contact with hot rocks, one of the steam explosions occurs which at intervals of perhaps more than a hundred years break the monotony of rhythmic flow and ebb, or rising and sinking, of a Hawaiian lava column. To take a specific example: It has been inferred (for reasons already mentioned) that the surface of the lava column—or at least the pyromagma—must have sunk in the Halemaumau pit before the explosions of 1924 to a depth of about 300 feet below sea-level in order to permit a strong influx of ground water.

> The evidence was clear that the lava had escaped at such a low level [under the sea] that it let in water. . . . This made the steam blasts, . . . discharged like geyser jets at progressively greater intervals. The discharges were nearly pure steam, . . . bringing up broken wall rock, not fresh lava.[17]

Thus, in the course of the Kilauea cycle, submarine outflow of pyromagma probably plays an important part as a producer of steam explosions. Submarine outflow has been appealed to also as an explanation of rather frequently occurring collapses in the summit sink of the great dome Mauna Loa—not all of which have occurred immediately following the visible

[16] These fissures are mere chinks, and those from which pyromagma issues are not generally on the floors, and not always on the lower parts of the walls, of the sinks, where fissures previously existing have become sealed. A cascade of pyromagma poured into the Halemaumau pit at Kilauea in 1934 from a horizontal crack hundreds of feet above the floor.

[17] T. A. Jaggar, Kilauea's Last Lava Flow, *Paradise of the Pacific,* 46 (6), 5 pp., 1934.

outflows on the flanks of the mountain—and more particularly to account for the explosive eruption of 1877.[18]

Recently published records of great activity indicate that the cycle of Hawaiian volcanicity is quite as well exemplified on Mauna Loa as on Kilauea, and on a far grander scale. Far from being full-grown and even decadent, as some have thought, Mauna Loa is still in the prime of life and fully active. Not only do extensive flank outflows of pyromagma take place every few years, but summit flows tend to fill the great "sink" Mokuaweoweo on the crown of the dome. It is this great area of collapse, and the similar hollow in the flat crown of Kilauea, which become filled and overweighted so that they sink under the burden of lava flows, as pictured by Jaggar, and not merely the Halemaumau pit and other similar lava-lake-containing pit craters some of which are situated within the limits of the larger sinks. Flows of pyromagma thus overspread the floor of Mokuaweoweo in 1933, 1935, 1940, and 1942. That of 1940 was very extensive, covering many square miles—in fact the greater part of the floor of the sink—and produced an estimated bulk of 448 million cubic metres of lava.[19]

During the course of the cycle some superficial exposure of epimagma to atmospheric oxidation is a necessary preliminary to its engulfment, if much heat is to be generated by reactions between juvenile gases from the magma and oxygen carried down in combination as ferric oxide in engulfed bench lava. Exposure of frozen-lava benches takes place during the frequent episodes of low lava level in a pit crater, but on frequent occasions, and even when the Halemaumau pit at Kilauea has been full to the brim, high craggy peaks and islands of bench magma and clastolithic sediment

[18] T. A. Jaggar, The Crater of Mauna Loa, *The Volcano Letter*, 360, 1931; History of Mauna Loa, *Paradise of the Pacific*, 46 (2), pp. 5-8, 1934: "The lava of the summit crater had been lowered along Kona cracks below sea-level."

J. D. Dana, quoting Coan, has reported columns of steam three miles high over Mauna Loa on this occasion, and has recorded also a well authenticated submarine eruption off the Kona coast of Hawaii on "the day the light disappeared from the summit" (i.e. a lava lake disappeared) (*Am. Jour. Sci.*, 36, n. 29, 1888).

[19] H. H. Waesche, Mauna Loa Summit Crater Eruption 1940, *The Volcano Letter*, 468, pp. 2-9, 1940.

have been thrust high above the lava lake (Figs. 3, 4, 6).[20] It may be that oxidation of lava thus exposed to the atmosphere provides most of the oxygen for the volcanic furnace in lava volcanoes of the Hawaiian type, as Jaggar originally assumed, though a further supply of oxidised rock is derived from the perilith as rock slides (Chapter III).

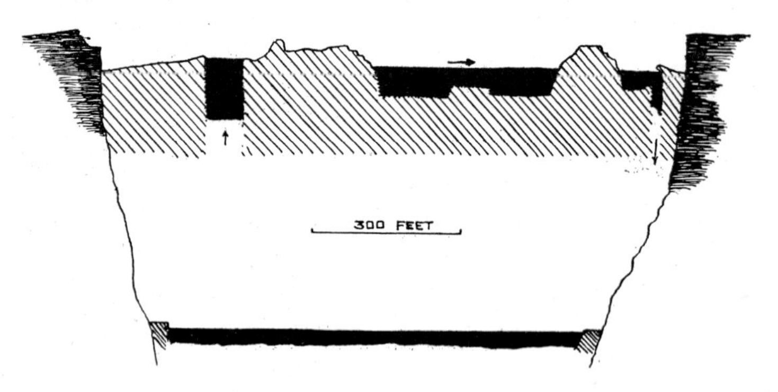

Fig. 6. Section of the Halemaumau pit, Kilauea, Hawaii, showing rapid rise of the lava-lake level from June, 1916, to January, 1917. Pyromagma black; epimagma shaded. True scale. (After Jaggar, 1917.)

Thus, in the top of the lava column at Kilauea as it existed in 1915, for example,

we must imagine a stiff paste [epimagma] rising with the old-talus breccia cemented in it, with a ring platform [bench lava] carried on its crest, the whole somewhat cylindrical in shape, and a shallow inside cup containing the lava lake fed by a well leading downward through the midst of the column of stiffer lava. Up the feeding well [compare Fig. 6] was coming the hot gas froth [pyromagma] . . . heated by gas reactions, . . . liquefied and vesiculated to a fluid condition very different from the fundamental [hypo]magma deep down, under pressure, where it is presumably stiff, rigid, and heavy.[21]

[20] It is obviously a mistake to assume, as a great many observers have done, that these crags or peaks are "floating" islands. Rebutting the "time-honoured fallacy" that such masses of bench lava float on the surface of deep lakes of pyromagma, Jaggar has remarked: "Solidified basalt has a higher specific gravity . . . than liquid basaltic melt. Much more is it heavier than gas-foamed liquid basaltic melt of the . . . lava lakes. . . . These solid crags grade downward into a red-hot, dense, heavy substance." (T. A. Jaggar, *The Volcano Letter,* 272, p. 1, 1930.)

[21] T. A. Jaggar, *The Volcano Letter,* 332, p. 2, 1931.

"Progressively until 1924 the upper part of this lava column was becoming more fluid as the talus fragments had in a measure been appropriated as part of the bench magma column." From time to time the lava column rose and sank hundreds of feet (Fig. 6), the lowering being "in just such proportion as there had been greater clearing out of the throat below." By these processes "the throat is cleared bigger and the pulsations of subsidence are permitted to go deeper." The throat was funnel-shaped (Figs. 3 and 6), and each uprise of lava made room for much pyromagma in a "ring channel" around the periphery ("wall cracks" in Fig. 3, section).

Ring Dykes

A complex of fissures in the form of a ring 3½ miles in diameter, from which lavas have issued in the present century surrounds the central sink or caldera on a low-angle basalt dome the summit of which emerges from the ocean as the volcanic island Niuafo'ou, near Tonga, in the South Pacific.[22] That these are sealed by dykes is indicated by the level of a caldera lake in the sink, which is 75 feet above the sea (Fig. 7), for it is well known that basalt-built structures will fail to hold up ground water to this height in such a situation unless intersected by dykes.[23]

Solidification in similar ring channels in ancient "Kilaueas" has been claimed by Jaggar as the origin of the ring dykes of Scotland.[24] Tyrrell[25] has made a similar suggestion and has pointed out that the trend of opinion among modern investigators of the Tertiary igneous rocks of the west of Scotland favours the view that there are among these some eroded stumps or basements of great volcanoes of the Hawaiian type.

What we now see at the present level of erosion is the final emplacement of magma in the ring structures; and while the rings break through the main mass of flood basalts, there is no reason

[22] T. A. Jaggar, Geology and Geography of Ninafoou Volcano, *The Volcano Letter,* 318, pp. 1-3, 1931.

[23] H. T. Stearns, *loc. cit.* (8).

[24] T. A. Jaggar, *The Volcano Letter*, 7, 1925.

[25] G. W. Tyrrell, Flood Basalts and Fissure Eruptions, *Bulletin Volcanologique,* 1, pp. 89-111, 1937.

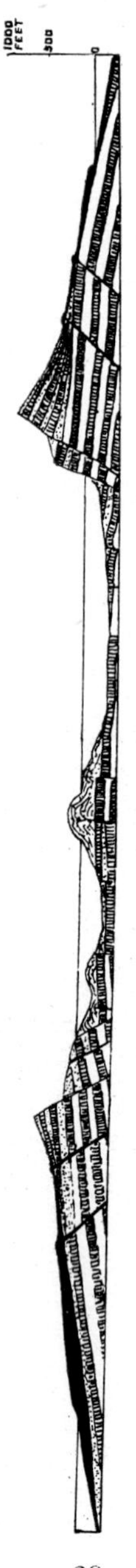

Fig. 7. Section of the Pacific island Niuafo'ou (basalt dome and central sink) showing structure of ring faults, fissures, and dykes and peripheral lava outflow. Vertical scale twice the horizontal scale. (After Jaggar. *The Volcano Letter*, 318, 1931.)

to disbelieve that they also represent the earlier foci of eruption and were the centres from which the earliest flood basalts were erupted.[26]

Preparation of Pyromagma for Outflow

Flow and ebb of magma in a funnel-shaped pit, as described above, must result in the development of a larger and larger sponge containing a progressively increasing quantity of the fluid pyromagma which has been heated up in place by chemical reactions. A cycle of activity has been observed to culminate with a tumultuous withdrawal of the magma column which has been likened to the escape of water down the waste pipe of a basin when the plug has been removed. The cause of such withdrawals seems to have been the escape of the gradually accumulated hot pyromagma to the surface by way of some crack in the mountain at a lower level than the lava lake—in the case of the high dome Mauna Loa this will be one of the rather frequent lava flows which gush out on the flanks of the mountain.

In the case of the lava lake at Kilauea, which is nearly ten thousand feet lower than that on Mauna Loa, the systematically observed and measured changes of the lake level have been found to take place mainly in sympathy with movement of the magma column under Mauna Loa,[27] although there is no relation even remotely resembling hydrostatic equilibrium between the columns of the two volcanoes (Chapter II). As already cited, Jaggar explains some ebbings of the column which cannot be definitely correlated with visible surface outflows, or with inter-adjustments between the adjacent columns, by postulating escape of lava somewhere on the ocean floor.

The heating up of magma by chemical reactions in lava lakes at the summits of basalt volcanoes considered as a stage in their cycle of activity seems thus to be an essential step in the preparation of the great gushes of lava that contribute to the upgrowth and outgrowth of basalt domes.

[26] G. W. Tyrrell, *loc. cit.* (25), p. 109.
[27] T. A. Jaggar, *The Volcano Letter,* 332, p. 2, 1931.

Very little has been published in the way of discussion of the significance of the high ferrous-oxide content of basalts of the plateau and mid-oceanic kind in relation to Jaggar's theory that iron is the vehicle for the supply of oxygen to support combustion of hydrogen in the volcanic furnace. These basalts contain in most cases from 7·5 to 10 per cent of ferrous oxide, whereas in Daly's average of all basalts the ferrous-iron content is only 6·34 per cent, in average andesite 3·13, in dacite 1·33, and in rhyolite 0·82 per cent.[28] If it be a fact that hot and mobile pyromagma is cooked up to that condition by a process such as has been described in this chapter just prior to its effusion, is it not permissible to surmise that the viscid condition and low temperature at which the more acid lavas commonly arrive at the surface may be in part the consequences of their low iron content?

It cannot be assumed that the process referred to in the foregoing section has been alone responsible for the preparation of eruptible pyromagma, for this seems to be supplied to some volcanoes directly from the depths—at Matavanu (Samoa), for example, and in Mexican volcanoes such as Jorullo and Paricutin. This recalls a remark of Fenner[29]: "Reactions in an ascending magma may at times be highly exothermic."

---

[28] R. A. Daly, Average Chemical Compositions of Igneous-rock Types, *Proc. Am. Ac. Arts and Sci.*, 45 (7), pp. 211-240, 1910.

[29] C. N. Fenner, The Katmai Magmatic Province, *Jour. Geol.*, 34, pp. 673-772 (p. 740), 1926.

CHAPTER V

Pumice Volcanoes

THE FOREGOING CHAPTERS DESCRIBE ONLY SOME OF THE VOLCANIC processes that are observed in and about the vents in which basalt lava ascends, giving rise to the lava-volcano phenomena which are of a relatively quiet, effusive, and generally non-explosive kind; and perhaps the hypothesis of volcanic cycles which culminate in basaltic outpourings will afford an explanation of the essential differences between "lava" volcanoes on the one hand and "explosive" or "pumice" volcanoes on the other.

The variable mineral composition of volcanic rocks indicates a very considerable range of solidification temperatures. The presence of high-temperature forms of crystallised silica, for example, not only tridymite but cristobalite as well[a], among the final products of crystallisation in some of the more acid effusive rocks may indicate volcanic temperatures in the vicinity of 1500° C.[b] It may very well be true that some such temperature is required to make these lavas flow freely.

In most cases of present-day volcanic action, however, the hypomagma seems to emerge at or to approach the surface at temperatures decidedly below those at which lavas are freely liquid—indeed below, rather than above, 1000° C. The mineral composition of many rocks formed from lava is such as to indicate that the temperature of their emission has been below 850° C. Yet the dacite lava of Lassen Peak, Cali-

[a] E. S. Larsen and others, Petrologic Results of a Study of the Minerals from the Tertiary Volcanic Rocks of the San Juan Region, Colorado, *Am. Mineralogist,* 21, pp. 679-701, 1936; C. S. Hurlbut, X-ray Determination of the Silicate Minerals in Submicroscopic Intergrowths, *Am. Mineralogist,* 21, pp. 727-730, 1936.

[b] This argument holds good, of course, only in cases where the associated minerals justify an assumption of a high temperature and where it is not necessary to assume (on account of association with low-temperature minerals) that the cristobalite present has been introduced by gaseous transfer, being formed by a "pneumatolytic" process at a low temperature, as was the case in the laboratory experiments of Greig, Merwin, and Shepherd (Notes on the Volatile Transport of Silica, *Am. Jour. Sci.,* 25, pp. 61-73, 1933).

fornia, when reheated to that temperature in the laboratory, has been observed to show only the first signs of a reduction of rigidity, being still several hundred degrees below the point at which it becomes liquid or even pasty.[1] Such lavas are commonly of andesitic or dacitic composition and are among those which habitually exude in a nearly rigid condition so that they pile up over the volcanic orifices as tholoids, or cumulo-domes. The actual temperatures of emission in such cases are estimated to vary from below 850° C. (at Lassen Peak) to several hundred degrees higher (at Mont Pelée, Martinique). (All observers agree that the lava which has exuded from Mont Pelée has been relatively very hot.) So greatly does a small variation in the gas-content of a magma change its properties, however—both its viscosity when in the condition of a glass and the temperature at which it will crystallise or pass in the reverse direction from a crystalline to a viscous glassy condition—that statements of actual volcanic temperatures give little information as to the condition of the lava.[2] The relation of viscosity to water-content has been investigated in the case of obsidian.

[1] A. L. Day and E. T. Allen (*The Volcanic Activity and Hot Springs of Lassen Peak*, pp. 50-51, Carnegie Inst., Washington, 1925) record that this rock when reheated begins to soften slightly at 850° C., but its mobility is very slight up to 1040° C. and does not become great even at 1260° C., though this seems to be above the "melting point" in the sense that crystals of constituent minerals have dissolved or are dissolving, and the lava tends to become homogeneous and to assume a condition of chemical equilibrium. Such material is still so viscous, however, that, according to E. S. Shepherd and H. E. Merwin (Gases of the Mt Pelée Lavas of 1902, *Jour. Geol.*, 35, pp. 97-116—see p. 115, 1927) its resistance to vesiculation is "comparable to that of a rigid body," and gas vesicles can expand rapidly in it only when they are so numerous already "that they need expand only a few volumes to make a foam."

[2] Experimental work by R. W. Goranson on remelted granite from Stone Mountain, Georgia (Notes on the Melting of Granite, *Am. Jour. Sci.*, 23, pp. 227-236, 1932) indicates that, while the melting (or crystallising) point of the gas-free material is about 1050° C., the corresponding point for the fused granite if it contains 6·5 per cent water in solution is about 700° C.

Great differences in the behaviour of New Zealand obsidians when subjected to heating (which have been recorded by P. Marshall, Acid Rocks of the Taupo-Rotorua Volcanic District, *Trans. Roy. Soc. N.Z.*, 64, pp. 323-366, 1935) may perhaps be attributable, on the other hand, to difference of chemical composition. Obsidians of ordinary rhyolitic composition became pumiceous only at 1000° C. and did not fuse until a temperature of about 1200° C. was attained; whereas more sodic (comendite) obsidian from Mayor Island swelled into pumice at 600-660° C., and fused at 900° C.

Small differences in the water-content (tenths of a per-cent) make great differences in the viscosity of such glasses, and for the very "dry" glasses it is difficult to imagine them flowing at any temperature yet observed around volcanoes.[3]

A notable feature of lavas that exude as cumulo-domes, according to a generalisation by Williams,[4] is that they are "erupted at temperatures below those of corresponding lava flows," which is equivalent to a statement that actual volcanic temperatures are commonly too low to give such lavas the mobility required for flow. They are indeed in an "under-cooled" condition to the extent of being too viscous even to allow of the free vesiculation which would result in an explosive formation of pumice and pumiceous ash.

Basaltic hypomagma also, according to Jaggar's interpretation of Hawaiian phenomena, is very viscous—even semi-rigid—when it is approaching the surface at a temperature somewhat below that at which it can become freely fluid. He pictures it as passing very readily, however, as a result of mere release of pressure, into liquid, frothing pyromagma which liberates gases and thus sets the volcanic-furnace mechanism in operation; whereas it appears that, in the general case, a non-basaltic magma fails to respond in this manner to release of pressure, perhaps because its viscosity diminishes with rise of temperature only at a much slower rate than is the case with basalt. As such magmas fail to effervesce freely they fail to develop the internal furnace effects which, it seems probable, prepare some basalts for quiet outflow.

When these viscous lavas do froth or foam the result is commonly an explosion, the viscous glass being torn to shreds by the expanding gases. Reactions between the gases must in this process be delayed until the gases escape from enclosing vesicles. In the case of extrusion of the lava in a semi-rigid

[3] L. H. Adams, "The Volatile Constituents of Magmas," in Annual Report of the Director of the Geophysical Laboratory, *Carnegie Inst. Wash. Year Book*, 37, p. 108, 1938.

[4] Howel Williams, The History and Character of Volcanic Domes, *Univ. Cal. Publ., Bull. Dep. Geol. Sci.*, 21 (5), p. 139, 1932.

condition to pile up over the vent much of the gas remains trapped and is still in solution in undercooled glass.[5]

Magmatic Explosion

Statistics collected by Sapper[6] show that during the last four centuries a very much greater quantity of material has been ejected explosively by volcanoes than has been poured out as lava. The two great lava-producing regions, moreover, have been Iceland and the Central Pacific (including Hawaiian volcanoes) which have yielded 15·5 and 11 cubic kilometres of lava respectively, together producing 40 per cent of the world output. In contrast with this the volcanoes of the Java belt, or Sunda arc, alone have produced 185 cubic kilometres of fragmentary material, as against 1·5 cubic kilometres from the Central Pacific volcanoes; and when the output of the volcanoes of the Sunda arc is added to the total from the other volcanoes which are ranged in lines that encircle the Pacific Ocean it is found that these together have produced 96 per cent of the world total of fragmentary material. It is worthy of note also that the volcanoes of the Sunda arc have yielded no more than 0·5 cubic kilometres of lava—i.e. only 0·27 per cent of their product has been permanently solidifying lava.

So striking is the regional contrast between the quiet outwellings of pyromagma in the lava volcanoes, which build in some cases domes of Hawaiian type and in other cases basalt plateaux, and the explosive phenomena associated with the uprise of magmas that yield mainly pumice and pumiceous ash that it suggests the possibility that the pumice-making lavas are fundamentally different, or have been generated in some different way, from basaltic pyromagma.

As is well known from petrographic studies, however, there is no sharp line of demarcation between explosive lavas and effusive basalts. Chemically the difference is only that

[5] Shepherd and Merwin (*loc. cit.* (1)) have recorded that a specimen of obsidian (containing 75 per cent silica) was observed to swell into a froth and lose its dissolved gases when heated to 850° C. Rocks which behave in this way—generally exploding, which perhaps is an indication that reactions occur among released gases—are the "live" rocks as defined by A. Brun (*Recherches sur l'exhalaison volcanique,* 1911).

[6] K. Sapper, *Zeits. Vulk.,* 11 (3), 1928.

between the basic and the rather less basic (grading into intermediate and acid) silicate magmas from which all igneous rocks are derived.

The explosive magmas are known to be in all cases more viscous than the free-flowing, or "effusive," type of basalt. A suggestion has been made, however, that it is not necessarily the silica-content that controls viscosity in the critical temperature-range at which volcanic explosions take place, but rather the ferrous-oxide content of the magma. Washington,[7] who has made this suggestion, would divide basalts generally into two main classes, "the plateau basalts, which are high in iron, and the cone basalts, which are low in iron."

> Chemically the plateau basalts [with which must be included the dome-building basalts of Hawaiian type] differ materially from other basalts in one or two features. . . . The chief difference is seen in the much higher amount of iron oxides, with ferrous oxide greatly preponderating over ferric oxide. . . .
>
> It is a matter of common observation that basalts generally are . . . more fluid when molten than are more feldspathic or more silicic rocks. . . . The experience of iron and steel workers and smelters bears testimony to the lower fusibility and greater fluidity of slags containing considerable iron.[8]

Those basalts—all definitely poorer in ferrous oxide than the more fluent plateau- and dome-building variety—which, according to Washington, are potentially explosive cone-builders, though "no hard and fast line can be drawn," may be classed for the present purpose with the non-basaltic magmas as basalts of an andesitic volcanic habit. Washington notes that flows of such basalt "do not extend very far, and are often found consolidated on steep slopes." As noted in Chapter IV, these basalts have a content of ferrous oxide far below the 7·5 to 10 per cent found in mid-oceanic and plateau basalts, though above that in lavas of intermediate to acid composition which contain only from 3 to less than 1 per cent ferrous oxide.[8a]

[7] H. S. Washington, Deccan Traps and other Plateau Basalts, *Bull. Geol. Soc. Am.*, 33, pp. 765-804, 1922.

[8] H. S. Washington, *loc. cit.* (7), pp. 709, 803.

[8a] Quantitatively such basalts are relatively insignificant. P. C. Putnam estimates they make up 10 per cent only of the lavas of Central America (Magma Mass Underlying Central America, *Jour. Geol.*, 34, pp. 807-823, 1926).

The Source of Magmas

It has been pointed out by Van Bemmelen[9] and many others that the pumice-making (generally andesitic) volcanoes, which are arranged in linear groups (Fig. 106), are so closely associated with mountain and island arcs consisting of recently folded geosynclinal sediments[10] that it must be assumed there is a genetic relation between the magma that feeds them and the folding process. The magma may originate perhaps by palingenesis of downfolded sedimentary material, as has been supposed by Hobbs[11], or by a large-scale fusion and assimilation in primary olivine-basalt magma. "Neither volcanism nor plutonism can be understood until we understand the formation of mountain chains" (Daly).

Much attention is directed now towards evidence of granitisation[12] of sedimentary and other rock material by various processes, chiefly metasomatic. Investigators of the origin of plutonic acid and intermediate rocks who favour the theory of granitisation do not generally claim or attempt to explain also the development of the bodies of magma that manifest themselves in volcanic phenomena; but a very close relationship between plutonic emplacement and volcanism in the orogenic belts has been demonstrated—notably by the observations on the geology of the Barisan Range (Sumatra) collected in Van Bemmelen's discussions of the problem. The argument, often urged, that typical igneous magma cannot be produced by simple fusion of crustal rocks of other than

[9] R. W. van Bemmelen, On the Origin of the Pacific Magma Types in the Volcanic Inner Arc of the Soenda Mountain System, *Ingenieur in Ned. Indie,* 5, series iv, pp. 1-14, 1938; The Volcano-tectonic Origin of Lake Toba (North Sumatra) *De Mijningenieur,* 6, pp. 126-140, 1939.

[10] R. T. Chamberlin has distinguished thin- and thick-shell mountain uplifts. "On theoretical grounds the thin-shell mountains should be accompanied in their growth by relatively little volcanism, while the growth of thick-shell mountains should be attended by relatively greater outpourings of lava" (The Building of the Colorado Rockies, *Jour. Geol.,* 27, pp. 225-251, 1919).

[11] W. H. Hobbs, Some Considerations Concerning the Place and Origin of Lava Maculae, *Gerlands Beitr. zur Geophysik,* 12, pp. 329-361, 1913; *Earth Evolution and its Facial Expression,* pp. 28-71, 1921; also The Growing Mountain Ranges of the Pacific Region, *Proc. II Pan-Pacific Cong.,* I, pp. 746-757, 1926: "Maculae of magma are formed as a result of the local fusion of pelitic (shaly) sediments due to local relief of load. . . . The magmas associated with arcuate mountains are in all earlier stages of the process intermediate in composition—andesites." (p. 747).

[12] As defined by M. MacGregor and G. Wilson, On Granitisation and Associated Processes, *Geol. Mag.,* 76, pp. 193-215, June 1939.

igneous origin loses its force when allowance is made for change in chemical composition brought about by the introduction of elements by volatile fluid "emanations" from the depths. Metasomatic alteration, it is claimed, can produce magmatic mobility and may be followed by more or less perfect fusion of granitised material, so that the generation of the magmas of intermediate to acid composition responsible for volcanic phenomena in the folded belts may perhaps be thus accounted for. Van Bemmelen's conclusion is that "volcanic activity on the top of orogenic uplifts is caused by palingenetic, hybridic, or syntectic magmas. These magmas are saturated with emanations, and this explains the high explosiveness."

On the other hand, basic magma from the depths, undifferentiated and uncontaminated, is commonly regarded as the actuating hypomagma of the clusters of volcanoes that rise from the ocean floor, where it probably breaks through a thin crust. In the kratogens (non-folded regions) it may rise also where tapped by deep fissures which have fed plateau-building floods and some dome-building basalt volcanoes such as are present in east-central Africa.

A different view of the magmatic contrast is maintained by Kennedy and Anderson,[13] who would derive the andesitic fold-belt volcanoes from a primary basalt magma distinct from and of less deep-seated origin than that which provides the olivine basalt for oceanic and kratogenic volcanoes and plateau basalts. The differentiation products of the olivine-basalt (sometimes termed "crinanitic") magma are trachytes and phonolites. The magma termed "tholeiitic," on the other hand, which gives rise to andesitic volcanoes (as well as basaltic cone-builders of andesitic habit) differentiates in the direction of andesite and rhyolite. Its chief chemical peculiarity is oversaturation with silica.

All such views of magmatic origins are opposed to those of Bowen,[14] who insists on the possibility—and indeed the

[13] W. Q. Kennedy and E. M. Anderson, Crustal Layers and the Origin of Magmas, *Bull. Volcanologique*, 3, pp. 24-82, 1938.

[14] N. L. Bowen, *The Evolution of the Igneous Rocks*, pp. 75ff., 1928.

probability—of derivation of all igneous rocks from one primary magma of olivine-basaltic (plateau basalt) composition, and has shown how a more andesitic basalt may separate from this primary basalt and become the starting point for further differentiation such as will result in the production of all the magma types associated with non-basaltic volcanism.

Apart from the important question of defining the distinction between "lava" and "pumice" volcanoes, speculations and theories regarding the origin of magmas would be out of place in a work which sets out to treat of volcanoes as geomorphic forms. In the words of Bonney (who, however, preferred to think of magmas rather as "integral portions of the inner part of the earth" than as "the result of the local melting-down of sedimentary strata"):

> If we agree upon certain characteristics as denoting an igneous rock, the antecedent history of the rock (for our special purpose) becomes immaterial [and the same applies to magma as a volcanic agent]. For instance, I can think of a piece of glass as an (artificial) igneous rock even though I may have formerly seen a crucible full of the material from which it has been made.[15]

The Viscous Lavas

So far as information is available, there seems to be this one important difference in physical properties between the iron-rich olivine-making basic magma and all other magmas (including its own differentiates) that it becomes by far the most fluent of lavas at volcanic temperatures. Less iron-rich lavas are much more refractory and remain extremely viscous—in some cases almost rigid at such temperatures.[16] Petrographically this property is reflected in the generally hemicrystalline and even largely glassy character of the acid volcanic rocks. They remain in the condition of undercooled glass, crystallisation having been inhibited or delayed by the slowness of molecular diffusion in the very viscous glass when it has been at a temperature low enough for crystallisation to take place.

[15] T. G. Bonney, Presidential Address, *Proc. Geol. Soc. London*, p. 35, 1885.
[16] Compare Day and Allen, *loc. cit.* (1).

An important cause of variation in viscosity more potent even than a considerable change of temperature is variation in the content of potential volatiles. This is quite apart from the chemical composition of the silicate portion of the magma containing these. Though there is not very much quantitative information available on this subject beyond that afforded by some laboratory experiments which indicate that the melting point of silicates is lowered considerably by absorption of water, petrologists all assume that magmas which contain much dissolved gas (generally taken to be water) have relatively low viscosity. Volcanic phenomena indicate that explosive lavas are undoubtedly both viscous and rich in gases; but it seems probable that, within the range of temperature at which explosion of magma takes place, most hypomagmas—as a function of chemical composition—would, instead of being merely viscid, be practically rigid but for the fluxing effect which is due to the presence of volatiles as long as these remain in solution.

The Pyromagma in Explosive Volcanoes

The terms hypomagma, pyromagma, and epimagma can be applied only tentatively to phases of pumice-making lavas. Jaggar has suggested that "epimagma may be the substance of the cumulo-domes of such volcanoes as Mont Pelée, these being, as in Kilauea [crags], the slowly tumescent, semi-solid upper portions of the lava column."[17] Perret's observations, published later, have shown, however, that the substance of Peléan and other tholoids is by no means dead lava. On the contrary, its glassy base imprisons potential volatiles in abundance, the explosive emission of which results in *nuée ardente* eruptions (Chapter I), and the explosive disintegration of the lava of tholoids produces much of the fragmentary debris which accumulates as sandflows and volcanic breccias. "Explosion that gives rise to *nuées ardentes* is evidently the result of rapid vesiculation;"[18] and it is in the substance of the tholoid that such explosion commonly takes place.

[17] T. A. Jaggar, Seismometric Investigation of the Hawaiian Lava Column, *Bull. Seism. Soc. Am.*, 10, p. 164, 1920.

[18] F. A. Perret, *The Eruption of Mt. Pelée 1929-1932,* p. 88, Carnegie Inst., Washington, 1935.

Possibly this rock should be classed as very viscous pyromagma, or some of it may be unchanged hypomagma continuous with that of the column below. As regards the hypomagma column Jaggar has expressed the opinion:

The dormant lava of an explosive volcano, unrelieved by abundant pyromagmatic effervescence at high temperature as at Kilauea, is probably hypomagma, slowly honeycombing, heating, and expanding beneath the surface with accumulated gas. Disruption finally relieves pressure, effervescence is rapid and paroxysmal, and the gas reactions are explosive.[19]

Some preparation, possibly of the nature of a preliminary generation of pyromagma, has been observed to be in progress prior to a major outburst at Mont Pelée. Perret, interpreting observations made by Heilprin, has described the condition of the viscid mass of exuded lava at the summit of the mountain four hours before a great explosion took place, as follows:

The volcanologist sees the entire mass of lava rapidly heated by countless steam columns rising from underground conduits; conceives of exothermic reactions in the powerfully circulated liquid brought up from the interior of the volcano; and foresees the eventual attainment of a critical condition where the combined energy of the superheated vesicles shall exceed the resistance and the whole grand mass shall explode with cataclysmic violence. Cooling is not a factor in this great combination of causes; the total energy continues to rise. The mysterious interval of external calm which often precedes the great explosions . . . is explained by the temporary stoppage of the vents; the myriad steam holes in the carapace of the dome are closed by the expansion and swelling of the liquid lava, stopping the release of gas only to augment the final catastrophe.[20]

The remarks of Shepherd and Merwin[21] on a specimen of rock taken from the surmounting spine of the tholoid of Mont Pelée in 1902 have shown that this should be classed as pyromagma which has been checked by undercooling from degenerating into epimagma; and it may be assumed that the

[19] T. A. Jaggar, *loc. cit.* (17), p. 167.
[20] F. A. Perret, *loc. cit.* (18), pp. 88, 89.
[21] E. S. Shepherd and H. E. Merwin, *loc. cit.* (1), p. 110.

material of this specimen is very similar to that of portions of the tholoid which disintegrated from time to time so as to give rise to *nuées ardentes*. The specimen "is almost holocrystalline, but contains ramifying vesicles which give the rock a notably sugary texture. Such a rock while crystallising with every pore under very high pressure might shatter explosively if erupted into a region of low external pressure."

A general statement which these authors make regarding the Mont Pelée lavas is that "at about 1000° C. each cubic metre of issuing lava has at least 50 cubic metres of gas available; this means a minimum partial pressure of fifty atmospheres." They justifiably assume that "this represents a minimum of gas, since the more highly charged material exploded in the form of ash and lapilli."[22]

In cases of very explosive activity it appears highly probable, therefore, that the pyromagma phase is short-lived; or, in an explosive paroxysm, the stage characterised by development of pyromagma may be elided, the rather viscid pyromagma, itself generated with explosive rapidity, breaking down immediately into epimagma with the simultaneous ejection and explosive interreaction of enormous quantities of gas. Highly viscous magma inflated so that it has become a pumiceous foam may be disintegrated by the further dilatation of its vesicles into minute shreds and dusty fragments. Epimagma in such a condition is frequently ejected violently by Vulcanian eruptions as pumice and pumiceous ash. The solid materials in *nuées ardentes* are more of the nature of pyromagma, as they are still emitting gas. The cooled, or spent, residue of *nuées ardentes* may be regarded as epimagma, however.

Phreatic Explosions

Some volcanic explosions, such as occur occasionally in lava volcanoes, Jaggar has suggested, may take place if "conditions of accumulating epimagmatic cooling, dominating the free heating of pyromagma and quenching it, . . . lead a quietly-erupting liquid-lava volcano into an explosive phase."[23]

[22] *Loc. cit.* (1), p. 109.
[23] T. A. Jaggar, *loc. cit.* (17), p. 167.

As a general explanation of such explosions as occur in the volcanoes of Hawaii, however, this suggestion has been dropped by Jaggar, though retained by some other volcanologists.[24] The "steam-boiler," or "phreatic,"[25] explanation of Hawaiian explosions has been stated in Chapter IV (see also Chapter XIV). Dana has designated such explosive outbreaks, which emit no effusive lava and eject only fragmentary material derived from the wall rocks (perilith), "semivolcanic."[26] These might be considered to come into the category of "ultra-vulcanian"[27] phenomena or "indirect explosions,"[28] though such outbursts are not necessarily always of the simple steam-boiler type but may be due, it has been suggested, to reheating of potentially explosive glass,[29] or to retrograde boiling in a cooling glass perhaps in accordance with the theory of Morey, which will be outlined on a later page.

Revival After Dormancy

It has been observed, as noted on an earlier page, that the extremely stiff, semi-rigid lava composing a still hot tholoid (cumulo-dome) is liable to explosive disintegration with the emission of much gas. Such disintegration—that which gives rise to *nuées ardentes*—takes place probably at spots which have been reheated, and have thereby had their viscosity reduced, by gas-fluxing. In addition to the potential explosiveness of cool volcanic glass due to the presence of gases, and especially of unoxidised gases, in solution in it, there is a further reserve of energy due to its having failed to assume a crystalline condition (owing to its high viscosity) when its temperature has fallen below that at which minerals can separate from such glass by crystallisation. Release of a portion of the latter energy may be the cause of some of the phenomena of renewal of activity in dormant volcanoes which have been previously

[24] A. L. Day, *Some Causes of Volcanic Activity,* p. 19, Franklin Inst. Philadelphia, 1924.

[25] The description "phreatic" has been applied to all volcanic phenomena which may be attributed to the explosive conversion of water of superficial origin into steam (E. Suess, *The Face of the Earth* (English edition), 4, p. 568, 1909).

[26] J. D. Dana, *Characteristics of Volcanoes,* pp. 23-24, 1890; *Manual of Geology,* 4th ed., p. 292, 1894.

[27] G. Mercalli, *I vulcani attivi della Terra,* 1907.

[28] F. von Wolff, *Der Vulkanismus,* I, pp. 94, 546, 1914.

[29] See Chapter I.

plugged by stiff undercooled lava. The softening, if not actual melting, of a plug of this kind seems to be a necessary preliminary to its disintegration and explosive ejection, if a new eruption is to take place in an old crater instead of breaking out somewhere else along a line of greater weakness.

In a similar way, the upward thrusting of solidified lava in an old pipe, which seems to have occurred sometimes so as to cause its protrusion as a "plug dome" of the cylindrical kind (Chapter XI), implies that some softening by reheating of the plug has taken place. Thus

> the lava of 1915 [at Lassen Peak, California] does not seem to have been a new magma erupted through an old conduit. It seems more like an old conduit-lining [filling?] which had been fissured and reheated by a fresh influx of juvenile gases from below until finally it acquired sufficient mobility to allow it to be forced upward by pressure.[30]

Such a softening, but in this case softening to the extent of fully melting old lava, seems to be indicated (as Fenner[31] has assumed) by the appearance of lava in the crater of Katmai (Alaska) and by the injection of magma close to the surface under the adjacent Valley of Ten Thousand Smokes prior to the explosions which took place in that volcanic field in 1912. It "remained comparatively quiet for a certain period before explosive inflation" occurred. Not only did sufficient time elapse to allow the magma to assimilate a large quantity of still older rock of a different composition—a heat-consuming process—but the nature of some fragmentary ejected material shows it to be scoriaceous scum which had formed on a pool of liquid and "potentially explosive magma."

Possibly the lava which, it seems, thus sometimes floods a crater prior to an explosive outburst is actually cooling in contact with the atmosphere during its quiet interval until it reaches Roozeboom's "second boiling point," at which "retrograde boiling" with ebullition of volatiles previously in

[30] A. L. Day and E. T. Allen, *The Volcanic Activity and Hot Springs of Lassen Peak*, p. 52, Carnegie Inst., Washington, 1925.

[31] C. N. Fenner, The Katmai Region, Alaska, and the Great Eruption of 1912, *Jour. Geol.*, 28, pp. 569-606 (p. 603), 1920; A. L. Day, "Katmai Studies" in Ann. Rep. Geophys. Lab., *Carnegie Inst. Wash. Year Book*, 33, pp. 63-67, 1934.

solution will take place, explosively inflating the lava and causing, or at least initiating, an explosive ejection of pumiceous ash.[32]

In such a case it may be that volcanic reheating has generated the liquid, but a subsequent cooling off has preceded its explosion; and it is inherently probable that, under varying conditions of equilibrium, in some cases heating and in other cases cooling may release the trigger for explosive effervescence. Heilprin's observations which appear to indicate reheating at Mont Pelée during a quiet interval prior to an explosion have already been mentioned (p. 50). Actual reheating close to the surface to the extent of producing a visible glow in rock that had been solid for some years took place on the flank of the tholoid at Santa Maria volcano, Guatemala, several days before the ejection of a *nuée ardente* from the hot spot in 1929.[33]

Reheating and softening of lava plugs are credibly explained by Daly[34] and others as the results of gas-fluxing from beneath caused by the rise of juvenile gases. Morey's theoretical discussion of crystallisation in magmas and some of its effects,[35] however, suggests supplementary or alternative processes which may contribute to the renewal of some mobility in magma that has been near the surface for some time already, and to the generation of a vapour pressure such as will lead, if the magma is confined by gas-tight walls, to an explosive outbreak.[36] Crystallisation expels water vapour or other gas

[32] An experimental analogue to such an occurrence has been described by Morey (G. W. Morey, The Development of Pressure in Magmas as a Result of Crystallisation, *Jour. Wash. Ac. Sci.*, 12, pp. 219-230, see pp. 224-225, 1922): "If potassium metasilicate at its melting point [976° C.] be saturated with water at one atmosphere pressure it takes up about 1 per cent, enough to lower its melting point about 35°. If the saturated liquid be cooled quickly it becomes supersaturated; the molten aqueous glass remains liquid until cooled several degrees below its melting point. First a few bubbles begin to form within the glass, then suddenly the bubble-formation becomes rapid, the viscous mass swells into a pumiceous mass, increasing in volume many times and overflowing the crucible. This is an example of the second boiling point at atmospheric pressure; of a boiling, attended by a sudden liberation of vapour, taking place as a result of cooling."

[33] K. Sapper and F. Termer, Der Ausbruch des Vulkans Santa Maria in Guatemala vom 2-4 Nov. 1929, *Zeits. f. Vulk.*, 13, pp. 73-101, 1930.

[34] R. A. Daly, *Igneous Rocks and the Depths of the Earth*, p. 380, 1933.

[35] G. W. Morey, *loc. cit.* (32).

[36] A similar suggestion had been made previously by A. Lacroix, *La Montagne Pelée et ses éruptions*, p. 358, 1904.

which has been in solution in the glass and (under favourable conditions only—those which prevent the escape of gas and allow of its accumulation) a very high pressure will be thus generated, quite sufficient in some cases to lift a column of solid overlying lava or impel it upward. Investigations carried out and described by Goranson[37] indicate the possibility that pressures thus developed may reach the vicinity of 5000 atmospheres. "The data show that the pressures developed on crystallisation may not only comply with but actually exceed the pressures necessary to explain certain volcanic phenomena" (GORANSON).

A vulnerable point in the application of Morey's theory in this form to the explanation of volcanic phenomena may be the question of the existence of a gas-tight container or perilith. It is highly improbable that slow crystallisation will result in concentration of gas under pressure in any ordinary case of a body of magma close to the surface; but in the case of what has been termed an "avalanche of crystallisation" such as may occur in special circumstances (p. 56) the case may be different.

Rittmann[38] has based on Morey's theory a hypothesis to account for accumulation of enormous gas pressures such as are relieved, it has been thought, by Plinian explosions (Chapter I). In this case he pictures a long period of preparation during which crystallisation goes on in a body of basaltic magma. The process of differentiation by crystallisation produces residual acid glass rich in dissolved gases, and this glass will explode when retrograde boiling sets in. Thus, as Williams[39] remarks, "the kinetic eruptive energy of a magma depends on the extent to which it has crystallised."

Paroxysmal eruptions of Plinian intensity, which may possibly be generated by long preparation during dormancy in the manner suggested, can occur perhaps only if the magma

[37] R. W. Goranson, High Temperature and Pressure Phase-equilibrium in Albite-Water and Orthoclase-Water Systems, *Trans. Am. Geophys. Union,* 19, pp. 271-273, 1938: Silicate-Water Systems, *Am. Jour. Sci.,* 35, pp. 71-91, 1938.

[38] A. Rittmann, *Vulkane und ihre Tätigkeit,* p. 112, 1936.

[39] Howel Williams, Volcanology, *Geology 1888-1938,* pp. 365-390 (p. 382), Geol. Soc. Am., 1941.

chamber is at an unusually great depth, as Escher[40] and Sonder[41] maintain, and therefore contains magma under enormous pressure, a condition which permits it to retain much gas in solution. Some release of pressure eventually starts an eruption, which is thereafter self-propagated. "The explosions continue with increasing intensity as deeper and therefore more gas-rich layers are evacuated. In this way it is possible for great volumes of glassy pumice to be erupted."[42]

Morey's theory has been applied to some particular cases, notably to the great explosion of Bandaisan, in Japan, which occurred in 1888 after a thousand years of dormancy,[43] and also the activity which occurred in 1915 at Lassen Peak, California, described by Day and Allen.[44]

Should the viscosity of an undercooled glass or hemi-crystalline lava (in which crystallisation has been delayed by the viscosity resulting from low temperature) be reduced very considerably without change of temperature an "avalanche of crystallisation" may occur. A cause that has been invoked to account for such a lowering of viscosity in a rigid, though still hot, lava plug is absorption of water by the glass, which seems at Lassen Peak to have been at a temperature between 700° and 800°C. Morey[45] has suggested the possibility of absorption of ground water by such glass, and Day and Allen consider it very probable that the Lassen Peak lava has absorbed water vapour the origin of which has been rain or melt-water from snow. According to Morey,

> if an undercooled magma were to come in contact with percolating waters or the vapour generated thereby, . . . introduction of the water might of itself induce crystallisation in virtue of the lowered viscosity of the resulting magmatic solution, and it is conceivable

[40] B. G. Escher, On a Classification of Central Eruptions according to Gas Pressure of the Magma and Viscosity of the Lava, *Leidsche Geol. Meded.*, 6 (1), pp. 45-49, 1933.

[41] R. A. Sonder, Zur Theorie und Klassifikation der eruptiven vulkanische Vorgänge, *Geol. Runds.*, pp. 499-548, 1937.

[42] Howel Williams, Calderas and their Origin, *Univ. Cal. Publ., Bull. Dep. Geol. Sci.*, 25 (6), p. 337, 1941.

[43] H. Williams, *loc. cit.* (39).

[44] *Loc. cit.* (30).

[45] G. W. Morey, *loc. cit.* (32), p. 228.

that the result would be a sudden and violent outburst of steam and ash at a comparatively low temperature.

Liberation of latent heat when such an "avalanche of crystallisation" takes place must produce an actual rise of temperature in that portion of the lava in which an explosion is brewing, and may contribute appreciably to the result in some cases.[46]

The Origin of Magmatic Gases

Though one may dismiss without discussion the contention of Scrope that magmatic pressure, the force which causes magma to rise from the depths of the earth, "consists unquestionably in the expansive force of some elastic aeriform fluid struggling to escape,"[47] the problem of the origin of the volatile substances in solution in magma rising from a deep-seated source still remains urgent.

Some authorities have been satisfied to assume that these are of primitive, or strictly "juvenile" origin (Fourier, Tchermak, Reyer, Fisher, Suess, and the trend of modern opinion); others (Prestwich and Russell, for example) picture the introduction of water to a rising magma column already well on its way towards the surface, or (Shaler) derive it from that entrapped in marine sediments; while others again—and among these have been some of the strongest advocates of the theory that water is the cause of magmatic pressure—have attempted to explain a mechanism by means of which water may penetrate from the ocean or from the earth's surface into the hot sub-crustal rocks or magma (Humboldt, Lyell, Bischoff, Daubrée, Judd, Bonney, Arrhenius).[48]

Morey[49] touches only very lightly on this question, and the ancient controversy has not been reopened in recent years. It may be noted, however, that laboratory experimentation has thrown some light on the question of the gas-content of

[46] E. T. Allen (Chemical Aspects of Volcanism, *Jour. Franklin Inst.*, 193, pp. 29-80, see p. 6, 1922) touches on this matter. See also R. W. Goranson, Heat Capacity; Heat of Fusion, *Handbook of Physical Constants*, pp. 232-242, Geol. Soc. Am., 1942.

[47] G. Poulett-Scrope, *Volcanos*, 2nd ed., p. 30, 1862.

[48] See A. Geikie, *Textbook of Geology*, 4th ed., pp. 353-357, 1903; K. A. von Zittel, *History of Geology and Palaeontology*, p. 279, London, 1901.

[49] G. W. Morey, *loc. cit.* (32).

hypomagma, though the results do not seem conclusive and have not as yet definitely indicated the source of the volatile constituents. Goranson's work on obsidians,[50] as well as various new determinations of the gas-content of unaltered rocks, have led Adams[51] to the conclusion that "primitive, deep-seated magmas are not necessarily highly charged with volatile constituents."

On this question, opinions differ, however; and Gilluly[52] has been led by a study of reliable published analyses of rocks to the opinion that average magma contains certainly not less than 1·5 per cent and probably not less than 2 per cent of water. In the case of primary magmas he finds reasons for believing that the content may be 4 per cent in basaltic and perhaps as high as 8 per cent in granitic magma at depth, judging from their geological effects. As Allen[53] has pointed out, determinations of water in igneous-rock analyses must be fairly reliable for otherwise it would be "impossible to get an acceptable summation." Allen also has expressed the opinion that the volatile contents of a magma must have been much greater than what remains in a crystalline igneous rock derived from the magma.[54]

---

50 *Loc. cit.* (2).

51 L. H. Adams, The Volatile Constituents of Magmás, *Carnegie Inst. Wash. Year Book*, 37, p. 109, 1938.

52 J. Gilluly, The Water Content of Magmas, *Am. Jour. Sci.*, 33, p. 430, 1937.

53 E. T. Allen, *loc. cit.* (46).

54 A similar opinion has been expressed by A. L. Day, *Some Causes of Volcanic Activity*, p. 15, Franklin Inst. Philadelphia, 1924.

CHAPTER VI

Vesuvius and its Cycle of Activity

MOST EXPLOSIVE VOLCANOES HAVE ALTERNATING OR OCCASIONAL episodes of quiet effusion of lava. In many cases, however, the lavas of such volcanoes are very viscid and congeal as convex tongues on the flanks of a cone consisting in the main of pumiceous ash which has been ejected explosively; and throughout extensive regions, including the belts of most active volcanoes, the lava tongues are generally of very limited extent, so that, as has been noted in Chapter V, lava solidified as such makes up but a very small proportion of the total output of solids. In contrast with most of the volcanoes of such regions Vesuvius is rather exceptional in that, while it is very definitely at times an explosive, pumice-and-ash ejecting vent and subject to paroxysmal and voluminous escape of gas, a definite portion of its cycle of activity is devoted to the quiet emission of fluent pyromagma. This widely overspreads the flanks of the eruptive cone, solidifying as lava flows with surface forms resembling in detail those of Hawaiian volcanoes and of Etna. The resemblance of the lavas of Vesuvius to those of Etna suggests, indeed, that in the course of long ages of activity Vesuvius may pour out so much lava on its flanks that it will become a great lava dome, or shield, like Etna, through which a central funnel may be maintained, and above which a debris-built cone may rise, built by occasional explosive eruptions (pp. 78, 150).

In the intervals between major eruptions the funnel-shaped crater of Vesuvius is filled in by layer upon layer of fluent lava very much in the manner of the infilling of the Halemaumau pit at Kilauea (Chapter II) after its throat had been cleared by the phreatic explosion of 1924, though in the case of Vesuvius the pyromagma wells up from beneath the floor (Fig. 8) instead of entering through chinks in the side walls. The deep and extensive funnel blasted out of the summit of

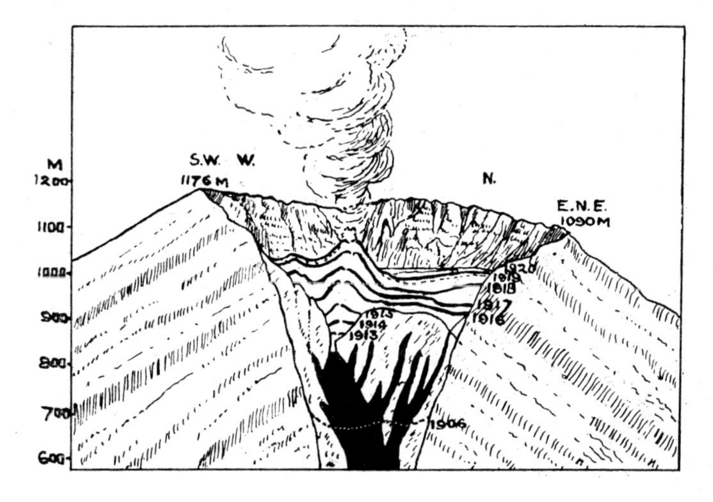

Fig. 8. True-scale section of the crater of Vesuvius showing progressive filling with solidified lava flows from 1913 to 1920. (After Malladra.)

Fig. 9. Vesuvius, 1924. The large crater, cleared out to a great depth in 1906, has been more than half-filled by successive lava flows that cover the wide floor. The active "conelet," built largely of "spatter," is emitting a thin stream of fluent pyromagma.

Vesuvius in 1906 was afterwards filled in this way[1] (Fig. 9), lava eventually overflowing through a gap in the rim. Outflow of lava took place in this way in several eruptions between 1930 and the major outbreak in 1944.

With the exception of its habit of changing to an explosive volcano (or assuming the Vulcanian type of activity) the cycle of activity of Vesuvius is, indeed, not unlike that of Kilauea. Some remarkable evidence of the occurrence of volcanic-furnace effects has been observed by Perret, who records that part of a solidified lava flow, on which it had been possible to walk, has been seen to remelt and flow; and undoubtedly reheating and remelting of pyromagna take place on a large scale in the lava column and in the crater. In the "lava phase" of activity this pyromagna accumulated at a high level escapes through a crack in the cone and flows out on the flank. Then the "gas phase" (p. 68) and the "dark or ash phase" (Fig 10) of the cycle observed by Perret[2] follow, for, when the crater has been emptied by outflow of pyromagma and pressure is relieved on the magma column, new (at first effervescing and later explosive) hypomagma has been exposed.

The new hypomagma is sufficiently viscous — possibly because of low temperature—to be subject to explosive vesiculation when the lid is removed from it or it rises into the region of pressure-release.

It is possible that the important difference in physical properties between Hawaiian and Vesuvian pyromagmas which results in the introduction of magmatic explosions and paroxysms of gas and ash emission into the Vesuvian cycle of activity is in some way a consequence of difference in chemical composition. It is perhaps related to the very highly potassic character of the Vesuvian magma.[3] It may be of some significance that the alkaline basalt of Vesuvius contains also less ferrous oxide than do most Hawaiian and plateau

[1] A. Malladra, Sul graduale riempimento del cratere del Vesuvio, *Atti VIII Cong. Geog. Ital.*, 2, pp. 67-75, 1923.

[2] F. A. Perret, *The Vesuvius Eruption of 1906*, p. 27, Carnegie Inst., Washington, 1924.

[3] For chemical composition of Vesuvian lavas see H. S. Washington, in F. A. Perret, *loc. cit.* (2), p. 145.

Fig. 10. Continuous emission of ash-charged gas during the dark or ash phase of eruption of Vesuvius in 1906 (April 16th). The ash cloud is "soaring upward in huge spiral volutes" (PERRET) because of some obstruction in the throat of the volcano.

basalts (Chapter V), though it is much richer in this possible "flux" than are explosive lavas of andesitic composition. It thus appears that the "basaltic" character of Vesuvian magma cannot be accepted unreservedly as evidence that the characteristics of a volcano—whether it is to be placed in the category of lava volcanoes or pumice volcanoes—are determined by causes quite apart from possible chemical control of the viscosity of the magma.

The lava of Etna also is a potash-rich basalt, and the activity of Etna is not very unlike that of Vesuvius (p. 78). The soda-rich basalt of the active Fogo, in the Cape Verde Islands, has produced a volcano almost as large as Etna which presents numerous points of resemblance to both Etna and Vesuvius[4] (Fig. 11).

Fig. 11. The active volcano Fogo, Cape Verde Islands. (After Stübel, redrawn.)

Recent examination of the lapilli (lava pellets) ejected by the eruption of Tarawera (New Zealand) in 1886[5] indicates that it was a basaltic magma that rose there in a fissure and erupted with the production of a great quantity of exploded fragmentary material, which is mainly fine. This differs but little in composition from the average of Hawaiian and plateau basalts except in its lower content of ferrous oxide. Unless this eruption is to be interpreted as basaltic "fire-fountaining" on a truly gigantic scale, the highly explosive nature of the magma as made manifest by the phenomena of the eruption points a warning, therefore, against too confident prediction of the eruptive properties of magma from chemical compo-

[4] F. von Wolff, *Der Vulkanismus*, II (2), pp. 1050-1052, 1931.

[5] L. I. Grange, The Geology of the Rotorua-Taupo Subdivision, *N.Z. Geol. Surv. Bull.*, 37, p. 79, 1937.

sition. It gives some support to the opinion of Perret[6] that what he calls the classic distinction between basaltic "effusive" lava and non-basaltic "explosive," or potentially explosive, lavas cannot be rigidly insisted on. "This distinction", he writes, "can be maintained only with great reserve, inasmuch as many of the more acid volcanic cones are built up of lava outpourings as well as of fragmental deposits, while in the basaltic Vesuvius and other volcanoes of similar nature the most violent explosive phenomena often accompany the molten flows."[7]

It may be noted further with regard to the Tarawera magma mentioned above that, though only fragmentary material was ejected, thoroughly molten pyromagma rose in the open rift. This is shown by the fact that many bombs ejected from the rift have cores of old rhyolite thinly coated with basalt (Fig. 53, 1). These closely resemble bombs of the same kind ejected from Vesuvius.[8] The basaltic nature of the Tarawera magma might be thought more destructive to the theory of the non-explosive character of basalt than the example of Vesuvius, in the case of which the peculiar properties of the magma concerned may possibly be related to its alkaline composition. Even in the case of the Tarawera eruption, however, the phenomena, though highly explosive, differed from typical explosive eruptions of andesitic, dacitic, and rhyolitic lavas. They seem, indeed, to have been closely similar to phenomena accompanying some great eruptions in the basalt field of Iceland. The Laki fissure eruption of 1783, for example, produced not only lava but also a great blast of lapilli (p. 188).

[6] *Loc. cit.* (2), p. 78.

[7] It is clear from statements made and opinions expressed from time to time by Jaggar that he is another volcanologist who regards chemical composition of magma as of little significance in determining the type of eruption to which the magma will give rise. Thus: "If we grant that engulfment and the admission of ground water is what leads to explosion, and that on the other hand the rise of primitive magma and exclusion of ground water is what leads to lava flow, then it would appear that only the latter magmatic happening can be construed as a measure of pure volcanicity. Any magmatic vent in the earth-crust may enter into an explosive phase. . . . On the other hand it is quite possible that some volcanic regions [i.e. those in which the volcanoes are explosive pumice-makers] are characterised by intrusive magma in just as great volume as anything poured out in Iceland or Hawaii." (T.A.J., *The Volcano Letter,* 355, p. 2, 1931.)

[8] Compare F. A. Perret, *loc. cit.* (2), Fig. 51, p. 85.

There still seems, therefore, to be no reason for doubting the value of the generalisation advanced by Russell in regard to the relation of composition of magma to type of activity, which is as follows: "It is variation in the composition of lavas and in the amount of water contained in them, rather than variation in temperature which controls the marked contrasts observed in their fluidity."[9] Variation in fluidity, in its turn, controls the type of eruption. The hypothesis that chemical composition influences the general nature of volcanic phenomena far more than any other single property cannot be lightly discarded; for a survey of the volcanoes of the world shows that the type of activity that will be associated with the rise of a particular kind of magma may be predicted with a fair degree of confidence.

Perret's Theory of Vesuvian "Mixed" Activity

Perret, who doubts the value—at least in its application to Vesuvius—of the hypothesis that chemical composition of magma exerts an important control over the volcanic phenomena, favours instead the theory that "mixed" phenomena—alternation of effusive and explosive phases of eruption—result from various degrees of closure of the vent. This theory is applied in particular to the case of Vesuvius, but it may be of great value with a more general application.

Perret's views with regard to eruption and the maintenance of volcanic activity agree closely with those of Jaggar, as outlined in earlier chapters, in that he affirms that the "uprise of magma" from the depths is continuous, but takes place "with extreme slowness"; "gas is the active agent, and the magma is its vehicle"; heating or reheating of lava takes place prior to or during eruptions; and there is a significant cyclical succession of volcanic phenomena between major eruptions.

As regards the first and fourth of these points, it must be noted that there is a remarkable similarity in the very gradual cyclic filling of the craters of Hawaiian volcanoes (Chapter IV) and of Vesuvius with solidified lava flows after

[9] I. C. Russell, *Volcanoes of North America*, p. 57, 1897.

c

they have been cleared out by explosions (Figs. 6 and 8). Not until the deep holes thus excavated have been filled up in this way with lavas does external outflow of pyromagma become possible.

The Vesuvian Cycle

The cycle of activity as exhibited by Vesuvius is described thus by Perret[10]:

> The filling of the crater with magma from the depths is a slow process. There results an accumulation of liquid material within and beneath the volcanic edifice, with gradually increasing gas-content and tension. The release and discharge of all or part of this accumulation constitute eruption and with consequent exhaustion give to the cycle of manifestation its quality of periodicity.

Perret agrees with Jaggar (Chapter IV) that liquid lava (pyromagma) is cooked up or generated "in the upper part of the column," while below "the magma is a paste" (hypomagma). According to Perret, however, this paste "permits the transfusion of gas," for he considers "the actual upward movement of lava" to be "too slow to permit of its being the carrier of all the gas which is emitted." He thus pictures an accumulation in the volcanic throat of gas derived from a greater depth.

Perret's statement that "the main function of the magmatic reservoir is the evolution and accumulation of gas" may be taken to apply to potentially explosive volcanoes only, and is perhaps too dogmatic even when thus restricted. It is too suggestive of adherence to a primitive theory of a volcano blowing up like a bursting steam boiler. The magma in a conduit or "hearth" may be thought of rather as accumulating gas *in solution,* which will be released when pressure is relieved (Chapter V). This is the case of the "closed conduit," a condition preceding the Vesuvius eruptions of A.D. 79 and 1631. Perret remarks: "As it is also normal for most volcanoes with acid lavas, the type of eruption resulting therefrom has become classic and the one whose mechanism is most generally recognised."[11]

[10] *Loc. cit.* (2), p. 58.
[11] *Loc. cit.* (2), p. 59.

More instructive is Perret's development of a theory of explosion in an open conduit. At the top of the column is effervescing pyromagma

with occasional irruption of great high-tension gas-bubbles producing mild explosive effect. . . .

But if this freedom of gas emission prevails throughout the upper layers, such is not the case below, where pressure of the superincumbent material progressively increases with depth until a condition is reached which, as regards the retention and accumulation of gas, is *almost equivalent to a closed conduit.* In other words, the open-chimney condition produces in time a potentially explosive magma in the *lower* portions of the conduit, which will be last in coming into external manifestation during the course of an eruption.

We have thus a sort of standpipe whose contents may be considered as divided into a number of zones with downwardly increasing gas-accumulation, each subjected to a pressure that imposes a state of equilibrium and renders its explosiveness latent and proportionate to its depth.

Under such conditions the magmatic column, although occupying an open conduit, is as a whole potentially explosive, and all that is required to cause a great eruption is the intervention of some factor that is capable of disturbing the existing state of equilibrium.

Such a disturbance might be brought about in various ways, but inasmuch as the volcanic standpipe has been built up to a considerable height above the surrounding surface and within comparatively fragile containing-walls, the simplest contributing cause would be such a fracturing of the volcanic edifice as would result in lateral outflow and drainage sufficiently rapid and copious to reduce materially the level of the liquid in the tube. This, by relieving the pressure upon subjacent zones, will result in reactive gas-release, with expansional rise of material to a higher level, which in turn extends downward the condition of relief, with corresponding reaction in zones of greater pressure and power of gas-concentration. This progressive process of expansion, rise, and expulsion [Fig. 10] will continue—with concomitant destruction of the now unsupported terminal cone structure—until the paroxysmal liberation of gas from the surcharged lower layers culminates in the outbreak and exhausts the accumulation of energy and material,[12] so that there supervenes the cyclical interval of repose and renewal.[13]

By way of illustration the actual succession of phenomena during the Vesuvius eruption of 1906 may be cited:

By some . . . means fissuring occurred on the morning of April 4 in the southern wall of the cone and at a high altitude (1200 metres) with at first a sluggish subterminal outflow. This was followed by downward extension of the fissure and a more rapid effusion, which effected the first true lowering of the lava-level. . . . There followed progressive lowering and augmentation of the lava outpourings, with corresponding reactive explosive manifestations, illustrating in the most convincing manner the mode of action here set forth.[14]

Perret's picture of the volcanic pipe standing full of liquid, reheated lava ready for eruption forcibly suggests comparison with a great geyser ready to erupt on the introduction of a bar of soap or *the removal of some of the cooler water from the surface.*

Rittmann's Cycle

In the opinion of Perret,[15] whose views agree closely with those developed by Jaggar in the course of his Hawaiian studies, an eruption exhausts the magma available in the throat or feeding channels of the volcano, and an interval of rest which follows the eruption is occupied by a very gradual upwelling of viscous hypomagma, after which a new cycle of activity is in full swing. According to some volcanologists, however, the stage is set for an eruption not so much by this gradual accumulation of lava in the crater and the remelting and preparation there of liquid magma in readiness for outflow in the manner that has been described but rather by a gradual development of a gas-saturated condition in magma in the upper part either of the magma column or of an underlying magma chamber or volcanic hearth. (Perret[16] has come to

[12] At Vesuvius in 1906 this took two weeks, comprising the "gas phase" and the "dark or ash phase" of the eruption.

[13] F. A. Perret, *loc. cit.* (2), pp. 60-61.

[14] *Loc. cit.* (2), p. 61.

[15] *Loc. cit.* (2), p. 59.

[16] "Lava upheld by the compressed gas constitutes the crater filling" (F. A. Perret, The Volcano-Seismic Crisis at Montserrat, p. 75, Carnegie Inst., Washington, 1939).

believe that there may even be a large pocket of separated gas, but this does not seem essential, or indeed probable.) Rittmann[17] has shown deductively that when such a gas-saturated condition has been arrived at any drop in the lava level, which may result either from outflow or from withdrawal of magma into some newly opened cavity, will reduce hydrostatic pressure below the vapour pressure in the upper part of the column. This will cause paroxysmal effervescence in magma now oversaturated with gas, and an explosive eruption will supervene and continue until the oversaturated magma is exhausted and its place is taken (according to Rittmann's postulate) by new magma less rich in gas. According to this hypothesis the interval of inactivity in the cycle, during which incubation of the next eruption is in progress, is not merely the period of preparation and accumulation of pyromagma but includes the time required for development of the gas-saturated condition.

The Eruption of 1944

The period of incubation of the great eruption of Vesuvius studied by Perret was from 1872 to 1906. Thirty-eight years later (in March, 1944) another cycle culminated in voluminous outflow of lava followed by an explosive outbreak. The phenomena of this eruption were similar to those of 1872, when twenty million cubic metres of lava poured down the north flank of the mountain.

---

[17] A. Rittmann, *Vulkane und ihre Tätigkeit,* pp. 135-139, Figs. 23, 24, 1936.

PART II

VOLCANIC LANDSCAPES

CHAPTER VII

Domes and Cones of Basaltic Lava

VOLCANIC ACTIVITY, CONSIDERED AS A GEOMORPHIC AGENCY, interferes with the course of the normal cycle of erosion in various ways. Outflow of lava even in limited quantities may make dams which divert rivers and impound lakes, thus putting in operation processes of degradation and aggradation controlled by new local base-levels; and a major volcanic outbreak may either destroy parts of a prevolcanic landscape by explosion or engulfment or may bury it beneath widespread sheets and piles of lava flows or showers of volcanic debris.

Such events as these are volcanic "accidents" — in the nomenclature used by W. M. Davis—in so far as they interfere with the orderly succession of landscape changes normally in progress, i.e. with the continuity of a geomorphic cycle which may be brought to a premature close by their agency.

Volcanic interference with the progress of the geomorphic cycle may be only local, affecting isolated districts dotted island-like over a region, though even in such a case the effects of aggradation due to increased loads of debris in rivers may be felt over a wide belt. In some regions, however, the accumulation of materials of volcanic origin has taken place on such a grand scale that it has furnished very extensive initial landscape surfaces made up of constructional forms, on which an entirely new cycle is inaugurated when rivers flow and erosion and redistribution of this material begin. Such new groups of landforms may rise out of the sea or may bury former lands.

Quite commonly the new constructional surface consists for the most part of a few large volcanic mountains—with which may be associated many mounds—these being built of lava and of the scoria and spattered debris derived from lava, but there may be also extensive low-lying fields of newly solidified lava associated with these.

Volcanic Mountains

Two distinct types of material are built into volcanic mounds and mountains, but one of these generally predominates over the other. They are solidified lavas of "effusive" (flow) origin and explosively ejected debris. Practically all large constructional volcanic forms are "composite" in the sense in which that term is usually applied to volcanoes. They are built, that is to say, of both effusive lavas and layers of fragments ejected as projectiles, these being stratified and interbedded; but the two contrasting kinds of constructional material are present in widely different proportions, the one or the other dwindling to little more than a trace in the extreme types, which are of rather common occurrence. These are lava-built domes on the one hand and "ash" cones on the other.

To the slopes of a cone built by the accumulation of heaped fragmentary debris about a centre of emission lava may be added in small quantities from time to time to solidify as tongues on the surface without modifying to any great extent the smooth conical slope, which is controlled by the angle of repose of the predominant grade of fragmentary

Fig. 12. Profiles of two great lava domes on the island of Hawaii (Mauna Kea at left; Mauna Loa at right) as seen from the flank of the Hualalai dome. The summits of Loa and Kea are both more than 13,000 feet above the sea. A broadly rounded ridge slightly mars the symmetry of the elongated dome of Mauna Loa. This is situated over a feeding fissure, or rift zone, the line of which is continued in a nearer row of scoria mounds of recent growth. (From a photograph by Emerson.)

debris continuously supplied at the summit (Chapter XIII). Large lava-built mountains, on the other hand, and those in which there is a great predominance of lava over fragmentary debris, assume forms controlled by the accumulation of lava. They are commonly domes[1] rather than cones (Figs. 12, 24).

Home Studios, Takapuna, photo.

Fig. 13. The Rangitoto Island basalt cone, Auckland, New Zealand, showing the concave profile on the flanks. At the summit a large scoria mound has been built within a broad crater with an irregular convex rim.

Basalt Cones

Though the largest and most perfect of lava-built mountains are domes, yet some of the smaller mountains that have resulted from accumulation of free-flowing basalt are conical, with long concave lower flank profiles. Of these Rangitoto, New Zealand, which has been described by Hochstetter,[2] may be taken as an example (Figs. 13, 14). Though slightly irregular in form

[1] The term "dome" has been long in use for the description of Hawaiian lava-built mountains (see, for example, A. Geikie, *Textbook of Geology,* 4th ed., Fig. 57, p. 328, 1903). As the type of a "lava dome" Mauna Loa, Hawaii, is cited by R. A. Daly (*Igneous Rocks and their Origin,* p. 135, 1914). See also S. Powers, Volcanic Domes in the Pacific, *Am. Jour. Sci.,* 42, pp. 266-267, 1916. In German nomenclature these are "shield volcanoes" as distinguished from the conical composite, stratified "stratovolcanoes."

[2] F. von Hochstetter, Geologie von Neuseeland, *Reise . . . Novara,* Geol. Teil, I Bd., p. 168, 1864.

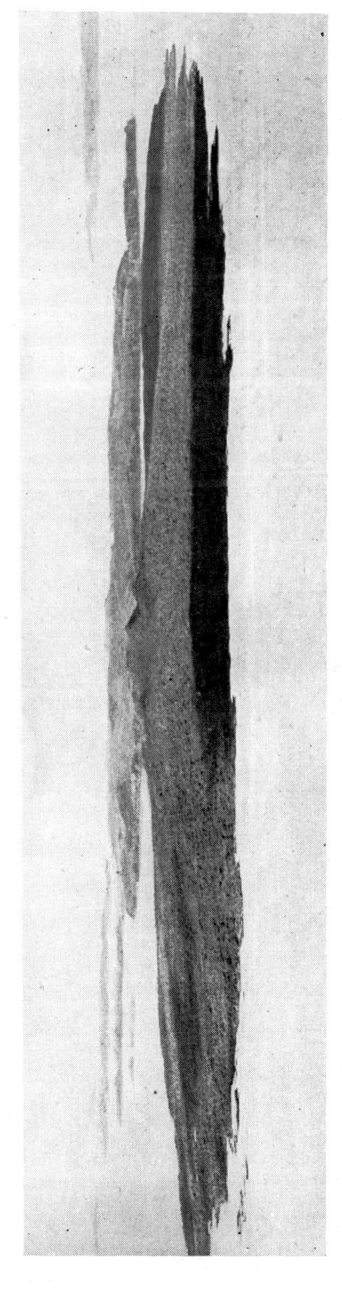

V. C. Browne, photo

Fig. 14. Rangitoto Island basalt cone (compare Fig. 13) from the air. This view shows an extensive low-angle basalt-lava fringe, above which the slope steepens.

near the apex Rangitoto is sufficiently symmetrical to present an appearance almost identical in profile from every point of view. The lava cone is surmounted by nested or concentric ash and scoria cones. The inner of these is an ordinary scoria mound (Chapter X). The outer, and older, may be built of more widely scattered debris. Beyond this somewhat uneven convex ring the profile closely resembles that of the Skjaldbreit, or "dyngja," type of cone in Iceland (p. 88). It is possible that the lavas of an outer flat-lying fringe have

Fig. 15. The cone of Batur, Bali, viewed across a lake in a caldera which surrounds it. (Drawn from a photograph by E. G. Zies.)

N. H. Taylor, photo

Fig. 16. Whatitiri basalt "cone," New Zealand. Relief, 850 feet above the surrounding lava plain.

flowed not from the summit but from fissures midway down the slopes,[3] comparable to those on the basalt island Niuafo'ou (Fig. 7).

The larger, still active cone of Batur (Batoer) volcano, Bali, East Indies, is also described as built of basalt flows, which however are only "fairly" mobile and may be considered to be of andesitic habit. Batur has more the appearance of a cone built partly of ash and has almost the steepness characteristic of an andesitic mountain (Fig. 15).

Some lava cones which have piled themselves up to very moderate heights over the sources of outflow of great sheet flows of basalt may well be described as embryonic domes. A craterless mound of this kind is Whatitiri, in the North Auckland district of New Zealand (Fig. 16). An example of such a basalt cone of small dimensions, which is of more symmetrical form, may be taken from the San Franciscan basaltic volcanic field, Arizona (Fig. 17). This cone buries a

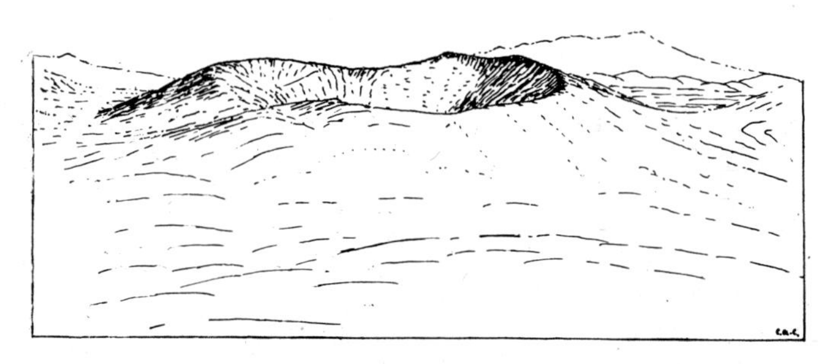

Fig. 17. A basalt cone with an enlarged crater, San Franciscan volcanic field, Arizona. (Drawn from a photograph.)

mound of scoria,[4] which had been built, no doubt, by fire-fountaining (Chapter X) prior to the emission of flowing lava. A final explosion, apparently, has cleaned out the crater.

Central and Fissure Vents

It is essential for the building of either a cone or a symmetrical dome of lava that the material for it shall be

[3] P. Marshall, *Geology of New Zealand,* p. 90, 1912.

[4] H. H. Robinson, The San Franciscan Volcanic Field, Arizona, *U.S. Geol. Surv. Prof. Paper,* 76, p. 88, 1913.

ejected or poured out from a central point or small area. Thus all such forms, whether they consist of fragmentary materials or of lava, must be the products of "central" eruptions as distinguished from those in which lava has welled out unrestricted from elongated fissures (Chapter VIII). Alignment of central vents in some cases—and especially alignment of numerous small mound-building vents—indicates that they are related to fissures in the underlying rocks; and some volcanic mountains have obviously (as indicated by their form) been built over moderately elongated fissures or systems of fissures which have extended to the surface. Where the vents that emit lava are opened from time to time at different points on such a rift zone a dome built over it is elongated instead of being radially symmetrical. Most Hawaiian basalt domes have some such elongation, and they thus tend to be elliptical or oval domed ridges (Fig. 24). In the case of the Haleakala dome (Maui Island) three such broad ridges lead up to the summit (Fig. 214). The underlying fissures (rift zones) are well marked on the Haleakala dome also by the crowding of numerous scoria mounds or "adventive" cones (Chapter X), which mark points of lava outflow along the radial ridges.[5] Other examples of elongated domes are the two

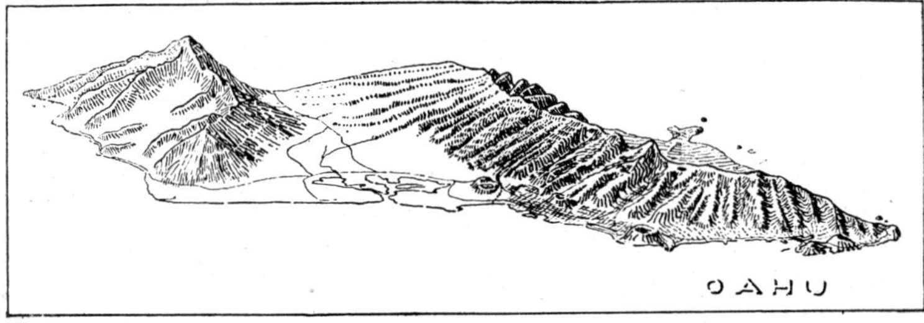

Fig. 17A. Diagram of the Hawaiian island Oahu, viewed from the south and drawn with the vertical scale somewhat exaggerated. The island consists in the main of two elongated basaltic domes, which have been dissected and partly destroyed by erosion. The western of these (left) forms the Waianae Range and the eastern the Koolau Range. (After C. K. Wentworth.)

main volcanoes, now both deeply dissected to form mountain ranges, which make up the island of Oahu (Hawaiian Islands) (Fig. 17A).

[5] H. T. Stearns, Origin of Haleakala Crater, *Bull. Geol. Soc. Am.*, 53, pp. 1-14, 1942; Geology and Ground-water Resources of Maui, *T.H. Div. Hydrog. Bull*, 7, 1942.

The intersection of major rift zones or fissure systems may be a main cause of the localisation of activity in a more or less central position which controls dome-building,[6] just as it is assumed to be in the case of more explosive cone-building volcanoes by many volcanologists (Chapter XIII). Emission takes place, or is localised for a period, however, at what Stearns terms "the most active place on the rift."

Flank Outflow of Lava

In the case of the great active dome Mauna Loa it has been suggested that at earlier stages of growth of the mountain, when the dome was lower, voluminous floods of lava poured from a central orifice on the summit. However this may be, outflow now takes place in greatest volume from radial fissures

Fig. 18. The central or summit cone on the basaltic dome of Etna. This contrasts in its large dimensions with one of the very numerous "adventive" scoria-built mounds on the dome. (From a photograph.)

which open from time to time on the mountain-sides; and some have thought that such outbreak of lava is a result of the great height to which the dome has grown and the consequently great hydrostatic pressure in the lava column at the level of these flank outflows. Lava still rises to the summit,

[6] H. T. Stearns, Geology and Water Resources of the Kau District, Hawaii, *U.S. Geol. Surv. W-S. Paper,* 616, p. 319, 1930.

however, and escapes from chinks there (Chapter IV), flooding the floor of the extensive "sink" on the summit.[6a]

Lateral escape of lava through radial fissures is a feature also of the activity of Etna (Sicily), which is a basalt dome similar in many respects to Mauna Loa though surmounted by a debris-built cone of considerable size (Figs. 18, 59). This cone may be a feature peculiar to the overgrown condition —and late phase of activity—of Etna, and be due to uprise of lava more viscous than that of former times; but it seems at least equally probable that the platform of lava flows of which the bulk of Etna is composed has been continuously erected by flank outflows around a core of fragmentary ejected material which is surmounted by the visible cone of similar construction. Whether or not Etna had this somewhat Vesuvius-like history, the great dome of lava is very similar to those of Hawaiian volcanoes. The edges of very numerous and very gently inclined lava flows which have built up the plateau around the central cone are visible in cliffs 3000 to 4000 feet high at the rear of a vast amphitheatre, Val del Bove,[7] east of the summit — a lateral "caldera" or "sector graben" (Chapter XVI)—from which a great bite has been taken out of the lava fields by some means so as to expose their structure.

The radial fissures from which lava flows issue on the flanks of both the Mauna Loa and Etna domes extend progressively outward—i.e. down the slope of the mountain—so that successive phases of the eruption from a fissure occur at successively lower levels on the mountain-side (Figs. 19, 20).

If one accepts the theory that there has been a change over from summit overflow (of "superfluent" lavas, as Dana[8] termed them) to flank outflow ("effluent" lavas of Dana) during the growth of a dome from small to large dimensions, then the stage at which the dome is no longer built higher but grows in girth only is one of maturity, bordering on senility, in the growth cycle. In the Hawaiian Islands no domes are known in the early stage of vigorous upgrowth

[6a] Since 1832 eruptions at the summit have gone on for 3 per cent of the time, and on the flanks for 3·3 per cent (G. A. Macdonald, *Am. Jour. Sci.*, 241, p. 242, 1943).

[7] Illustration in K. Sapper, *Vulkankunde*, Pl. 24, 1927.

[8] J. D. Dana, *Am. Jour. Sci.*, 35, p. 228, 1888; *Characteristics of Volcanoes*, p. 2, 1891.

with floods of lava pouring from the summits over the upper slopes; but if the Matavanu volcano (Savaii, Samoa) is an independent vent it furnishes an example. If such superfluent emission has been the rule in an early stage of growth, it is not necessary, however, to postulate for it the existence in all cases of a wide-open pipe-like vent or crater similar to that of explosive volcanoes like Vesuvius. Sinks like

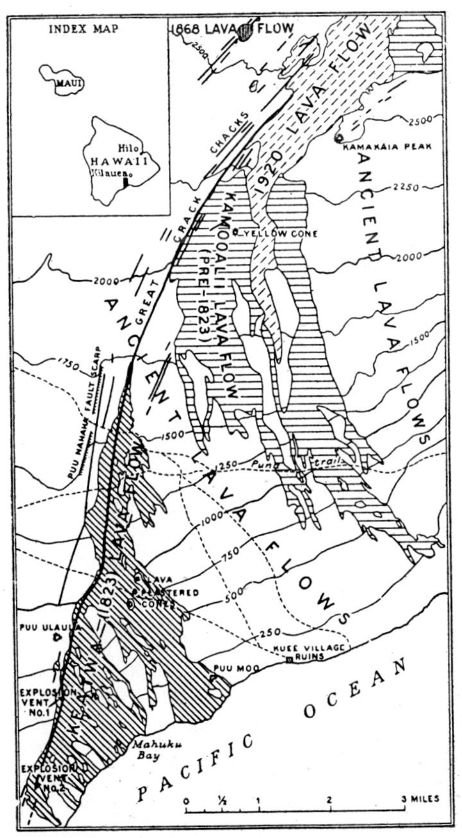

Fig. 19. The "Great Crack" on the flank of Mauna Loa; the extent of the lava flow of 1823 from it; and earlier flows. (After Stearns, *Jour. Geol.*)

those on the summits of Mauna Loa and Kilauea may have been opened by subsidence from time to time and intermittently filled in with lava during the superfluent stage as well

Fig. 20. Adventive cone, scoria mounds, and spatter cones built by fire-fountaining during the escape of voluminous lava flows from two radial fissures on the flank of Mauna Loa, Hawaii, the sources of the lavas of 1919 and 1926. (Drawn from a photograph.)

as later. Thus Jaggar[9] remarks of the Mauna Kea (Hawaii) dome: "The growth must have been pericentric about a sink now buried."

Superfluent Vents

Little is known, of course, of the actual summit vents from which superfluent emission of lava took place during the early stages of up-building of domes now fully grown; they must have been of ever-changing form, however, and not even constant in position.

Hualalai [an extinct or dormant dome on the island of Hawaii] preserves a small crater ring, a circular lava basin enclosed by a rampart of its own building. These are rarely solidified in Hawaii, but are genetically of great importance, for they represent profuse lava flooding such as must have taken place when the volcanoes were superfluent—overflowing from their central craters. . . . It is an annular ridge, thirty-seven metres in diameter, enclosing a floor

[9] T. A. Jaggar, Seismometric Investigation of the Hawaiian Lava Column, *Bull. Seism. Soc. Am.*, 10, p. 187, 1920.

two metres deep and rising on the outside from four to five metres. . . . These ring ramparts are rarely preserved because of the habit of engulfment. . . . If small pools so obliterate the traces of their rims of overflow, the large craters must necessarily do so by the same process.[10]

The broadest, or lowest-angle, lava domes known are some mounds which are apparently of purely superfluent origin and centred over the sources of voluminous basalt flows which have fairly recently (though not in historic times) built up the surface of the Snake River plains in Idaho (Chapter VIII); these have marginal slopes of generally less than 100 feet per mile. It is uncertain whether the higher and much more convex (and certainly more complex) domes of the Hawaiian type have grown up over such infantile forms or have been in their infancy cones like that of Rangitoto (p. 72) or of Skjaldbreit in Iceland (p. 88).

Basalt-dome Structure

All high basalt domes are composed of very numerous thin flows of lava. Once the dome form has been assumed it may be that convexity of profile can be maintained by accretion of tongues and sheets of lava which flow from vents at or near the summit, provided that these streams are sufficiently voluminous and flow far across the nearly level summit before losing their initial free-flowing liquidity. If they have become more viscous and sluggish by the time they reach the steeper slopes on the periphery they may congeal somewhat more thickly there, thus building the dome outward more rapidly than upward.

Alternatively it may be supposed that a growing cone becomes a dome as flank outflows congeal on it. Superfluent lavas may later build up the broadened summit, solidifying on it as nearly horizontal sheets concurrently with further broadening of the base by accumulation of effluent lavas on the flanks.

Lack of definite knowledge of the cause of the dome form is demonstrated by the following quotation from Zies[11]:

[10] T. A. Jaggar, *loc. cit.* (9), pp. 188-189.

[11] E. G. Zies, Surface Manifestations of Volcanic Activity, *Trans. Am. Geophys. Union,* 19, pp. 10-23, 1938.

The slopes of the basaltic volcanoes could maintain a very low gradient if the lava would always come from the central orifice, but flank eruptions often alternate with eruptions from the crater and therefore pile up material below the upper part of the cone and alter what might be called the normal gradient.

The view that a low gradient would be maintained if lava came always from the summit is diametrically opposed to that expressed by Dana[12], which is as follows:

> Since a cone diminishes in diameter upward, a flow of lava from the summit region, having like width throughout, would cover a much larger part of the circumference in the upper part than in the lower. The part of the cone below would require in fact a great number of ordinary lava streams to make one coat over the surface. The consequence of this condition is that such discharges add to the height and make the cone steeper above, and give it also a concave outline.

In Dana's opinion, "the flattened summit which gives so vast bulk" to a basalt dome is attributable to "discharges over the upper slopes and not over the summit." These

> tend to widen the upper part and flatten the summit so as to produce a convexity in the profile. . . . [At Mauna Loa] the outflows of the century have had the distribution required to produce the actual form. Part are basal; another large part start just below the summit, and none of much size from the vicinity of the summit crater.[13]

The upper slopes of Mauna Loa, and also the dips of the lava beds near the summits of domes that have been dissected so as to exhibit the internal structure (Fig. 23), are of the order of 250 to 350 feet per mile, and both the surface slopes and the dips of lavas increase to 1000 feet per mile (11°) midway down the flanks.[14] Similar in a general way in surface

[12] J. D. Dana, *Manual of Geology,* 4th ed., p. 275, 1894.

[13] J. D. Dana, History of the Changes in the Mt Loa Craters: Part 2, *Am. Jour. Sci.,* 36, pp. 81-112 (p. 111), 1888.

[14] H. T. Stearns, Geology and Ground-water Resources of the Island of Oahu, *T. H. Div. Hydrog. Bull.,* 1, p. 13, 1935.

Structural studies made by Stearns in the dissected domes of Oahu (see also Geology and Ground-water Resources of Lanai and Kahoolawe, Hawaii, *T.H. Div. Hydrog. Bull.,* 6, 1940) show that the nearly horizontal lavas on dome summits have accumulated for the most part within central "sinks"—not true "craters" (Chapter XVI)—the development of which is in part responsible for the conversion of cones into domes.

form to the domes of the Hawaiian Islands are those of other regions—Nyamlagira and Nira Gongo in Central Africa, for example—and it is a fair inference that their internal structures are similar. In all cases the lava flows composing domes are thin[15] and extremely numerous. Few of the flows are of great lateral extent, though some are very long.

A fallacious and misleading diagram which has found its way into a popular textbook of geology[16] represents the structure of a basalt dome as a pile of horizontal pancake-like lava flows of progressively dwindling diameter. The only field observer who has, even temporarily, entertained this hypothesis of dome structure is Daly,[17] who has based upon it some conclusions as to the order of succession of lavas on Mauna Kea, Hawaii. Cross,[18] however, has firmly rejected the hypothesis as applied to this particular dome.

Basalt domes built over a land surface are flanked by long slopes, broadly concave in profile, which are the surfaces of gently inclined lava flows. Thus they exhibit to some extent the concavity of flank profile characteristic of all volcanic piles of "central" construction.

The Domes of the Island of Hawaii

The Mauna Loa volcano has often been described as a vast pile built up to a height of nearly 30,000 feet from the floor of the Pacific Ocean and comprising the whole, or greater part, of the island of Hawaii and its foundations.[19] Jaggar has described Mauna Loa, however, as merely the youngest member of a complex of domes—or, as some would call them, shield volcanoes—which together have built the island. Clustered domes (Fig. 12) crown a complex pile, of structure unknown below sea-level, which has grown up in the course of long ages from the ocean floor. The base of the Mauna

[15] Average thickness 15 feet in a typical area (H. T. Stearns, *loc. cit.* (6), p. 65).

[16] Longwell, Knopf, and Flint, *Textbook of Geology,* I, 2nd ed., Fig. 188, p. 280, 1939.

[17] R. A. Daly, The Nature of Volcanic Action, *Proc. Am. Ac. Arts and Sci.,* 47, pp. 47-122 (see p. 103 and Fig. 11), 1911.

[18] Whitman Cross, Lavas of Hawaii and their Relations, *U.S. Geol. Surv. Prof. Paper,* 88, pp. 37-38, 89, 1915.

[19] Though adopted in *Volcanoes,* p. 103, 1931, this view has since been rejected by G. W. Tyrrell (Flood Basalts and Fissue Eruptions, *Bull. Volcanologique,* 1, pp. 89-111, 1937) following the publication of the work of H. T. Stearns, *loc. cit.* (6).

Loa dome itself is, however, two or three thousand feet above sea-level. "Both Kilauea and Mauna Loa appear to be over an older topography."[20]

Kilauea, taken by some volcanologists to be an adventitious outgrowth on the flank of Mauna Loa[21] which has become a "parasitic" centre of activity, is now generally regarded as an independent dome centred over a distinct rift zone or swarm of fissures in the underlying structures. Though 9000 feet lower Kilauea is thus, equally with Mauna Loa, a member of the Hawaii-Island complex of interfingering and imbricating domes. According to some Kilauea is older,[22] though according to others it is younger,[23] than Mauna Loa. The lower dome is at present, however, in some danger of burial beneath lavas which flow occasionally from fissures on the flank of its high neighbour, for Mauna Loa is growing more vigorously than is Kilauea, as was pointed out long ago by Dutton.[24]

Fig. 21. Mauna Kea dome, Hawaii, as seen from Hilo. Height, 13,825 feet. Adventive cones, or scoria mounds, on the summit and flanks together with a certain amount of dissection on the windward (north-east) side, somewhat mar the symmetry of the dome.

The three other basalt domes which, with Mauna Loa and Kilauea, make up the superficial part of the island are Mauna Kea, Kohala, and Hualalai (Fig. 24). These are extinct, with the possible exception of the last. Mauna Kea is by far the largest (Figs. 21, 22), being actually higher than Mauna Loa. Its height is augmented, however, by a debris-built cone on its summit. This cone (Fig. 22), which is of unusually large size as compared with most of the adventive mounds on lava

[20] T. A. Jaggar, *loc. cit.* (9), p. 194; *The Volcano Letter,* 344, p. 1, 1931.

[21] R. A. Daly, *loc. cit.* (17), pp. 109-117.

[22] T. A. Jaggar, *loc. cit.* (9), p. 196; *The Volcano Letter,* 344, 1931; H. T. Stearns, *loc. cit.* (6), leaves the question open.

[23] J. B. Stone, The Products and Structure of Kilauea, *Bishop Mus. Bull.,* 33, 1926.

[24] C. E. Dutton, Hawaiian Volcanoes, *U.S. Geol. Surv. 4th Ann. Rep.,* pp. 75-219, 1884.

Fig. 22. Mauna Kea dome, Hawaii, as seen from Kilauea. Adventive, debris-built cones and mounds are seen to attain considerable dimensions on the summit and flank. (Drawn from a photograph.)

domes, gives the mountain a decided resemblance to Etna (Figs. 18, 59) when seen from some points of view. This pile may, however, be merely the product of an expiring burst of explosive activity—the last phase (p. 148).

Another great dome similar to those on Hawaii Island rises to a height of 10,000 feet as Haleakala, forming the eastern part of the adjacent island of Maui. It retains the smoothness of the constructive-dome form on the dry leeward side, though long ago growth as a dome has ceased and the mountain is considerably dissected in other parts (Chapter XVIII). On the sides of deep valleys cut in the dome by erosion, and especially around a great erosion caldera excavated across its centre,

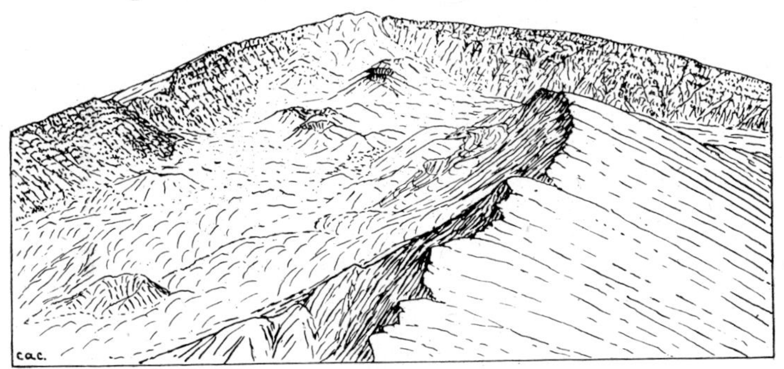

Fig. 23. Profile and surface of the Haleakala dome, Island of Maui, showing lava-flow outcrops on the walls of the erosion caldera in the summit of the dome. Eroded valleys have been partly filled in by the lavas and fragmentary products of a volcanic episode of late date. (Drawn from a photograph.)

may be seen the exposed edges of the very numerous basalt flows of which the dome has been built, all bedded parallel to the dome surface (Fig. 23) in the same manner as those exposed on the wall of Val del Bove, Etna.

Fault Scarps on Hawaiian Domes

Not only are many of the domes which form the Hawaiian Islands much dissected (Chapter XVIII), but a number of them have lost something of their symmetry as a result of flank subsidences along faults.[25]

Dana attributed the faulting of Hawaiian domes to gravity subsidence resulting from development of actual voids in the interior due to drawing off of lava from the column to feed flank outflows at low levels (perhaps submarine).[26]

Sliding on the steep oceanic slopes flanking domes seems a highly probable cause of subsidence of sectors along gravity faults. Jaggar has expressed a doubt as to whether the submarine slope at the base of the Kohala dome, Hawaii (where down-faulting has taken place), is "so steep as to occasion slip", but this opinion may have been influenced by a preference for a theory that all Hawaiian faults are related to the underlying rift system.

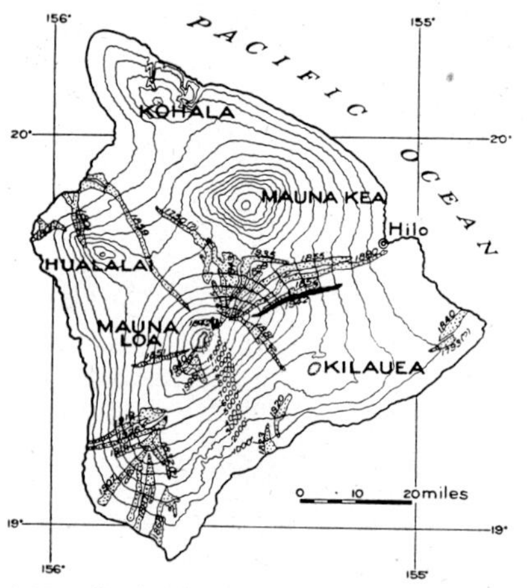

Fig. 24. The domes of Hawaii Island, and recent lava flows. (After G. A. Macdonald.)

Jaggar[27] has given a detailed description of the down-faulting on the north-east side of the Kohala dome (Fig. 24). The sector makes a rectangular coastal re-entrant twelve miles

[25] J. D. Dana, *Am. Jour. Sci.,* 36, p. 111, 1888.
[26] J. D. Dana, *Am. Jour. Sci.,* 37, p. 95, 1889.
[27] T. A. Jaggar, *loc. cit.* (9), pp. 181-184.

long and two miles deep, which might perhaps be classed correctly as a "sector graben" (Chapter XVI). Jaggar writes:

> The straight-shored embayment is the portion with the many and deep canyons, and between the mouths of the canyons the inter-stream ridges terminate each at the coast in a clean precipitous facet, all the facets lying in one plane inclined at a very high angle seaward. There are many great waterfalls over the facet cliffs from hanging gullies which have not yet trenched the spurs to sea-level.

Jaggar remarks also: "This chapter in the physiographic history of Kohala probably typifies a common happening in the Hawaiian Islands, for the islands show many signs of faulting."

Though not all the suggestions that have been made of marginal-faulting features on Hawaiian domes have stood the test of detailed investigation, this explanation seems unques-

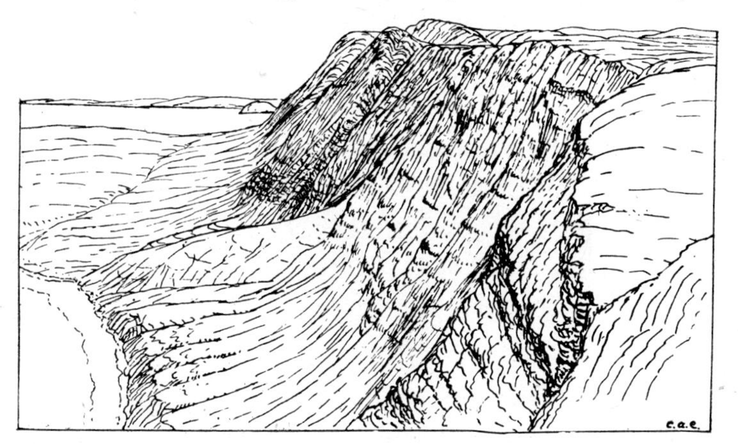

Fig. 25. View eastward along the fault scarp down which the northern half of the East Molokai dome (Hawaiian Islands) has subsided. The low coastal salient at the left appears to be a step or splinter between fault scarps above and below. The latter descends into very deep water as a steep submarine slope. (From a photograph.)

tionably correct for several great lines of cliffs facing the ocean. In addition to the Kohala scarp (described by Jaggar) there is a similar great cliff facing northward along the eastern half of Molokai (Fig. 25); and coastal cliffs of north-western Kauai and eastern Niihau are probably developed by retrogradation from fault scarps.[28] In the case of East Molokai less than half of the original basaltic dome has survived the collapse, the northern part having dropped, apparently to a great depth below sea-level (with the exception of a small splinter). In the centre (Fig. 25) the scarp rises 3500 feet above sea-level, and it seems to descend as far below. The great depth of water off shore leads to the abandonment of an alternative hypothesis of marine retrogradation to account for the great cliff.

Icelandic Cones

In Iceland some volcanoes, which have emitted very fluid basalt, have built true cones of very symmetrical form and

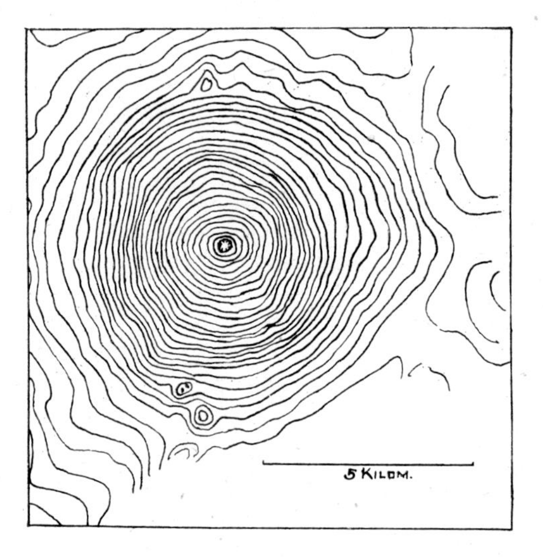

Fig. 26. The Skjaldbreit lava cone, Iceland, from an official map. Contour interval, 20 metres.

composed, apparently, of lava throughout. These are all of moderate dimensions (Fig. 26) and are no longer active. They have been classed by some observers—and probably justly from the point of view of genesis, if not of form—with Hawaiian

[28] N. E. A. Hinds, The Relative Ages of Hawaiian Landscapes, *Univ. Cal. Publ., Bull. Dep. Geol. Sci.,* 20 (6), pp. 143-200, 1931.

basalt domes under the head of "shield volcanoes." Von Knebel and Reck's ideal sketch of an Icelandic "shield volcano" figures it as a broad-crowned dome,[29] but, on the other hand, von Wolff[30] selects the symmetrical cone Skjaldbreit (Fig. 26) as exemplar of the Icelandic form, and indeed, according to his view, of "shield volcanoes" in general. He distinguishes the Icelandic variety of "shield volcanoes" as the "dyngja" type from the "Hawaiian type," which amounts merely to a separation of basalt cones from basalt domes.

The steepness of the slopes of Skjaldbreit is about 1000 feet per mile, which is the same as that of the flank slopes of Mauna Loa. Thus the difference in form between domes and the most perfect cones is nothing more than a difference in the breadth of the summit convexity; and it is conceivable that a Skjaldbreit cone might grow into a broad dome if outpouring of superfluent lava (such as has been the last phase of activity at Skjaldbreit) were to give place to such effluent emission as would build out the flanks of the young cone in the manner described by Dana (p. 82).

An explanation on very similar lines is adopted without reservation by von Wolff[31] for the difference between Icelandic cones and Hawaiian domes. Ignoring the fact that Hawaiian lava volcanoes are of all dimensions from the small, almost undissected, and very distinctly dome-shaped Kahoolawe (Fig. 27), 1429 feet high, up to the giant Mauna Loa the summit of which is more than 10,000 feet above its base, he states that, whereas Icelandic cones are not full-grown specimens, the Hawaiian domes have reached their limit of height and have afterwards extended only laterally.

Skjaldbreit cones have quite small, circular apical craters (Fig. 26), which are surrounded by distinctly raised rims of the kind recognised by Jaggar around a lava vent on the Hualalai dome (p. 80). The slopes of these raised rims are about 30°. Such rims may be an indication of overflow of voluminous lava all around the crater, but alternatively it is

[29] W. von Knebel and H. Reck, *Island,* 1912; F. von Wolff, *Der Vulkanismus,* I, Abb. 132, 1914; II, Abb. 22, p. 882, 1931.
[30] *Loc. cit.* (29) (1931), pp. 882, 895.
[31] *Loc. cit.* (29) (1914), p. 454.

suggested that they are built without overflow by spatter from lava "fountains" (vertical spurts of gas) in the crater.[32] According to Dana[33] and Daly[34] such rims which have been built around the lava lake at Kilauea when the pit containing it has been full to the brim have consisted in part of spatter but have been raised by overflow, though this has not been voluminous. One such rim, which was in existence in 1894, with lava in the act of spilling over it, is figured by Daly.[35]

Fig. 27. The northern half of the little-dissected basalt dome Kahoolawe, Hawaiian Islands. (From a photograph.)

Dome Form Characteristic of Basalt Volcanoes

The eruptions in the Samoan Islands resemble those of Hawaii. The domed shape of the little dissected Savaii Island

[32] H. Reck, Ueber die Entstehung der isländischen Schildvulkane, *Zeits. Vulk.*, 6, p. 72, 1921-22.

[33] "The small overflows, lapping in succession over the borders, often make them steep" (J. D. Dana, *Am. Jour. Sci.*, 35, p. 221, 1888).

[34] R. A. Daly, *Igneous Rocks and their Origin*, p. 135, 1914; *Igneous Rocks and the Depths of the Earth*, p. 156, 1933.

[35] R. A. Daly, *Our Mobile Earth*, Fig. 86 and p. 152, 1926. The lava lake is "rimmed in by a wall . . . of its own congealed substance. Note the little overflow tongues of liquid basalt pouring over and solidifying so as to increase the height of the retaining wall."

has been remarked on by Thomson,[36] and its elongated-dome form is well shown on a map by Friedländer.[37] The active volcano Niuafo'ou (Fig. 7) is a low basalt dome with a collapsed crown,[38] and probably most of the basaltic mountains, many of them thoroughly dissected by erosion, which make the numerous high islands of the South Pacific Ocean, notably Rarotonga (Fig. 28) and the high-peaked Society Islands (Fig.

C. A. Cotton, photo

Fig. 28. Rarotonga Island, a dissected basalt dome.

217) have originated as domes similar to those of Hawaii. The same appears to be true of large basaltic islands in other oceans, and there are basalt domes in east-central Africa. The twin volcanoes, ruins of which now form the unique feature Banks Peninsula, on the east coast of the South Island of New Zealand (Fig. 218), are also basalt domes and there

[36] J. A. Thomson, The Geology of Western Samoa, *N.Z. Jour. Sci. and Tech.*, 4, pp. 49-66, 1921.

[37] I. Friedländer, Beitrage zur Geologie der Samoainseln, *Abh. Kgl. Bay. Ak. Wiss.*, 2 Kl., 24 (3), pp. 514-527, 1910.

[38] T. A. Jaggar, Geology and Geography of Niuafoou Volcano, *The Volcano Letter*, 318, pp. 1-3, 1931.

are two others, Karioi and Pirongia near the middle of the west coast of the North Island of New Zealand. Pirongia (Fig. 29), which has a relief of 3000 feet, is eight miles in diameter. Karioi has a relief of 2400 feet and a diameter of five miles. Both these mountains, which are in an area mapped geologically,[39] are, though long extinct, of recent formation. From their large diameter and moderate height it must be assumed that they had initially the form of domes rather than of cones; for, though they are seamed by ravines, they have not been very much reduced in height by erosion. Little is known of their internal structure, however.

Fig. 29. Pirongia, New Zealand, a somewhat dissected basalt dome. (Outlined from a sketch by F. von Hochstetter.)

---

[39] J. Henderson and L. I. Grange, Geology of the Huntly-Kawhia Subdivision, *N.Z. Geol. Surv. Bull.*, 28, 1926.

CHAPTER VIII

Lava Plateaux and Plains

In various parts of the world, though in most cases in rather remote times, piles of basalt sheets with aggregate volumes of the order of 100,000 cubic miles have overspread former landscapes of varied and even strong relief, completely burying them. Such "basalt plateaux" of regional extent are thoroughly dissected for the most part (Fig. 30). Less extensive basalt "plains", generally due to more recent outpourings, are known also. Their survival as plains may be due either to the resistance which the basalt terrain offers to erosion or to their low-lying position, with little available relief. Mature dissection has overtaken some lava fields of comparatively modern origin, however (Fig. 31).

Ignimbrite Plateaux

Not only basaltic effusive lavas, but also rhyolitic materials ejected as incandescent "sands" have built plateaux in this way, becoming agglutinated to make the firm rock "ignimbrite" which has been commonly mistaken for effusive lava (Fig. 32). The whole volcanic field of the central part of the North Island of New Zealand consists of an ignimbrite plateau except where low-lying parts of the thick sheet of this material have been buried under accumulations of younger volcanic origin. The sheet is broadly warped and flexed up in parts, and is broken to some extent into fault blocks. The origin of the ignimbrite sheet of New Zealand and some of the surface forms assumed by it will be discussed in later chapters. It is mentioned here along with plateaux of basalt lava because of the resemblance of its material to rhyolite lava and the lingering suspicion in some minds that such is its origin.

Basalt Plateaux

Extensive dissected basalt plateaux in which structural-bench landscape features are widely developed are known in

Fig. 30. A dissected basalt plateau, Isafjord, north-west Iceland.

Fig. 31. A maturely dissected lava field at the base of the Sierra Nevada fault scarp, Owens Valley, California. *C. A. Cotton, photo*

H. T. Ferrar, photo

Fig. 32. Escarpment of an ignimbrite plateau, Piopio, near Te Kuiti, New Zealand.

India, South Africa, and western North America. The Indian basaltic terrain (Fig. 33) is lava-covered to a depth of nearly two miles near Bombay though the covering thins eastward.

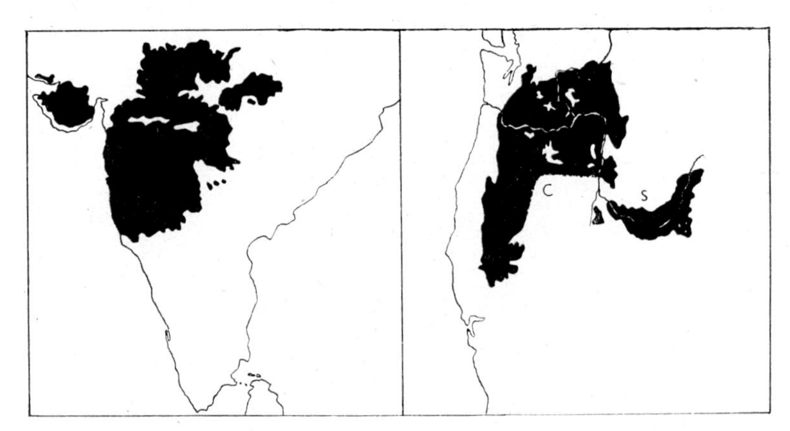

Fig. 33. Left: Deccan traps. (After H. S. Washington.)
Right: Columbia River basalts (C) and Snake River basalt plain (S). (After Salisbury.)

These basalts are known to geologists as the Deccan "traps".[1] In south-eastern Africa a basalt plateau breaks off in the great eastward-facing escarpments of the Drakensberg Mountains. In western North America (Fig. 33) the Columbia River basalts comprise more than 60,000 cubic miles of piled-up lava sheets spread over a region considerably more than 100,000 square miles in extent. The basalt plateau thus defined, which is thoroughly dissected, does not include the adjacent Snake River basalts, in southern Idaho, which are younger and differ in geomorphic expression from those of the Columbia River region, forming a little-dissected plain 20,000 square miles in extent (Fig. 33).

Landscapes of very strong relief are buried under both the Columbia River and Snake River lavas. Peaks thousands of feet high project through the piles of basalt sheets, which themselves have an aggregate thickness of thousands of feet; and the Snake River, trenching down to the undermass through the basalts, has revealed rugged profiles of the buried landscape.[2]

Another basaltic plateau is now represented by marginal remnants only, but is believed to have been of vast extent in very recent times when it bridged the northern part of the North Atlantic Ocean, joining Britain, across Iceland, to Greenland. Of this "Thulean" plateau[3] the best known surviving portions include the deeply dissected basalt terrain of the west of Scotland and the plateau which is cut into broad-shouldered subdued forms by the mature glens of Antrim.

In these and also some more ancient and equally large plateau-basalt regions partial survival of plateaux and of the basalts that compose them is attributable, at least in part, to the very great thickness, extent, and homogeneity of the accumulations. A far advanced stage of erosional reduction of a land surface once mantled with lava may witness the complete removal of the lava, though not usually in such a

[1] The term "trap," applied to basalt, is derived from the stepped landscape profiles developed on dissected plateau basalts (descending structural benches). (Compare German *Treppe*.)

[2] I. C. Russell, *Volcanoes of North America,* 1897.

[3] G. W. Tyrrell, *Volcanoes,* pp. 194-195, 1931.

manner as to permit the resurrection of a buried, or fossil, landscape; for lava sheets, unless much decomposed by chemical weathering, consist of too resistant a material to be easily stripped away from an undamaged undermass.[4]

Complete removal of the basalts will be possible, of course, only if the available relief has been sufficient to allow of it, which generally will be the case if the base of the lava cover is high above sea-level. The floor on which the lavas rest may have been bowed down, however, beneath the enormous load, or may have subsided owing to withdrawal of magma from beneath it, so that the base of the lava pile, even if originally high, may be now below the local base-levels and even below the level of the sea.

Should plateau basalts be not only dissected but also completely removed by erosion, the lava landscape will be replaced by one developed on the undermass. This may be composed perhaps in the main of subsequent features determined by the materials and structures of the stratified and other rocks of the pre-basaltic land. Stencilled on such a landscape by superposition, however, much of the river pattern of the former plateau may survive. Moreover, the landscape forms will generally include some which are derived by subsequent erosion from the exposure of intrusive igneous bodies injected more or less contemporaneously with the upbuilding of the plateau of basalts now destroyed. Dykes and sills may stand out as prominent hogbacks and mesas. Such are the landscape forms of some districts in South Africa, where great mesas of intrusive dolerite are left standing in relief. The injection of this dolerite as sills is believed to have occurred at the same time as the outpouring of the plateau basalts of which the Drakensberg is a remnant.

Fissure Eruptions

It is rarely possible to ascertain what vent supplied a particular lava flow among the many of which ancient plateau-making piles of basalt are built, but the general nature of

[4] A buried landscape of low granite hills has been exhumed locally, however, from under the Columbia River basalts (see N. E. A. Hinds, *Geomorphology*, photo on p. 294, 1943).

such vents is indicated by the "dyke swarms", as they have been called in Scotland, which intersect the underlying rocks. These are seen at places where the basalts have been removed by erosion. The basalt sheets themselves are intersected by many dykes, and these no doubt fill fissures through which lava has risen to feed the later flows.[5] Dyke swarms have been mapped in the surveys of the west of Scotland[6] and of the state of Washington.[7]

Almost certainly outflow of basalt took place quietly from such fissures—i.e. without violently explosive phenomena—for fragmentary volcanic products are but rarely interbedded at all abundantly with the plateau-building lavas; and there is every reason for believing that the mode of eruption closely resembled the effusion of lava from fissures on the flanks of the modern basalt domes Etna and Mauna Loa (Chapter VII), though the average volume of lava outflowing in an eruption may have been considerably greater.

From the Laki fissure in Iceland, which is regarded as affording a modern example more or less closely resembling the vents from which ancient basalts were emitted, three cubic miles of basaltic lava flowed out in 1783. The fissure was twenty miles long. Lava welled up and flowed over both sides of it. This is the only large-scale fissure eruption that has been witnessed in a plateau region; but the traces of another which must have occurred very recently have been found at the Craters of the Moon, Idaho,[8] where a fissure emitted the latest lava flood contributing to the construction of the vast Snake River lava plain (Fig. 33). The line of

[5] In Antrim "the basalt itself is often traversed by dykes, each of which is probably the feeder from which some overlying bed of basalt was poured out. These dykes are still more numerous, or are more readily observable, in the Chalk and other beds below the basalt, which crop out around the basaltic area." (J. B. Jukes, *The Student's Manual of Geology,* 2nd ed., p. 330, 1862.)

In India also "dykes of large size . . . are observed at a number of places in the neighbourhood of the trap area around its boundary," where the underlying terrain is exposed. (D. N. Wadia, *Geology of India,* p. 199, 1919.)

[6] G. W. Tyrrell, *Volcanoes,* pp. 194-195, 1931; Flood Basalts and Fissure Eruption, *Bull. Volcanologique,* 1, pp. 89-111 (p. 98), 1937.

[7] R. A. Daly, *Igneous Rocks and the Depths of the Earth,* p. 140, 1933.

[8] H. T. Stearns, Craters of the Moon National Monument, Idaho, *Geog. Rev.,* 14, pp. 362-372, 1924.

the fissure from which pyromagma here flowed is marked by a row of spatter cones and scoria mounds (Fig 34) as has been the case also on Etna and the Hawaiian domes (Fig. 20) and also along the Laki fissure in Iceland. Though such features may be present during the whole of an eruption, voluminous lava, pouring at places over perhaps both rims of the fissure, may eventually bury all the minor forms with the exception of those built up during the concluding phase.

Plateaux have been built only by the accumulation of many flows which have not always followed one another in quick

Fig. 34. Spatter cones aligned on a fissure from which a flood of basalt lava has been emitted very recently, making the latest addition the surface of a portion of the Snake River lava plains, Craters of the Moon, Idaho. (From a photograph.)

succession. Though plateau-making flows seem to be on the average thicker and probably of greater areal extent than those which go to the making of Hawaiian domes the difference is not very great. Averages calculated from a limited number of observations give plateau flows a thickness of forty feet, however.[9] When the fact is taken into consideration that such flows have spread out nearly horizontally there is no reason to suppose that the individual volumes of flows of such thickness have been exceptionally great or differed essentially from those of present-day flows from basalt volcanoes.

[9] G. W. Tyrrell, *loc. cit.* (6) (1937), p. 101.

The flows have certainly been extremely fluid, however, and have spread very widely.

Against the interpretation of dykes as indicators of the positions of the multiple feeding fissures for plateau-basalt eruptions objections may be based on the observed fact that dykes penetrate the basalt sheets as well as the underlying rocks and on the absence of visible continuity in most cases between dykes and sheets. Several examples of such continuity are known and have been described, however[10]; and it has been shown that

> the same fissure may repeatedly emit lava with long intervals between the successive flows. The fissure and the dyke which has served all the flows must, therefore, break through the succession of lavas up to the last one emitted, and thus penetrate the whole series. The fissure is ultimately filled with the latest magma to be delivered, but it may nevertheless have served as the feeder of many of the earlier flows. . . . Multiple dykes may yet prove to be the strongest evidence of fissure eruption.[11]

Individual fissures are not large.

> The width of an active fissure rarely exceeds five feet, and may be less. Large volumes of lava seem to have no difficulty in finding their way to the surface through narrow fissures. Exposures on Oahu and on Hawaii indicate that these feeders do not widen downward, at least in the first 4000 feet.[12]

The only rivals of the generally accepted fissure-eruption theory in explanation of voluminous plateau basalts are the two "areal-eruption" hypotheses, this name having been used in two different senses. "Areal eruption" as understood by von Wolff[13] is Daly's hypothesis of the "deroofing" of a batholith,[14] and von Wolff has copied a diagram from Daly in illustration of the hypothesis. Overflow of magma welling up from a deroofed area might produce lava floods around its

[10] G. W. Tyrrell, *loc. cit.* (6) (1931), p. 198; (1937), p. 105; R. E. Fuller, The Closing Phase of a Fissure Eruption, *Am. Jour. Sci.,* 14, pp. 228-230, 1927.

[11] G. W. Tyrrell, *loc. cit.* (6) (1937), p. 106.

[12] H. T. Stearns, Geology and Water Resources of the Kau District, Hawaii, *U.S. Geol. Surv. W-S. Paper,* 616, p. 140, 1930.

[13] F. von Wolff, *Der Vulkanismus,* I, pp. 435-436, 1914.

[14] First stated as "eruption through local foundering" (R. A. Daly, The Nature of Volcanic Action, *Proc. Am. Ac. Arts and Sci.,* 47 (3), pp. 60-67, 1911).

periphery as implied by Daly.[15] A thick sheet of ancient basaltic rock ("diabase") at Neurode, in Germany, is ascribed by von Wolff[16] to such areal eruption. This rock passes laterally into gabbro, and a similar relation of lavas to plutonic rocks in other examples is pointed out by Daly.[17] The whole hypothesis is highly speculative, however, and is not supported by many observed facts.

"Areal eruption" as understood by Reck[18] is entirely different. This is a theory that plateau-basalt outflow has taken place from clusters of vents in "central" areas instead of from fissures, and is based on the occurrence at a few Icelandic localities of scoria mounds closely clustered over small areas in contrast with the linear arrangement of such features along outflow fissures which is more often seen. Von Wolff[19] notes that such clusters are not numerous. He mentions four of them, of which one (situated seven miles south of Rejkjavik) is typical. This group of "raudholar" (red craters) consists of about 100 mounds, of heights up to 300 feet, all crowded into an area half a mile square. As these craters are in a swampy situation von Wolff suggests that they have been formed as a consequence of contact of magma with ground water, the magma being injected probably as a thin sill under a recent lava flow—that is to say the whole group of craters appears to be the result of a local accident and cannot be regarded as marking a main source of outflow of lava. There are similar but smaller clusters in northern Iceland.

"Multiple-vent" basalts (Tyrrell[20]) contribute a quota to "plateau" basalts, but are generally of small extent as compared with the output of fissure eruptions. Such output from isolated vents is, on the other hand, the origin of smaller basalt plains such as those of the western districts of Victoria (Australia) and the North Auckland district of New Zealand.

[15] See also R. A. Daly, *Igneous Rocks and the Depths of the Earth,* Fig. 60, 1933.
[16] *Loc. cit.* (13), p. 439.
[17] *Loc. cit.* (15), pp. 145-146.
[18] H. Reck, Die Masseneruptionen unter besonderer Würdigung der Arealeruption in ihrer systematischen und genetischen Bedeutung für das Isländische Basaltdeckengebirge, *Deutsche Islandforschung,* Breslau, 1930.
[19] F. von Wolff, *Der Vulkanismus,* II (2), p. 896, 1931.
[20] G. W. Tyrrell, *loc. cit.* (6) (1937), p. 92.

Acceptance of the theory that fissure eruptions are the chief source of plateau basalts does not imply support to the conception of vast regional floods of lava issuing from "fissures hundreds of miles in length and a mile or more broad," as Tyrrell has expressed it, an idea at one time associated in some minds with that of fissure eruption.[21]

Geikie[22] is regarded as the originator of the current hypothesis of fissure eruption, though he himself has credited the idea to Richthofen, and it was clearly in the mind of Jukes in 1862.[23] Geikie's brief statement of it, which embodies all the essential points, is as follows:

> Extensive eruptions of lava, without the accompaniment of scoriae, with hardly any fragmentary materials, and with, at the most, only flat-domed cones at the points of emission, have taken place over wide areas from scattered vents, along lines or systems of fissures. Vast sheets of lava have in this manner been poured out to a depth of many hundred feet, completely burying the previous surface of the land and forming wide plains or plateaux.[24]

Geikie's familiarity in the field with the geomorphic features of the Snake River lava plains enabled him to picture the mechanism of fissure eruption in a way which geologists conversant only with older lavas as seen in section had failed to do.

Volcanic Rifts

In addition to the indications of fissures given by lines of cones and scoria mounds in Iceland ("linear" as contrasted with "areal" eruptions) some open fissures are found. These have developed in basalt fields apparently as a result of tension. Some have emitted lava and scoria, but others are no more than gaping cracks miles in length and generally containing water (Fig. 35). Depths as great as 600 feet are recorded

[21] Even I. C. Russell, as late as 1897, has pictured the effusion of lava from "intersecting fissures having an aggregate length of several hundred miles" as "the most truthful conception of fissure eruption" (*Volcanoes of North America*, p. 38).

[22] A. Geikie, *Geological Studies at Home and Abroad*, p. 271, 1888.

[23] *Loc. cit.* (5).

[24] A. Geikie, *Textbook of Geology*, 4th ed., p. 343, 1903.

Fig. 35. Flosagjá, a deep, water-filled tension crack, or rift, at Thingvellir, Iceland.

in such rifts in Iceland.[25] They are characteristic features of the lava-field landscape, and are called in the Icelandic *gjá*.

One of these rifts, the Eldgjá, has been enlarged by explosions to form a continuous broad trench, which, according to Sapper,[26] however, is a unique feature. It cuts indifferently

[25] A. Geikie, *loc. cit.* (24), p. 342.

[26] K. Sapper, Ueber einige isländische Vulkanspalten und Vulkanreihen, *Neues Jahrb. f. Min.*, 26, pp. 1-46, 1908.

across the country for twenty miles, stretching as far as the eye can reach. Such rifts,[27] or portions of them, as have not emitted effusive lava have been sites of fire-fountaining which has piled up bordering walls of scoria, notably along the Eldgjá.

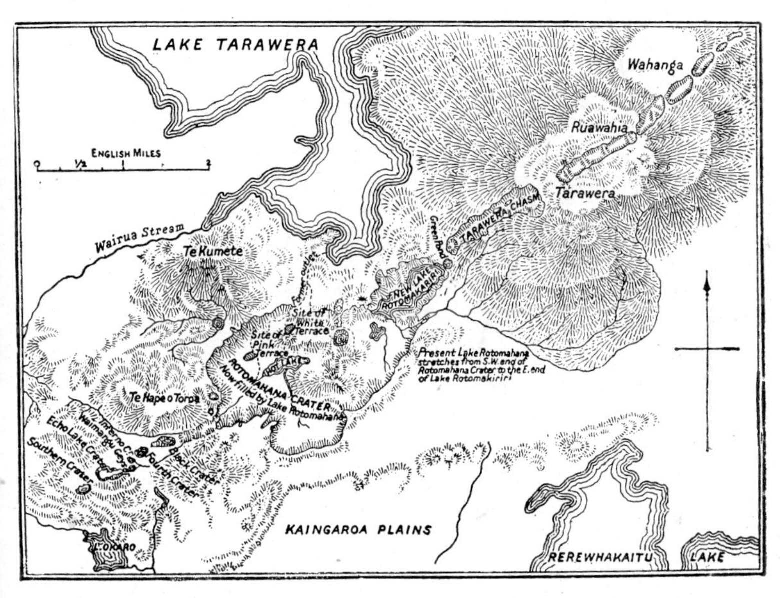

Fig. 36. The Tarawera volcanic rift (New Zealand), which crosses the Tarawera group of rhyolite domes and extends south-westward under lake Rotomahana, as it was shortly after the eruption of 1886. (After Bell.)

The Tarawera rift, in New Zealand, from which a vast quantity of basaltic fragments, chiefly lapilli, was emitted in 1886 (Chapter VI), may be placed in the same category. Though not in a basalt terrain, it intersects only rocks of volcanic origin (Figs. 36, 37).

Basalt Plateaux and Domes

Increasing knowledge of the mode of accumulation of Hawaiian domes, and diminished emphasis on the description of their summit sinks as craters identical with or comparable

[27] "Rift" does not necessarily imply a graben produced by trough faulting. The term is more suitable for an open crack which is the result of tension.

N.Z. Geol. Survey, photo

Fig. 37. View north-eastward along the rift opened by the eruption of 1886 across the Wahanga dome (part of the Tarawera cluster), New Zealand. Basaltic debris, ejected in 1886, overlies white rhyolitic pumice of a prehistoric shower, which in turn covers and somewhat smooths the spinous prominences on the rhyolite dome.

in function with the craters of explosive, cone-building volcanoes, have led to the recognition of close similarity in origin between dome complexes of Hawaiian type and lava plateaux. The phenomena of fissure eruption are common to both, and the petrographic similarity of the basalt in the two cases (suggesting that such basaltic magma is derived in bulk without differentiation from a sub-crustal source) supports the analogy. It remains to explain the development in the one case of a lava pile consisting of domes or lenses of inclined flows and the prevalence of more nearly horizontal structure in the other. It does not appear that the differences can be due to much greater volume in individual plateau-building than in dome-making flows, though the latter are characteristically thin and very numerous and have followed one another generally at quite long intervals. The building of the domes of the island of Hawaii has been a secular process.

Stearns has offered as an explanation of the difference between dome- and plateau-building that the rift zones, or systems of fissures that supply magma, under domes have remained open in the same place for very long periods, whereas in plateau-basalt emission activity shifts from time to time from one to another of many rift systems.

> In the Snake River plains [taken as an example] it has been the migratory character of the volcanic rifts that has caused the basalt extruded to spread itself over a wide area, forming a plain, rather than to remain stationary and build one or more lava domes. In Hawaii the permanency of the rifts has caused great domes to be built.[28]

One might expect the structure of a plateau as revealed by dissection to expose a multitude of small domes or cones interfingering and overlapping one another, if the analogy with dome-building is of value. Instead of this, however, individual flows seem to extend for many miles. Tyrrell[29] has suggested the hypothesis that lava plateaux originate as a result of levelling up of the surface of a dome complex by infilling of the hollows between individual domes, which have perhaps been quite large. Dips as steep as those characteristic of the flanks of domes are unusual among plateau basalts, however.

It has also been suggested that the building of plateaux and of domes may be distinct and successive phases in a fissure-eruption period. Thus, in the west of Scotland, it is believed that basalt domes (though they have since been destroyed by erosion and are recognised now only by their roots, which are ring-dyke systems) were once piled up above the plateau-making lavas "as a great line of shield-volcano complexes of Hawaiian type."[30]

Very low domed forms of the surface mark the most recent centres of effusion of basalt on the Snake River plains. These broad swells are "frequently 8 to 10 miles or more in diameter at the base, and perhaps 200 or 300 feet high. In fact their

[28] H. T. Stearns, *loc. cit.* (12), p. 139.
[29] G. W. Tyrrell, *loc. cit.* (6) (1937), pp. 109, 110.
[30] G. W. Tyrrell, *Volcanoes*, p. 195, 1931; *loc. cit.* (6) (1937), p. 109.

bases merge so gradually with the surrounding plain that no eye can recognise where the ascending surface actually begins."[31] In all probability such broadly convex forms have been characteristic of all lava-plateau surfaces during their growth, though it would be difficult to recognise them in the structures of dissected basalts. Russell remarks that these Snake River domes "owe their height mainly to the cooling and thickening of the lava about the place of emergence, so as to form the necessary slope down which it could continue to flow." He notes that the domes are not aligned on fissures. That some true fissure eruptions took place in the Snake River region is shown, however, by the features of the Craters of the Moon (Fig. 34).

The Snake River Basalt Plain

When the building of great lava plains was in progress outpourings of basalt (though separated in some cases by long intervals during which rivers excavated deep canyons) must individually have been rapid and soon completed.[32] Even in the case of relatively small flows which freeze over near the source the output of lava has amounted to millions of tons hourly, as estimated by Nichols[33]; but some flows of the Snake River field give the impression of having flowed out rapidly in even greater volume. They have made widespread sheets very thick in parts, though elsewhere thin, which have spread over uneven ground with apparently a lakelike surface "essentially horizontal," so as to merit Russell's description, "in truth lakes of molten rock." Stearns mentions that one of the upper flows of the Snake River pile—the Sand Springs basalt—has filled a former canyon of the Snake River for about fifty miles and to a depth in places of 500 feet, as well as spreading widely out above the filled canyon as a thin sheet over the plain.[34]

[31] I. C. Russell, Geology and Water Resources of the Snake River Plains of Idaho, *U.S. Geol. Surv. Bull.*, 199, 1902.

[32] Thus "fissure eruption" is described by R. A. Daly as a "sudden act" (*Igneous Rocks and the Depths of the Earth,* p. 140, 1933).

[33] R. L. Nichols, Viscosity of Lava, *Jour. Geol.*, 47, pp. 290-302 (p. 302), 1939.

[34] H. T. Stearns, Origin of the Large Springs and their Alcoves along the Snake River in Southern Idaho, *Jour. Geol.*, 44, pp. 429-450, 1936.

Observations such as those of Russell on the lake-like character of the flows of Snake River basalt emphasise a contrast between the flatness of great basalt plains and the slopes and convexities of surface of the Hawaiian dome complex. One of the most recently emitted of the voluminous Snake River flows must have spread and

> flowed with almost the freedom of water. On the plain it spread out and formed what may be termed a lake of liquid rock. The western shore of this lake for . . . 30 miles . . . is formed by steep mountain slopes, and . . . the margin of the lake is approximately a contour line. . . . No eye can observe that it is not a perfect plain. The liquid rock entered the mouths of the valleys in the mountains . . . and extended up them various distances, depending on their gradients. Where the ascent was gentle it flowed up them 5 or 6 miles. . . . The ridges between mountain valleys became capes and headlands; . . . peaks were converted into islands.[35]

A portion of such a lava flood, which occupies a valley marginal to the Snake River plain, is 300 feet thick, and locally a single flow of basalt covering the floor of a valley in Arizona is reported to be 750 feet thick.[36]

It is difficult to picture the mode of emplacement of such great single flows unless it be supposed that the great volume, and especially the great thickness, of lava in them have delayed cooling and solidification until the vast mobile flood has become a lake. This mode of accumulation of very voluminous flows very rapidly emitted, as compared with that of thinner, rapidly-congealing Hawaiian flows, may perhaps in itself explain why a vast plain resulted from the accumulation of the lavas instead of the up-building of a mountain or a range of lava domes.

Ponding of lava on a smaller scale has been observed on various occasions in Hawaii. The voluminous basalt which was poured out on the first day of the eruption of 1940 at the summit of Mauna Loa, for example, spread out with a level

[35] I. C. Russell, *loc. cit.* (31), pp. 102-103; see also A. Geikie, *Geological Sketches at Home and Abroad,* pp. 337-338, 1882.

[36] W. T. Lee, *U.S. Geol. Surv. Bull.,* 352, p. 54, 1908.

lake-like surface over an area of about 3 square miles in the summit "sink." The maximum depth of this pool, according to Schultz,[37] was about 50 feet.

Phonolite Plains

Few examples are known of wide-spreading sheets of mobile lava other than those of basaltic composition. One of the few exceptions among the non-basaltic lavas is perhaps phonolite, which in some cases has apparently flowed out almost as freely as basalt. Thus lava plains of phonolite are reported in Kenya, and phonolite sheets are interbedded with basalt flows in the Dunedin district of New Zealand.[38]

---

[37] P. E. Schultz, Some Characteristics of the Summit Eruption of Mauna Loa, Hawaii, in 1940, *Bull. Geol. Soc. Am.*, 54, pp. 739-746, 1943.

[38] P. Marshall, The Sequence of Lavas at the North Head, Otago Harbour, *Quart. Jour. Geol. Soc.*, 70 pp. 382-408, 1914.

CHAPTER IX

Lava Fields

MOST GREAT LAVA PLAINS ARE CHARACTERISTICALLY WITHOUT ANY considerable minor relief. Where a tumultuous aggregation of relief forms is found locally on a basalt surface, as is the case in some parts of Victoria, Australia, it is conspicuous, as Skeats and James have pointed out,[1] by contrast with the more usual "monotonous basalt plains" such as occur widely in western Victoria. It was Russell's observation of the absence of relief on great parts of the Snake River Plains (Idaho) that led him to the conclusion that the basalt of which they are composed had spread out with the smooth dead-level surface of a lake (Chapter VIII).

PONDED LAVA

According to this interpretation, "when a lava sheet cools without motion neither a characteristic pahoehoe nor an aa surface is produced. Many of the older sheets of Snake River lava illustrate this; they are simply plane surfaces."[2] The "pahoehoe" and "aa" surfaces (described below in this chapter) which are developed on Hawaiian and many other basalt lava flows, especially those which have flowed down and solidified on gentle slopes, both owe the characteristic features of their minor relief to continued movement after partial solidification of a flow has taken place. Small bodies of ponded lava which have been observed to congeal without disturbance by movement have assumed a smoother surface. Evidence of such freezing of a lava pool has been cited by Stearns,[3] and a perfect small-scale example of lava-lake solidification is seen

[1] E. W. Skeats and A. V. G. James, Basaltic Barriers and other Surface Features of the Newer Basalts of Western Victoria, *Proc. Roy. Soc. Victoria,* 49, pp. 245-278, 1937.

[2] I. C. Russell, Geology and Water Resources of the Snake River Plains of Idaho, *U.S. Geol. Surv. Bull.,* 199, p. 98, 1902.

[3] H. T. Stearns, The Keaiwa or 1823 Lava Flow from Kilauea Volcano, *Jour. Geol.,* 34, pp. 342-343, 1926.

on the lava floor, frozen in 1877, of the Kaneakakoi pit crater, adjacent to Kilauea, Hawaii, which Jaggar has described as the "smoothest kind of pahoehoe." He records that a long three-inch-wide crack in this floor has a measured depth of fifteen feet, which indicates that the basalt is a "columnar slab" of that thickness at least.[4]

Such columnar jointing, which reaches to the cooling surface, is not confined to lava fields that are free from surface relief, but is found conspicuously in the basalt of undulating and ridged forms such as those of parts of the Snake River lava field,[5] of the "stony rises" of Victoria,[6] and of the Penrose flow at Auckland, New Zealand.[7] It seems to be the indicator of a solidified crust of considerable thickness, however, which had not been subject to deformation until thus thickened.

Some lava sheets in Victoria have been described[8] as each a single flow unit ponded behind a solidified marginal barrier which has acted like a coffer dam, and these lake-like bodies must have been frozen over in such a way as at first to make smooth plains. At some stage of partial solidification of ponded lava escape of some liquid generally takes place, however, through such a marginal barrier. Thus, not only have marginal barriers come to stand out as ridges, but quite strong relief has also been developed on the lava-lake crusts behind them as a result of irregular subsidence. Such an explanation has been proposed for the rugged forms of basaltic relief termed "stony rises" in Victoria, which are described below (p. 123).

Pahoehoe and Aa

There are two distinct and contrasting modes of solidification which have been observed on those flows of basalt that spread out thinly and continue in motion during congelation, and with these are associated forms of minor relief on such

[4] T. A. Jaggar, *Bull. Haw. Volc. Obs.*, 13, Fig. 3, 1925.

[5] A Geikie, *Textbook of Geology,* 4th ed., pp. 344-345, 1903. Geikie notes also the absence of any "layers of slag or scoriae" and of any "scoriform"—i.e. aa—surface.

[6] Skeats and James, *loc. cit.* (1).

[7] Professor J. A. Bartrum has guided the author to this locality.

[8] By Skeats and James, *loc. cit.* (1).

flows. These are dermolithic and clastolithic solidification, as defined by Jaggar (Chapters II-IV). The former results in the formation of a glassy skin, elastic at first, beneath which movement of the lava generally continues, leading to the development of the characteristic surface termed "pahoehoe" (Fig. 38). In clastolithic solidification, on the other hand, cavernous, spiny-surfaced masses of solidifying lava[9] are

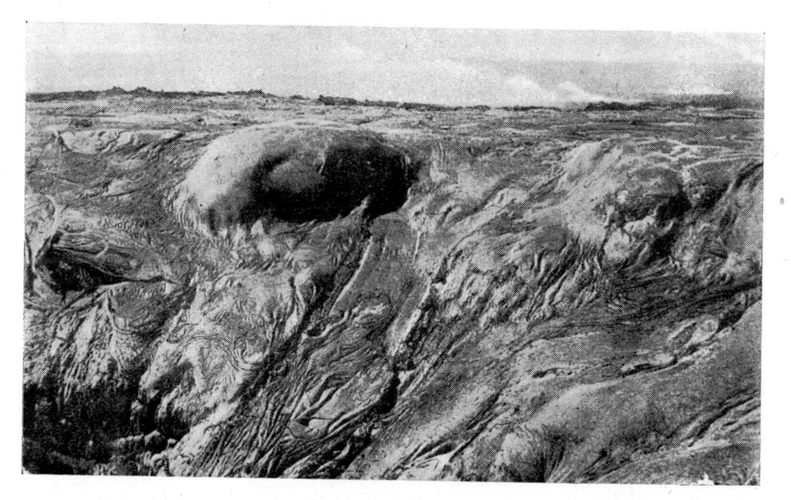

Fig. 38. A pahoehoe surface on basalt lava flowing from Matavanu volcano, Savaii, Samoa (1905-11).

rejected, as it were, by the pyromagma, giving to the flow the rugged "aa" surface. Solidified pahoehoe is described as "a 'live' lava, its gases trapped in solution; . . . its vesicles are distended and spherical," whereas "perfect aa should be a 'dead' lava; its vesicles are deformed and its gases largely replaced by air."[10]

The escape or expulsion of expanding bursts of gas from the still liquid lava may be regarded as a cause of rapid

[9] Aa masses may be termed "scoria," or "scoriaceous," but they are not, as Dana has pointed out, scoriaceous in the sense of being inflated by complete or closed vesicles. "The rock constituting the body of the mass is the ordinary solid lava, usually little vesiculated, not the scoriaceous; but the exterior surface is roughly cavernous and horridly jagged." (J. D. Dana, *Am. Jour. Sci.*, 34, p. 362, 1887.)

[10] T. A. Jaggar, Seismological Investigation of the Hawaiian Lava Column, *Bull. Seism. Soc. Am.*, 10, pp. 167-168, 1920.

cooling of aa and also of a sudden increase of viscosity; these together cause the solidification of the scoriaceous masses. "Great blocks appear to have cooled in this way so rapidly that no opportunity was given for the suddenly projected and rapidly expanding lava outbursts to 'heal' and resume a liquid flow. The projected masses are cooled almost instantly throughout the mass and remain discrete blocks of the roughest and most rugged outline."[11]

Many observers have maintained that there is no initial difference either in magmatic composition or in gas-content between the basaltic lavas that solidify as aa and pahoehoe. On Hawaiian flows, indeed, the character is said to change from pahoehoe to aa or from aa to pahoehoe as the flow progresses. Green,[12] as a result of observation of pahoehoe solidification on a small lava stream from under the great aa flow of 1859, has adopted the view that the lava forms pahoehoe "when it is not in too great quantity and runs out quietly."

According to the theory of Jaggar, the chief factor determining whether a lava shall form aa or pahoehoe is variation in gas-content. He argues that after emission flowing lava gradually loses gas, and finds in such gas-empoverishment an explanation of the most commonly observed change of solidify-

[11] A. L. Day and E. S. Shepherd, Water and Volcanic Activity, *Bull. Geog. Soc. Am.*, 24, pp. 573-606, 1913.

An alternative but less satisfying explanation of aa formation given by H. O. Wood, ascribes it to "the granulation by motion of a stiff overcooled fluid on the point of solidifying." (Notes on the 1916 Eruption of Mauna Loa, *Jour. Geol.*, 25, p. 335, 1917.)

P. E. Schultz (in G. A. Macdonald, The 1942 Eruption of Mauna Loa, Hawaii, *Am. Jour. Sci.*, 241, pp. 241-256, see p. 253, 1943) has watched solidification of lava issuing from the side of an aa flow and describes it as follows: "The lateral advance was made by yellow-hot lava in small amount which oozed from the loose steep wall in a highly fluid condition. It was apparently at such low viscosity solely on the strength of its great heat. I was unable to detect any effervescence or other indication of escaping gas. . . . After two or three seconds the fluid material very abruptly lost its high mobility and became friable, forming fragments that rolled down the steep, rough flank. During the next four or five seconds . . . the fragments still remained soft enough so as not to make any appreciable sound, yet of such a consistency that the descending pieces fragmented as they bumped down the slope. Then the fragments cooled and hardened to the point that the characteristic clinking sound was made, and at this point the yellow-orange glow was gone, the fragments being quite red in colour with greyed and darkened edges and projections."

[12] W. L. Green, *Vestiges of a Molten Globe*, 2, p. 172, 1887.

ing habit, that from pahoehoe to aa. He notes that "the swift and long Hawaiian flows are pahoehoe at the source and aa below, as might be expected."[13]

Washington[14] agrees that gas-content controls the mode of solidification. In contrast with the opinion of Jaggar, however, he maintains that aa is derived from a lava which contains more gas than does pahoehoe.

> The pahoehoe magma, by simmering in the throat of the volcano, has lost a large part of the vapour in solution in it, and issues thus with little internal mobility and consequent high viscosity, so that it cools rapidly from the surface inward and the viscosity increases rapidly to the point at which crystallisation is no longer possible and the lava solidifies as a whole, very largely as glass. On the other hand, aa lava issues at a lower temperature than pahoehoe but very highly charged with gas, giving it great internal mobility, so that it flows *en masse* with high velocity and crystallisation takes place very rapidly throughout and continues almost or quite to the point of complete solidification, the internal molecular mobility being maintained chiefly by the constant state of gas-saturation brought about by the continuous crystallisation and the maintenance of a high temperature caused by the same process. The two types of vesiculation that are characteristic of aa and pahoehoe are caused by the two different kinds of solidification.[15]

This does not seem at first sight to account particularly well for the common habit of lava flows, which change from "pahoehoe at the source" to "aa below". Possibly, however, lava which makes its way down the mountain-side in tunnels under the first-formed pahoehoe crusts is more highly charged

[13] T. A. Jaggar, *loc. cit.* (10). This explanation is favoured by most observers in Hawaii. "Pahoehoe is emitted with much included gas. If the gas is stirred out rapidly by flowing, cooling, or violent fire-fountaining, so that crystallisation starts, the lava changes to aa. . . . It is definitely established . . . that aa cannot revert to pahoehoe. It may appear to do so where pahoehoe emerges from a tube under aa" (H. T. Stearns, Geology and Ground-water Resources of the Island of Maui, Hawaii, *T.H. Div. Hydrog. Bull.*, 7, p. 25, 1942). G. A. Macdonald writes: "There seems to be no possible way in which the aa could be richer in gas than the pahoehoe" (*in litt.*, 1943). We may anticipate further discussion of the question.

[14] H. S. Washington, The Formation of Aa and Pahoehoe, *Am. Jour. Sci.*, 6, p. 409 ff., 1923.

[15] H. S. Washington, in F. A. Perret, *Vesuvius*, p. 146, 1924.

Without discussing the question A. Rittmann has adopted Washington's theory that lava which solidifies as aa is richer in gas than that which becomes pahoehoe (*Vulkane und ihre Tätigkeit*, 1936).

with gas and perhaps at the same time cooler than the first outflow of pyromagma, for it may be derived from a lower zone in the lava column. Lava of such deeper origin may be required to make the aa that comes to rest as the distal portion of a flow.

Other observers, who have not adopted Washington's explanation of the congelation of the pahoehoe skin as due to development of a viscous condition at a high temperature, but assume, on the other hand, that the skin is formed only after considerable chilling of the surface has taken place, ascribe to this lava a "capacity for retaining a high degree of liquidity through a relatively long period of cooling."[16] Lewis remarks also that it develops "a notable degree of viscosity only within a limited range of temperature as it approaches rigidity." This is a property possessed especially by basaltic as contrasted with the more siliceous lavas (Chapter V). It permits the formation of a pahoehoe skin over and around a mobile stream of pyromagma differing from it but little in temperature. Pahoehoe lavas grade into "pillow lavas", most of which have been chilled very rapidly under water (Chapter XV).

The Pahoehoe Surface

Pahoehoe lavas spread out freely as thin, and in some cases extremely thin-edged, sheets. Their chief characteristic, however, is the skin (Fig. 38), "smooth and billowy, satiny, glistening" (N. D. Stearns), which is so tough and elastic before it becomes finally a hard glass or slag that it assumes inflated, bulbous forms (Figs. 39, B; 40), is wrinkled into "tapestry", or twisted and rolled in a "ropy" surface by differential currents flowing under it (Fig. 39, A).[17] The "toes"

[16] J. V. Lewis, Origin of Pillow Lavas, *Bull. Geol. Soc. Am.*, 25, p. 646, 1914. The faculty of developing a skin on the surface, which is essential to the formation of pahoehoe (and also of "pillow" lava) is regarded by G. W. Tyrrell (*Principles of Petrology*, p. 38, 1926) as peculiar to basalt, being possessed, according to his statement only by "freely-flowing basaltic lavas which retain a high degree of liquidity through a lengthy period of cooling but nevertheless develop considerable viscosity as they approach the solidifying point."

[17] For an excellent photographic illustration of such development of a ropy pahoehoe surface on new lava from Nyamlagira see L. C. King, *South African Scenery*, Fig. 160, 1942.

shown in Fig. 39, B are composed of lava which "welled up a crack" in a previously solidified skin. They were "progressively formed by swelling and rupturing fronts, and the long one in the foreground is double and exhausted the lava available. When incandescent . . . these toes . . . resemble a bag of red jelly. They are a foot or two in diameter."[18]

Fig. 39. A: Ropy surface of pahoehoe lava. Wrinkles developed on the skin over a tongue of pahoehoe have been rolled into the semblance of ropes by the drag of movement from left to right, which has been most rapid in the middle of the lava stream under the thin skin. (Drawn from a photograph by Werth.) B: Toes of pahoehoe formed by budding of bulbous forms one from another, lava of 1919, Kilauea. (From a photograph by Maehara.)

Frozen cascades are formed also. Such details of the pahoehoe surface are developed obviously while the skin is still very thin. It thickens progressively, however, assuming greater rigidity.

Forms of a larger order which may diversify a pahoehoe surface are the results of bulging up of a somewhat thickened crust into mounds which Daly has termed "tumuli."[19] Some of these, both small and large, appear in Fig. 40. They are almost invariably cracked along the crest, and quite commonly rather viscous lava has welled out along such clefts in small

[18] T. A. Jaggar, Lava Stalactites, Stalagmites, Toes, and "Squeeze-ups," *The Volcano Letter,* 345, pp. 1-3, 1931.

[19] R. A. Daly, *Igneous Rocks and the Depths of the Earth,* p. 155, 1933. These features are referred to by J. D. Dana as "billows" or "hummocks." They are the "schollen-domes" of I. Friedländer.

Fig. 40. A pahoehoe lava surface, showing tumuli, Kilauea, Hawaii, 1921.

amount so as to form miniature convex flows and "squeeze-ups."

Tumuli are not gas blisters, as has been assumed by Dana[20], Sapper[21], and Tyrrell[22]. Jaggar[23], in agreement with Daly, explains them as follows:

In the filling of basins with pahoehoe lava are built [tumuli], "schollen-domes," or hillocks of swelling crust. What was a puddled flat with ropy lava shell begins in the course of hours to swell up. A laccolith of basalt paste is rising inside because of some equation between resistance to further spreading and resistance to upward lift. If upward lift is the easier for the onflow of the feeding stream which is pouring through a tunnel to feed the flat, there will be no further escape of "toes" pushing out from under the skirt of the crust. When the swollen dome, 50 to 100 feet across, lifts a shell three feet thick, there finally arises a star-shaped opening between sectors in the top of the dome. The dome gets to be 10 or 15 feet high. Then the [lava] paste "squeezes up" through the opening on top. It either trickles down and skins over, or it sputters up and builds a spatter cone.... It may rise as a stiff plug or spine or "squeeze-up."

Squeeze-ups of larger size have been formed notably at Sunset Crater, Arizona. It was for these that Colton[24] introduced the descriptive name "squeeze-up," with the synonym "anosma." In this case semi-solidified basalt from a main flow of considerable depth, which has cooled slowly, has exuded through cracks in the thick crust of the flow. The largest of the cracks is a mile and a quarter long and is of varying width up to seventy feet. The basalt squeeze-up from this fissure projects in places several feet.

The sides of the protruding basalt tongue are grooved, conforming to the walls of the fissure, and slickenside surfaces are usually present. . . . The more plastic inner layers of the mass have slid over the outer, less plastic plates so that we get a condition of a series of vertical layers. . . . The plates of basalt, as they have been thrust into the air, have bent under their own weight to form graceful arches in some places (Colton).

[20] J. D. Dana, *Rep. Geol. Wilkes Expl. Exped.*, 1849; *Am. Jour. Sci.*, 34, p. 356, 1887.

[21] K. Sapper, *Vulkankunde*, p. 214, 1927.

[22] G. W. Tyrrell, *Volcanoes*, p. 52, 1931.

[23] *Loc. cit.* (18).

[24] H. S. Colton, Lava Squeeze-ups, *The Volcano Letter*, 300, p. 3, 1930.

A smaller but similar squeeze-up at Auckland, New Zealand, has been described by Bartrum[25] as a slickensided dyke with surface protrusion.

These are "linear" squeeze-ups as defined by Nichols, who has described small examples in New Mexico; some of these are hollow as a result of draining-out of a portion of the lava. Nichols[26] describes and figures also "bulbous" squeeze-ups, which are of various sizes up to ten or fifteen feet across.

Among other minor features of pahoehoe surfaces are some "spiracles" or "hornitos"—openings through which gas has escaped from lava as it flowed beneath the pahoehoe crust, carrying up liquid clots so as to build spatter cones similar to those associated with the fissure vents from which lava flows issue but generally of very small dimensions as compared with these. In what is perhaps a rather exceptional case, some quite large mounds of scoria have been interpreted by Daly[27] as relief forms built in this way on the back of a lava flow

Fig. 41. Large cones of scoriaceous spatter on the surface of a basalt lava flow, Ascension Island. (From a photograph by R. A. Daly.)

(Fig. 41). Some moraine-like mound fields in New Zealand which are now interpreted as the hummocky surfaces of arrested lahars (Chapter XIII) have been mistaken in the past for somewhat similar adventive cones or mounds built by gas explosions and clustered on the surfaces of lava flows.

[25] J. A. Bartrum, Lava Slickensides at Auckland, *New Zealand Jour. Sci. and Tech.*, 10, pp. 23-25, 1928.

[26] R. L. Nichols, Squeeze-ups, *Jour. Geol.*, 47, pp. 421-425, 1939.

[27] R. A. Daly, The Geology of Ascension Island, *Proc. Am. Ac. Arts and Sci.*, 60, pp. 3-80, 1925. Large mounds of blocks discovered on a basalt flow at Mangere (N.Z.) by Professor J. A. Bartrum (*in litt.*, 1944) may be of such "hornito" origin, or are perhaps due to explosions made by contact of advancing lava with water.

Pressure Ridges and Residual Ridges

Of a larger order than the detail forms of pahoehoe surfaces such as have been described above are some grouped features of strong local relief found on parts of some lava flows. Of somewhat exceptional occurrence, these occupy in most cases only relatively small portions of lava fields which are elsewhere level-surfaced, as was noted by Russell in the Snake River region, though in that region Geikie[28] found relief present rather generally on the parts of the lava fields he traversed. Among these are some prominent ridges ascribed to pressure (though somewhat doubtfully in certain cases) if aligned at right angles to the direction in which the development of compression in the lava crust can reasonably be explained. Other similar ridges, however, differently aligned or without regular orientation, have been ascribed to the effects of either crumpling or irregular gravity subsidence of a thick pahoehoe crust after the withdrawal of lava from beneath portions of it. The forms in both cases are elongated domes traversed along the crest by a gaping crack or, in some cases, a subsided strip (Figs. 42, 43).

Great pressure ridges fifty feet high were rucked up on a lava flow on the summit of Mauna Loa (Hawaii) in 1940, and in association with them transported blocks of pahoehoe crust were piled into an impassable block field.[29] Such lava jams sometimes result from compression developed by landsliding of a thickened crust on an appreciable slope before the lava under it has become rigid.

Ridges on the lavas of the San Jose valley, New Mexico, are attributed to pressure and described by Nichols,[30] who regards them as distinct in origin from other ridges in the same locality which are residual, due to gravitational subsidence. The pressure ridges are found on the last mile of a great forty-mile-long flow, the surface of which is in most

[28] A. Geikie, *loc. cit.* (5).

[29] H. H. Waesche, Mauna Loa Summit Crater Eruption, *The Volcano Letter,* 468, pp. 2-9, 1940.

[30] R. L. Nichols, Pressure Ridges and Collapse Depressions on the McCartys Basalt Flow, N.M., *Trans. 1939 Am. Geophys. Union,* pp. 432-433, 1939.

parts smooth. The ridges are from 130 to 1200 feet long, up to 100 feet wide, and from 10 to 25 feet high.

In transverse cross-section the sides of the pressure ridges are steep, reminding one of the gable of a house or the cross-section of a broken anticline. They have a medial crack running along

Professor J. A. Bartrum, photos

Fig. 42. Elongated clefts in basalt fields. Above: Longitudinal cleft in the crest of an elongated tumulus or ridge of uncertain origin on a basalt flow at Penrose, Auckland, New Zealand. Below: Trench opened by collapse of a pahoehoe lava tube at Byaduk, Victoria, Australia.

the crest of the ridge which may be as much as 15 feet in width but which is usually much less. They are almost without exception lined up parallel with the flow and are in general close to its margin. The crust must have been several feet thick when folding of these ridges began, and Nichols estimates that the rucking up of each ridge absorbed five to ten feet of compression. Unconsolidated lava welled up into the crest-line clefts.

Nichols rejects the theory (Russell's) that the compression manifested by the arching of thick lava crusts has ever resulted from the drag of slow-moving viscous lava under them in much the same way as "tapestry" folds are formed by the wrinkling of a thin pahoehoe skin. He holds that this drag would be quite insufficient to buckle a crust several feet thick. These particular ridges at McCartys, moreover, are parallel to the general direction of movement of the lava. He suggests that the ridges are indirectly the result of collapse of a crust which was originally arched but buckled in compression when let down by outflow of lava from beneath it. Consistently with this collapse theory, which explains a "period of relaxation" following the folding, the summit clefts show signs of having been opened from below. It is found that they have been filled, and perhaps enlarged, by the injection of wedges of viscid lava. These are still present in a solidified condition, bearing on each side the scorings, or slickensides, inflicted by the enclosing walls of the crevice on the semisolid wedge during its injection. Together with linear squeeze-ups these afford some of the best examples of "grooved" lava.[31] Some further collapse of the surface, which has caused the crest-line clefts to gape, has led to the exposure of the solidified wedges (Fig. 43).

The ridges in New Mexico which Nichols ascribes to corrugation of a collapsing arch or dome are very similar in a general way, and are similarly aligned, to the "stony rises"[32] of Victoria, Australia. These, however, though attributed to collapse, have been regarded as the residual portions of a thick solidified crust left standing after the draining-away (perhaps out of pahoehoe tubes) of lava from beneath intervening strips in such a manner as to allow these to subside. Incompetent anticlines, if formed thus between subsided crustal

[31] R. L. Nichols, Grooved Lava, *Jour. Geol.* 46, pp. 601-614, 1938.

[32] So named by E. S. Hills (*Physiography of Victoria,* 1940); the inappropriate term "barriers" is applied to these forms by Skeats and James (*loc. cit.* (1)).

Fig. 43. Crest of a pressure ridge on McCartys basalt flow, New Mexico, showing the cleft containing a wedge of lava injected from below. (After R. L. Nichols, *Journal of Geology.*)

strips prior to the final hardening of the lava under them, would be in tension, and consistently with this the stony rises have clefts—gaping cracks and subsided strips—along the crests. The stony rises of the type locality are crowded near what has been a temporary front of the lava flow of which they form a part.

The term "barrier" is appropriately applied to the coffer-dam-like ridges which mark temporary halts of some lava-flow margins. Several of these have been described among the Victorian lavas. The solidified fronts have become ridges as a result of a draining-away of lava from beneath crusts upstream from them, which has caused subsidence there. These barriers must, therefore, be included among features due to collapse. In both the stony rises and the marginal barriers of Victoria the thickness of the anticlinal carapace is very considerable, as is indicated by the very general presence of columnar shrinkage structure in it. The columns are in all cases at right angles to the surface—of radial pattern, therefore, in the arched forms—which is a clear indication that the ridges are original forms, i.e. not formed by erosion. The same is true of the ridges on the Snake River basalt fields and of those on the Penrose lava flow (p. 112) near Auckland, New Zealand, one of which is shown in Fig. 42.

Among forms obviously attributable to collapse of the roofs of caverns are numerous trenches on basalt surfaces (Fig. 42). The caverns have originated as the tunnels, or tubes, through which lava continues to flow under a pahoehoe crust. Many such tunnels, which have been evacuated by escape of lava after the supply from the source has ceased, are of large dimensions. Owing to thickness and strength of the pahoehoe crust over them tunnels may remain long in existence; but Nichols has described some pits due to partial collapse of tunnels taking place while they still contained some liquid lava.[33]

Block Fields of Pahoehoe

Broken pahoehoe crust may be rafted or swept along, piled up like ice jams, and stranded as block fields, some of which have been mistaken for aa surfaces. The blocks show signs of partial engulfment. Examples of such block fields have been described at Ascension Island[34], in Réunion[35], at McCartys flow, New Mexico[36], and on a great scale on the lava flow of 1940 on the summit of Mauna Loa, Hawaii[37]. Daly describes the lava-block jams of Ascension as follows:

> The more rigid surface shell downstream tends to be folded and broken by transverse shears. Back of the frontal scarp, itself largely caused by the forward thrust of the surface shell, are found

[33] *Loc. cit* (31).

[34] C. Darwin, *Geological Observations,* 3rd ed., p. 41, 1891; R. A. Daly, The Geology of Ascension Island, *Proc. Am. Ac. Arts and Sci.,* 60, pp. 3-80, 1925.

[35] A. Lacroix, *Le volcan actif de l'île de la Réunion et ses produits,* p. 96, Paris, 1936.

[36] R. L. Nichols, *loc. cit.* (31).

[37] H. H. Waesche, *loc. cit.* (29).

isolated pressure ridges, grouped anticlines and synclines, anticlinoria, synclinoria, overthrusts, and underthrusts.[38]

In the case of the New Mexico example,

it seems likely that the ruptures are only superficial and that below three or four feet the basalt of the flow is compact. The average thickness of these upturned fragments indicates that the break-up must have occurred when the crust was about one foot thick.

Flat under-surfaces of some such blocks of pahoehoe crust, in common with other lava surfaces subjected to differential movement before they are quite solid, are grooved, or slickensided, if they have been thrust over rough edges of hard lava.[39] Such grooved-lava forms were observed by Russell on the under surfaces of blocks of broken crust erroneously termed aa.[40]

Lava Blisters

Yet another rather prominent relief feature which may be found on pahoehoe lava is a "lava blister." This term has

Professor J. A. Bartrum, photo

Fig. 44. Lava blisters at Byaduk, Victoria, Australia.

been restricted in Australia to steep-sided domes raised as blisters on thin sheets of lava which have spread over marshy flats. They are distinct in nature from tumuli. These blisters,

[38] R. A. Daly, *loc. cit.* (34), p. 18.

[39] R. L. Nichols, *loc. cit.* (31).

[40] I. C. Russell, *Volcanoes of North America,* p. 60, 1897.

of which there is a swarm at Byaduk, Victoria[41] (Fig. 44), have resulted apparently from a true blister-like lifting of the thin lava sheet by pockets of steam generated under it from the water in the ground. In some cases at Byaduk the steam seems to have burst explosively through so as to convert a blister into a cratered cone, but on most blisters collapsed summit hollows are quite shallow. Some at least of the blister domes are traversed by a few radial cracks, and there are some concentric fissures. Where gaping cracks are found they give the impression that the domes are hollow.

Lava Caves

Uncollapsed portions of lava tunnels such as have been already mentioned still exist abundantly in most basalt fields even where activity has been long extinct. Many have been explored—for example, at Auckland, New Zealand, at Byaduk, Victoria, and in the basaltic volcanic fields of Idaho, as well as in the lava flows at Kilauea, Hawaii (Fig. 45), and on Savaii and Upolu Islands, Samoa.

Among the features of tunnel interiors are crusts on the walls and miniature lava benches (compare the bench lavas around Hawaiian lava lakes) which mark temporary surface levels of the subsiding lava at the time when the supply was no longer sufficient to keep the tube full. Wall crusts that have partly peeled off have arched over under their own weight while still plastic.[41a]

> At this time, when the river of melt bubbling along the tube lowers so that the upper half of the tube is full of gas or air, the walls are of bright yellow incandescence and the gases escaping from the lava are continually burning to maintain a very high temperature on the ceiling of the cavern. With this temperature above 1200° C., the air being sucked in below as the hotter gas escapes through the cracks and windows in the roof, there is set up a blast-furnace condition. . . . This . . . melts the tunnel walls to a glaze of quite different crystallinity from normal lava. . . . On these walls [hang] . . . stalactites, some like currants, some like

[41] Skeats and James, *loc. cit.* (1).

[41a] Examples of this have been pointed out by Professor J. A. Bartrum in the Onehunga lava-tube cavern, Auckland, New Zealand.

grapes, some like walking-sticks, and some like worms. . . . They are the material of the gas-melted glaze. . . . These stalactites are sometimes two feet long.

Fig. 45. Lava-tube cavern, Kilauea, Hawaii.

Stalagmites formed on the floor "must be made by drip in some early high-temperature stage of the stalactite formation above them. . . . The drip is melted rock.[42]

Emplacement of Pahoehoe Lava

Pahoehoe flows, which chill at the surface and freeze over, continue to flow beneath the crust so formed. Their mode of emplacement is very different from that of lake-like and ponded flows. The pahoehoe lava occupies the available space strip by strip as streams of pyromagma pour out and feed the margin of the flow after passing by way of tunnels

[42] Quotations from T. A. Jaggar, *loc. cit.* (18).

under the skin and through the body of that portion of the flow that has already been arrested by partial solidification. Small flows congeal on slopes and terminate with thin edges, but more abundant lava spreads until its surface slope is gentle. The method of advance by outbreak from lava tubes to form new flow units, which soon develop skins and become tubes themselves, has been described by Stearns as "projecting one toe after another." Not only do the flow units accumulate side by side and one in front of another, but they pile over one another also. Thus sections through thick flows—those, for example, of the San Jose valley, New Mexico[43]—reveal that the lavas have arrived in place as successively congealed thin lobes or tongues, each with a somewhat convex surface, though the ultimate surface of the compound sheet is not far from level.

---

[43] R. L. Nichols, Flow Units in Basalt, *Jour. Geol.*, 44, pp. 617-630, 1936.

CHAPTER X

Aa and Block Lava; Scoria Mounds

AA LAVA, THE SOLIDIFICATION OF WHICH HAS BEEN DISCUSSED IN the foregoing chapter, has a bristling "scoriaceous," or "clinkery," surface. It has often been inferred that there is a superficial scoriaceous layer which floats on an aa flow, because such lavas when solidified contain compact, stony, and largely crystalline basalt in the body of the flow.[1] Russell has cited as evidence that aa scoria floats the discovery of "grooves and striations" on the under surfaces of "blocks of lava";[2] but this observation refers apparently to grooved lava in a jam of broken pahoehoe crust (Chapter IX) such as has often been reported as aa.

As they are formed on a moving aa flow superficially formed "granules and crusts" must sink, however, according to the principles of clastolithic sedimentation (Chapter IV), and the belief that the aa flow carries a jostling cargo of floating scoria seems just as mistaken as the fallacy that crags projecting above the surface of a lava lake are floating islands. An aa flow with scoriaceous surface is on the other hand a sponge of clinkery epimagma permeated by pyromagma and generally traversed also on the surface by rushing torrents of pyromagma (Fig. 46).

The whole flow presents a certain resemblance to the bed of an aggrading river; the pyromagma streams are in braided courses; and as they deposit solidifying aa this is built into moraine-like levees which border them and remain as prominent ridges after the whole flow is solid, making considerable relief. The surface streams feed the front of the flow in much the same way as the pyromagma escaping from tunnels feeds the front of a pahoehoe flow. The body of aa lava advances however, glacier-like, at a slow pace (Fig. 47) "moving like

[1] See, for example, H. S. Washington, Santorini Eruption of 1925, *Bull. Geol. Soc. Am.*, 37, p. 273, 1926.

[2] I. C. Russell, *Volcanoes of North America*, p. 60, 1897.

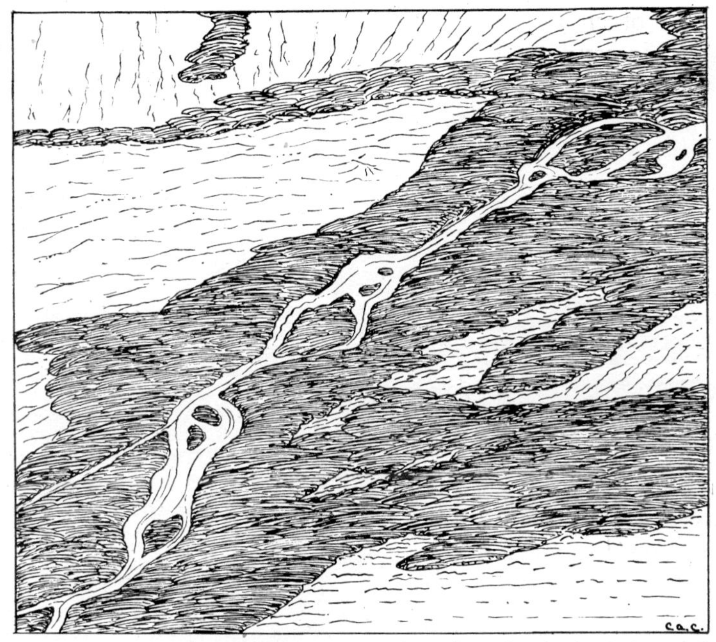

Fig. 46. Lava torrents in braided courses on the surface of a slowly-moving aa flow, Hawaii, 1926. (From a photograph.)

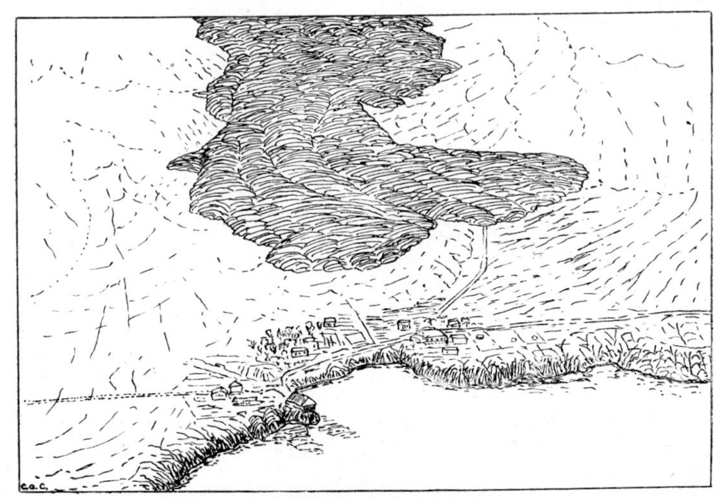

Fig. 47. The doomed village Hoopuloa, Hawaii, 1926. The aa lava, advancing 3 feet per minute, afterwards buried the village. The lava stream was here guided in its course towards the sea by the broad consequent depression between the convexities of two older tongues of lava, so that its emplacement resulted in an inversion of relief. (Drawn from a photograph.)

a wall of clinker" (N. D. STEARNS). Mobility is maintained in the interstitial pyromagma by both increasing gas-concentration and the liberation of heat as clastolithic sedimentation and crystallisation proceed. Solidified aa is generally scoriaceous on the under surface as well as above.

R. H. Finch, photo

Fig. 48. An aa lava flow, Mauna Loa, Hawaii. (Reproduced by permission of the Hawaiian Volcano Observatory.)

The surface of genuine aa lava is "roughly scoriaceous, horridly jagged, with projections often a foot or more long that are bristled all over with points and angles."[3] Typical Hawaiian aa is shown in Fig. 48.

A good example of an aa flow is one which issued as a great flank outflow from a lateral fissure in the cone of Vesuvius in 1906, draining away the pyromagma from the crater and lowering the lava column to such an extent, apparently, as to initiate the "gas phase" of the great eruption (Chapter VI).

[3] J. D. Dana, *Characteristics of Volcanoes*, p. 241, 1890.

Perret[4] has described this flow, which closely resembles Hawaiian lavas, as

> a comparatively homogeneous but fully scoriaceous layer covering a mass whose very considerable degree of liquidity even at a distance from the source permitted it to ascend to an almost surprising distance. . . . The fluent material followed even insignificant natural depressions, bifurcating and reuniting . . . , following sunken roads and railways. . . . There was no accumulation of material around the vents or any attempt at cone-building.

Perret terms the surface of this flow "medium aa" and remarks as follows on the great variety presented by different aa surfaces:

> We may have, for example, a mass almost wholly of detached fragments; or a flow presenting the appearance of a ploughed field, so earthy and incoherent are its materials, and with "séracs" like the flexed portions of a glacier; a coherent, massive core from whose advancing front fall blocks like bituminous coal, black without and red within, the centre of the stream bearing masses weighing tons; or an exceedingly rough and all but impassable flow, whose asperities are, however, united in one continuous monolith of granitic firmness.[5]

Block Lava

As used by many authors "block lava" is synonymous with "aa". Washington,[6] however, and more recently Finch,[7] have pointed out a contrast between the typically spinous "clinkers" forming a true aa surface and the blocks of angular form of which some lava fields, notably those consisting of the more acid lavas, are superficially composed (Figs. 49, 50). Finch suggests that the descriptive term "block lava" should be reserved for these latter.

Washington, though he regards block lava and aa as having something in common, has noted the distinguishing character of the former as follows: Block lava is composed of "loose, . . . angular blocks with sharp edges and smooth faces"

[4] F. A. Perret, *The Vesuvius Eruption of 1906*, pp. 75, 76, Carnegie Inst., Washington, 1924.
[5] *Loc. cit.* (4), p. 75.
[6] H. S. Washington, *loc. cit.* (1).
[7] R. H. Finch, Block Lava, *Jour. Geol.*, 51, pp. 764-770, 1933.

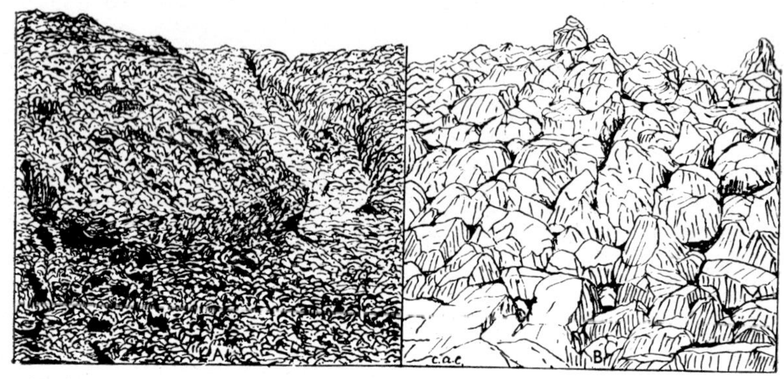

Fig. 49. Block lava of extreme type contrasted with aa. A: Alika aa flow, Hawaii, 1919 (from a photograph by Kanemori). B: End of a block-lava flow, Santorini, 1925 (from a photograph by H. S. Washington).

Professor R. Speight, photo

Fig. 50. Aa or block lava—a flow of andesite lava from the Red Crater on the summit of Tongariro, New Zealand.

(Fig. 49, B), whereas, in contrast with the looseness of surface on block lava, a scoriaceous basaltic aa surface is more or less agglutinated. "The partially agglutinated condition of typical aa is due," he remarks, "to the comparative fluidity of the basaltic lava, which also determines the characteristically rough and jagged surfaces of the aa." As Finch remarks, however, the cargo of loose blocks on a block-lava surface must have under it a stream of lava with greater continuity.

The advance of a block-lava flow, moving with a steep front 20 feet high, at Paricutin, Mexico, in 1943, has been described as follows: "At its front the flow broke off hard chunks, tossed them forward, and then occupied the gap. Peeling off, the fragments exposed a glowing, molten interior."[7a]

Block-and-Ash Flows

Some supposed block lavas may not be *lava* flows at all. Observations in Martinique and at Merapi (Java)[8] have shown

Fig. 51. Block-and-ash flows (such as are termed ladus in Java) from Mont Pelée, Martinique. Some ash may have been washed by rain from among the blocks. (Drawn from a photograph by F. A. Perret.)

that some extensive accumulations of angular blocks in glacier-like form which resemble some block-lava flows are in reality products of *nuée ardente* eruptions and have contained no liquid. These are "block flows" or "ladus" (ladoes). Block flows in the strict sense (Fig. 51) originate as rock slides of glowing but probably "dead" lava feebly ejected by minor *nuée ardente* eruptions from the carapaces of tholoids (and plugs of hot solidified lava) such as those on the summits of Mont Pelée and Merapi (Chapters I and XI), from which also many more mobile *nuées ardentes* consisting largely of fine ash have been expelled, or from the steep margins of coulees of viscid lava such as exude and congeal on the upper slopes of Merapi and other East Indian cones.

[7a] J. A. Green, Paricutin, *Nat. Geog. Mag.*, 85, pp. 129-136, 145-156, 1944.

[8] F. A. Perret, *The Eruption of Mt. Pelée 1929-1932,* Carnegie Inst., Washington, 1935; M. Neumann van Padang, De uitbarsting van den Merapi 1930-1931, *Vulk. Meded.*, 12, p. 22, 1933; C. E. Stehn, Merapi, *Bull. Neth. Ind. Volc. Surv.*, 70, pp. 120-131, 1935.

These block flows are mixed with a varying proportion of finer ash. Besides such "flows" there are gravitational rock slides, which accumulate also as ladus, for "gas-free lava of the Peléan type on leaving the vent will break up, separating into blocks without ash, forming a slide of great brilliancy, it may be, but without a single characteristic of the *nuée ardente*" (PERRET). In addition to these some piles of blocks are derived from hot monoliths fallen from the spines which are squeezed up through the carapace of the tholoid. These crack to pieces during cooling, as Perret has observed.

SPATTER OR DRIBLET CONES

Closely associated with basaltic lava fields and rising in many cases so abruptly from lava plains as to present a strong contrast with the insignificant relief of their surroundings, are piled-up forms which mark the positions of many more or less temporary lava vents. These are much more conspicuous than the features of subdued relief—low cones and domes (Chapter VII)—which mark the points of emission of less viscous, non-explosive, and freely effervescing lava in greatest volume. Such steeper forms—cones and mounds—are built by the piling-up of clots and fragments ejected forcibly in association with escape from the pyromagma of voluminous gas under considerable pressure. Over some such "lava fountains" only low mounds or domes are built of spatter—i.e. clots which are still liquid when they fall and are agglutinated like a multitude of miniature lava flows.[9] In many cases these domes are hollow, and they may be lined with lava stalactites as a result of high temperature developed by combustion of escaping gas.

FIRE FOUNTAINS

As temporary phenomena at lava vents in Hawaii high jets of escaping gas are of common occurrence which are known from their appearance as "fire fountains," because they carry up with them much incandescent pyromagma as large clots and abundant finer spray. Fire-fountaining to a height of 500 or

[9] Termed "driblets" by J. D. Dana (*Am. Jour. Sci.*, 35, p. 32, 1888).

600 feet or even more has been frequently observed both at points on the lateral fissures from which lava issues to flow down the flanks of Mauna Loa and from vents on the summit of the dome.[10] Around the orifices at which fire-fountaining activity is concentrated mounds or cones of material derived from solidification of the projected lava accumulate. The term "pyro-explosion," which has been used as a synonym for "fire-fountaining," seems less appropriate as the phenomenon is more continuous than even a series of explosions. In the Strombolian type of volcanic activity, however, as manifested frequently in minor eruptions within the crater of Vesuvius, the phenomena are intermittent and explosive, and the products are similar in a general way to those of fire-fountaining. It seems in such cases that large bubbles, or swarms of large bubbles, rise from time to time in the pyromagma and the gas expands explosively as it escapes, projecting liquid lava clots and spray.

In the case of continuous fire-fountaining ejected material when it falls cannot re-enter the orifice, but in Strombolian pyro-explosion it may do so—only to be re-ejected, however. The "conelet" usually present over the pyromagma vent within the crater of Vesuvius (Fig. 9) is a "spatter" cone. Through such a chimney most of the gas escaping from the lava column must pass.

The fine spray of pyromagma shot out to heights which may approach 1000 feet by violent fire-fountaining includes a little finely divided froth, or pumiceous ash, which is blown far away, but there are more abundant pellets of nut and pea size ("lapilli") consisting of vesicular lava; some of these also may be carried to a considerable distance by a strong wind. It is impossible on data available to distinguish all such material with certainty from the fairly fine products, generally described as "tuff," which have had their origin in explosive eruptions such as occur when basaltic pyromagma makes contact with surface water. Some tuff-built rings and cones which are attributed to such an origin may have been in reality

[10] During the eruption of 1940, for example. (H. H. Waesche, Mauna Loa Summit Crater Eruption 1940, *The Volcano Letter*, 468, pp. 2-9, 1940.)

built by fire-fountaining. These are generally spread over wider bases than are the mounds at present under consideration, however, and may be described most fittingly along with the products of explosive eruptions (Chapter XIV).

Lapilli and also coarser scoriaceous material and bombs derived from fire-fountaining and pyro-explosion fall in the immediate vicinity of the vent, and thus build a steep-sided mound around it. The close relation of such mound-building of discrete fragments to the building of spatter cones is shown by the occasional incorporation of large splashes of lava spatter with the loose materials of the walls.[11] A change of

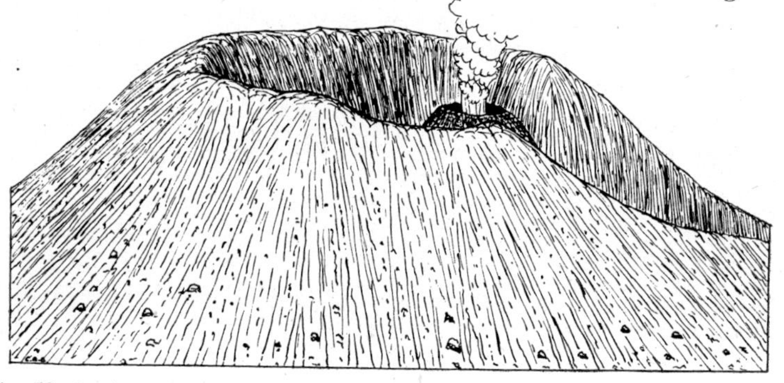

Fig. 52. Scoria mound built by Strombolian eruption within the main crater of Vesuvius in 1930. It has been breached by lava outflow, and a newer conelet has been built of spatter within it. (Drawn from a photograph.)

habit from spatter-cone to scoria-mound building which took place in the crater of Vesuvius in 1930 in a phase of vigorous activity illustrates the community of origin of these forms. The ejected lava clots, being thrown to a greater distance than usual, piled up a scoria mound of typical form (Fig. 52), within which a new spatter-built conelet later appeared.

Scoria

Among the products of fire-fountaining and pyro-explosion, and intermediate in size between lapilli and masses of liquid lava spatter, is scoria in large to nut-size lumps, mixed in varying proportion with more or less symmetrical bombs. This

[11] H. T. Stearns, Geology and Ground-water Resources of the Island of Oahu, Hawaii, *T.H. Div. Hydrog. Bull.*, 1, p. 14, 1935.

"scoria"[12] consists of masses of more or less vesicular pyromagma which have consolidated in mid-air so as to present a certain resemblance to the rough lava on the surface of an aa flow, though generally there has been less thorough ejection of gas and less crystallisation. Though vesicular, this scoria is much denser and less foamy than "pumice," which is typically formed from the more acid magmas and is rarely found, except in small particles, among the products of basaltic fire-fountaining and pyro-explosion.

The "cindery," or sometimes clinker-like, appearance of basaltic scoria has led some writers to designate it "cinders," and, following Dana, they term scoria cones "cinder cones."[13] Thus Stearns writes: "Heavier, scoriaceous masses are usually termed 'cinders'."[14] These writers refer to the surface rock of aa lava as "clinker." Clinkers are said to differ from cinders "in being stony instead of glassy, less vesicular, and generally spiny."[15]

Bombs

In a sense scoria lumps of irregular form are a variety of volcanic "bombs."[16] This name is best reserved, however, for bodies derived from pyromagma which have assumed more definite shapes during projection; though fragments of old rock (perilith) which have fallen into a crater and have later

[12] "Scoria: An Italian word meaning dross or scum, applied to the coarsish broken material ejected by a volcano . . . Less cellular than pumice, and rough in aspect like cinders." (T. G. Bonney, *Volcanoes,* 2nd ed., p. 342, 1902.)

As used by J. D. Dana "scoria" may include pumice as well as highly vesicular basalt as found in pahoehoe crusts. But he defines "ordinary scoria" as "such as is common about cinder cones outside of the crater, mostly stony in texture; the vesicles 65 to 95 per cent of the mass." (*Am. Jour. Sci.,* 35, p. 223, 1888.)

Typical scoria is derived from basalts and closely allied lavas. The close association of scoria ejection with basaltic activity has been recognised by G. Mercalli (*I vulcani attivi della Terra,* p. 119, 1907) who has bracketed together as a single type of activity the "Hawaiian," characterised by effusive basalt, and the "Strombolian," which projects basaltic scoria.

[13] "Scoria" and "scoria cone" are standard terms with I. C. Russell, however. (*Volcanoes of North America,* pp. 74-75, 1897.)

[14] H. T. Stearns, *loc. cit.* (11), p. 14; J. D. Dana, however, makes "cinders" a synonym of "lapilli."

[15] H. T. Stearns, *loc. cit.* (11), p. 18.

[16] Volcanic bombs have been described by H. Reck, Physiographische Studie über vulkanische Bomben, *Zeits. Vulk. Ergänzungsband,* 1915; see also F. von Wolff, *Der Vulkanismus,* I, pp. 383-384, 1914; and L. F. Brady and R. W. Webb, Cored Bombs from Arizona and California Volcanic Cones, *Jour. Geol.,* 51, pp. 398-410, 1943.

been ejected with thin skins of new lava adhering to them are also generally classed as bombs (Fig. 53, 1); and large nodules of sugary crystallised olivine, such as are found as inclusions in basalts, are sometimes thrown out thinly coated as "olivine bombs."[17] Spheroidal bombs from a few inches to a foot or

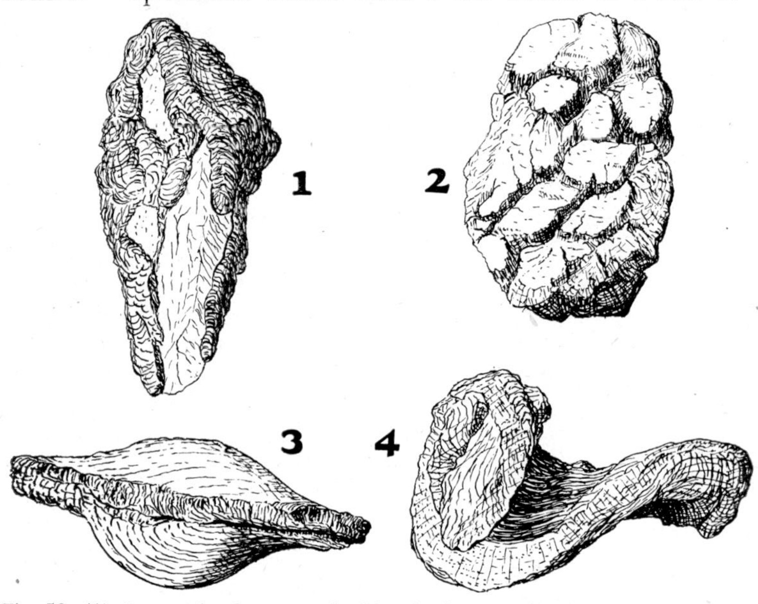

Fig. 53. (1) An angular fragment of white rhyolite coated with black basalt lava and ejected as a bomb from Tarawera (N.Z.) in 1886. (2) A "breadcrust" bomb ejected from Mont Pelée (Martinique) in 1902. (3) An "almond" bomb (length 4 inches), and (4) the twisted end of a much larger bomb of the same kind. (3) and (4) are from Auckland, New Zealand.

more in diameter, which have solidified rapidly on the surface, exhibit a netted pattern of wide-open cracks attributable to stretching of the skin due to swelling of the gas-rich core after the external crust has been formed. These are "breadcrust" bombs (Fig. 53, 2).[18] Other bombs, which generally are smaller, being commonly only a few inches in length, have obviously rotated very rapidly during their passage through the air and have thus spun into a spindle shape (Fig. 53, 3). These have been termed "almond" bombs because of the flattened form

[17] A. Lacroix, *Les enclaves des roches volcaniques,* p. 484, 1893.

[18] This descriptive term was applied by Johnston-Lavis to the bombs ejected from Vulcano in 1888.

they assume, apparently as a result of impact with the ground, when they fall while still in a semi-liquid condition. Large bombs of this kind have the thin ends of the spindle strangely twisted (Fig. 53, 4). This "may be due to differential motion during flight when chains of bombs are formed, especially when individual beads on the chain differ in size. . . . Twisting . . . of the ribbons occurs between beads" (BRADY and WEBB).

SCORIA MOUNDS

Such are the materials that accumulate as "scoria mounds," "scoria cones," or "cinder cones." The presence in addition of fragments torn from underlying rock formations (perilith) is not to be expected among the products of fire-fountaining and pyro-explosions. Cones or mounds which contain these fragments must be, in part at least, the results of explosions such as may be caused by contact of rising pyromagma with surface water. Fire-fountaining may supervene, however, and build its typical scoria mound of purely magmatic material over a pile of mixed debris that has been ejected by such an explosion, as seems to be the case in some examples near Auckland, New Zealand. The cores of some of the bombs built into basaltic scoria cones in Arizona and California consist of rock inclusions derived from perilith, though many bomb cores are olivine nodules and a few are large crystals of other minerals probably derived, like the olivine, from basaltic magma.[18a]

It is obvious that continuously showered coarse fragmentary materials may build a mound very rapidly around an active fire fountain (Figs. 54, 55); but such mounds rarely grow to very large dimensions because of the temporary and shifting nature of the vents in which basaltic pyromagma rises. They vary in height up to a few hundred feet, with about 500 feet as a limit rarely exceeded. These mounds and cones diversify basalt plains.

The mounds are uniformly steep externally (Figs. 56-58), for the material lies on their slopes at the angle of repose, which approaches 35° for coarse scoria. If added to gradually by rather quiet (i.e. non-paroxysmal) emission of scoria a

[18a] Brady and Webb, *loc. cit.* (16).

T. A. Jaggar, photo

Fig. 54. A fire fountain projecting scoria-making lava clots to a height of 200 feet, Alika flow, Mauna Loa, 1919. (Reproduced by permission of the Hawaiian Volcano Observatory.)

mound may grow into a *cone* surrounding an axial pipe in which lava has continued to rise, gently fountaining (Fig. 57). If, on the other hand, fountaining is rather violent, the coarse ejected materials fall in a wider ring and the common form of mound is produced, with a broadly convex summit (Figs. 56, 58). In this case the central crater—a feature which is, however, not always distinctly developed, or, at any rate, not always preserved in the final stage of growth—will be broadly bowl-shaped with centripetal slopes, like those of the exterior, at the steep angle of repose for coarse fragments of irregular shape (Fig. 56).

"Breached" craters (Fig. 52) may result from overflow of lava, which erodes and breaks down the weak scoria-built wall over which it has spilled, but some seem to be incomplete

G. A. Macdonald, photo

Fig. 55. Lava fountain 200 feet high building a mound of scoria on the flank of Mauna Loa, Hawaii, May 3, 1942. Darker (cooler) masses are falling.

Fig. 56. Mt Eden scoria mound, Auckland, New Zealand, a view showing the crater. The smooth constructional slopes on the exterior have been replaced by artificial terraces, most of which are old Maori fortifications.

V. C. Browne, photo

C. A. Cotton, photo

Fig. 57. Te Ahuahu scoria cone, North Auckland, New Zealand. Relief 400 feet. Note the single erosion gutter on the cone.

N. H. Taylor, photo

Fig. 58. A scoria mound of exceptionally large size, Mt Maungatapere, Whangarei, New Zealand.

rings (crescents) which have been thrown up around vents from which lava has continuously flowed during the eruption.

Though scoria-built mounds are generally minor features, there are a few quite large volcanoes which might be described as overgrown—in some cases multiple and composite—scoria cones. The great mound Izalco, in San Salvador, has been rapidly built of such material by Strombolian eruptions. Since its first appearance in 1770 it has grown to a height of 3000 feet. A multiple and composite example (incorporating lava flows) is the volcano Stromboli, in the Lipari Islands. This volcano has grown in ancient times to a height of more than 3000 feet above sea-level, or 6000 feet above its base on the floor of the Mediterranean Sea. Its flanking slopes are 30°. The activity in historic times has not been vigorous fire-fountaining but only mildly explosive ejection of scoria bombs and spatter from a group of vents in a lateral crater nearly 1000 feet below the summit. Early in the present century explosions of greater intensity occurred, but after a short period of rest the volcano has since resumed its typical "Strombolian" activity.

Some Mexican volcanoes also which rapidly build high cones by spectacular large-scale fire fountaining should perhaps be included here—e.g. Jorullo and Paricutin (p. 225).

Adventive Cones

Some basalt domes carry on their backs hundreds of scoria mounds and cones. Some of these on Etna (Fig. 59), on Haleakala (island of Maui), and on the three domes Mauna Loa, Hualalai, and Mauna Kea in Hawaii (Fig. 60) are so

Fig. 59. Summit cone and numerous adventive cones on the flanks of the Etna dome, as seen from Catania, Sicily.

large that if it were not for the fact that they are dwarfed by the great domes on which they are "parasitic," or "adventive," they would be known as independent volcanoes. Most, if not all, of these are developed, however, in relation to fissures from which effusion of lava, adding to the bulk of

Fig. 60. Adventive cones on the crown of the great basalt dome Mauna Kea, Hawaii. (From a photograph.)

the parent domes, has taken place. Stearns[19] has shown that those on Haleakala, in particular, are clustered along three convex ribs of the dome which mark the positions of rift zones through which magma has risen during its growth.

Vast numbers of adventive cones, along with scoria mounds and spatter cones, have been buried under lava flows during the growth of great domes and, no doubt, also during the up-building of lava plateaux. In some dissected basalt terrains abundant traces of such buried scoria-built features are found, though in others they appear to be rare or absent. According to Stearns,[20]

> large cinder cones are not commonly built on low basaltic domes with . . . fluid lavas. They are abundant on Mauna Loa, where . . . fluid lavas occur; hence the height or age of the dome probably has a direct bearing on the production of cinder cones. Evidently lava must be very frothy to rise to great heights above sea-level.

Fire-fountain deposits in the Waianae Range, in Oahu, which has been a relatively low dome, are associated with lavas of a special type, however. Thus "other factors besides height of the volcano are evidently important in the formation of cinder cones." A departure from the simple effusive plateau-basalt type of magma takes place in some cases, apparently,

[19] H. T. Stearns, Origin of Haleakala Crater, *Bull. Geol. Soc. Am.*, 53, pp. 1-4, 1942.

[20] H. T. Stearns, *loc. cit.* (11), p. 94.

after a long period of dome-building, and this seems to be of greater importance than the height to which the dome has grown. It may be that some intercrustal pocket of the primitive magma when nearly exhausted has yielded a residue of magma differentiated to approach andesite in chemical composition. Such, at least, is an explanation sometimes offered to account for excessive fire-fountaining and perhaps also for pyro-explosive activity such as may have become the dominant phase during the final stage of the active life of the Mauna Kea volcano, when effusion of lava had ceased and the building of unusually large cones of scoria was in progress on the summit and flanks of the dome (Figs. 22, 60).

In the most recent phase of activity on the Haleakala volcano[20a] also some lavas have approached andesite in composition, and scoria cones of considerable size have been built (Figs. 23, 61), though these have been of considerably less volume than effusive lavas.

In the case of the adventive-cone-building phase of activity on Mauna Kea some observers have seen what they regarded as evidence not merely of violent fire-fountaining but rather of a complete change to the ash-producing habit of andesitic volcanoes. Such a supposed change has been attributed by Dana[21] to "declining heat"—i.e. a cooling-off of the magma column. "A volcano", he writes, "often gasps out its life in cinder ejections; for this is the meaning of the summit cinder cones of Kea, Hualalai, and Haleakala." An examination of a number of adventive cones on Mauna Kea made by Jaggar[22] has revealed, however, that these have been built not of pumiceous ash but of the typical fire-fountaining products scoria and almond bombs. Some of their slopes, indeed, are steeper than thirty-five degrees,

> being evidently made of plastered lava fling, and steeper than the angle of rest of loose material. . . . The rock fragments other than the bombs are of broken pahoehoe lava, not pumiceous, and having

[20a] H. T. Stearns and G. A. Macdonald, Geology and Ground-water Resources of the Island of Maui, Hawaii, *T.H. Div. Hydrog. Bull.*, 7, 1942.

[21] J. D. Dana, *Am. Jour. Sci.*, 35, p. 31, 1888; 36, p. 108, 1888.

[22] T. A. Jaggar, Geological Notes on Mauna Kea, *Bull. Haw. Volc. Obs.*, 13, pp. 75-76, 1925.

Fig. 61. Lavas and scoria cones marking the final outburst of activity within a great eroded valley on the old basalt-built dome Haleakala, Maui, Hawaiian Islands. View looking south-east.

the appearance of being the broken ejectamenta of rather violent lava fountains in a pasty melt.

One explanation that may be suggested for the presence of a high-built cone on the summit of Etna is that differentiation has impaired the capacity of the lava column under it to supply pyromagma of an effusive or non-explosive nature. Effusive basalts are still poured out abundantly on the flanks of the Etna dome, however, progressively enlarging its diameter. The supply of magma is by no means approaching exhaustion. Actually the explanation of the complex activity of Etna seems rather to be that it resembles Vesuvius more closely than Hawaiian volcanoes, differing from Vesuvius only in that in the course of long ages it has poured out a vastly greater bulk of lava—sufficient to build a dome (pp. 59, 78).

Undissected Scoria Mounds and Cones

Scoria cones and mounds tend to escape dissection; they are a monument to the paradoxical aphorism "gravel is a resistant rock." But for their being clothed with vegetation mounds and cones which must be many thousands of years old might have have been built yesterday. The porosity of a scoria pile, which is generally without any appreciable amount of fine ash in the interstices, prevents the formation of surface streams even in the heaviest rains. Basaltic lava cones, equally permeable, are equally resistant to erosion. Rangitoto Island, for example (Figs. 13, 14), though in all probability it is thousands of years old, is without water courses. Te Ahuahu scoria cone, in northern New Zealand, is traversed by but one ravine, and that is only a narrow gutter a few feet deep (Fig. 57) draining a summit-crater hollow which has a soil-covered floor and collects some water in heavy rains. Though stripped of their natural vegetation to make way for the growth of grass pasture, the soil-covered slopes of the cone show no signs of the development of any soil-eroding gullies.

Most of the numerous scoria mounds, or adventive cones, on the summit of the Mauna Kea dome, in Hawaii, which has experienced four glaciations, were built, according to Wentworth and Powers,[23] prior to the last glacial stage. Yet these scoria-built features are scarcely, if at all, dissected. They must have endured thus for tens of thousands of years.

Neither scoria mounds nor basalt cones are everlasting, however. Weathering and the formation of soil gradually seal the surface. Rill formation and dissection are inevitable, however long-delayed they may be. It is probable, indeed, that under hot-and-wet climatic conditions weathering is so rapid that what has been said above of delayed dissection may be applicable only to these forms in regions with dry climates and in the cool zones.

---

[23] C. K. Wentworth and W. E. Powers, Multiple Glaciation of Mauna Kea, Hawaii, *Bull. Geol. Soc. Am.*, 52, pp. 1193-1218, 1941.

CHAPTER XI

Viscid Lavas; Coulees and Tholoids

In general the forms assumed by non-basaltic volcanic accumulations differ remarkably from those of basaltic lavas and the associated products of basalt eruption. It is true that no sharp line of distinction can be drawn between the effusive, or lava-making, and the explosive, or pumice-making, volcanoes and types of volcanism; but, judged by their products, the contrast is striking nevertheless.

Convex Lava Flows

The non-basaltic (that is, in general, the explosive) magmas furnish chiefly an abundance of debris of the kind that results from the explosive disintegration of magma (Chapter V), but some viscid lava streams proceed from such volcanoes, and also some lava masses are squeezed out in a condition too viscid to flow. Flows of viscid lava generally become tongues of small or moderate dimensions, which solidify and come to rest as distinctly convex features (Fig. 62), some of them congealing on very steep slopes, though in exceptional cases they advance over level ground. From steep-fronted tongues congealing on upper slopes of volcanic peaks great boulders break away and red-hot talus streams down; and ladus make extensive accumulations of breccia at lower levels, as has been observed during recent eruptions in Java and Flores.[1]

Flows of this kind are relatively thick and steep-sided and are of sufficient volume to add considerably to the bulk of some volcanic mountains by plastering on convex ribs and even by adding projecting spurs. Thus bulky flows of andesite descend the slopes and project from the broad base

[1] C. E. Stehn, Flores: Eruption of Lewotobi Lakilaki, *Bull. Neth. Ind. Volc. Surv.*, 58, 1932; Merapi (Java), *ibid.*, 70, 1935.

Fig. 62. Glacier-like trachyte lava flow, Ascension Island, South Atlantic. (After R. A. Daly, *Proc. Am. Ac. Arts and Sci.,* 1925.) Compare photograph in *National Geographic Magazine,* 85, p. 631, 1944.

C. A. Cotton, photo

Fig. 63. The mesa-like extinct rhyolite volcano Ngongotaha, New Zealand. Spurs in front and at the left have been interpreted as coulees.

of Tongariro volcano (New Zealand) extending out as headlands into the lake Rotoaira. Of a similar nature, apparently, are high spurs of rhyolite reaching out from the Ngongotaha mountain at Rotorua (New Zealand) (Fig. 63).[2] A fresh spur- or ridge-forming flow of rhyolite at Mono Craters, California, is illustrated in Fig. 64. For such very convex flows Putnam[3] has revived the name coulee. (In the French

W. C. Putnam, photo

Fig. 64. Front of a coulee of obsidian, Mono Craters, California. (Courtesy of the *Geographical Review*, published by the American Geographical Society of New York.)

the term *coulée* is applied to any lava flow.) Descriptions of the formation of a coulee of rhyolite at Vulcano in 1771 indicate that this lava moved with extreme slowness. The steep-sided tongue attained a length of only a quarter of a mile. Coulees at Mono Craters (Fig. 64).

advanced with a precipitous front having a height of from two to three hundred feet. . . . The edges of the coulees . . . must have been approximately perpendicular or perhaps overhanging. Even at the present day, after many blocks have fallen and the formation of a talus slope has commenced, the climber finds it extremely diffi-

[2] L. I. Grange, The Geology of the Rotorua-Taupo Subdivision, *N.Z. Geol. Surv. Bull.*, 37, Pl. 4, 1937.

[3] W. C. Putnam, The Mono Craters, California, *Geog. Rev.*, 38, pp. 68-82, 1938.

cult to scale these rugged and broken escarpments of glassy fragments.[4]

Such flows as these, though small, have sufficiently strong relief to block valleys and divert rivers (Chapters XVII, XVIII) if favourably placed, and they may produce inversion of the landscape form if they flow into and occupy valleys. Doubt has been cast, however, on the classic explanation given by

A. S. Reid, photo

Fig. 65. The Scuir of Eigg, western islands of Scotland, a ridge which stands where perhaps a river once flowed. At the right and in the foreground are structural terraces of the basalt terrain.

A. Geikie of the Scuir of Eigg (Fig. 65) as an example of such inversion. This prominent landmark is a ridge formed by a tongue (or a dyke) of glassy rhyolite or pitchstone. As its history is interpreted by Geikie,[5] this rock originated as a coulee which was guided westward by a ravine previously cut by erosion in the Thulean basalt plateau; but according

[4] W. C. Putnam, *loc. cit.* (3), p. 79.

[5] A. Geikie, *The Scenery of Scotland,* pp. 278-282, 1865.

to an alternative explanation of the history of the feature given later by Harker[6] the pitchstone is intrusive, and an outcrop of gravel containing petrified wood, which has been supposed to be part of the floor of the ancient ravine down which rhyolite flowed, is otherwise interpreted. If this is correct, the Scuir, though a monument to the resistant nature of the pitchstone as compared with the surrounding basalt flows and dolerite sills, becomes merely an example of differential erosion.

Convex flows of andesite (Fig. 194) commonly become divides on the slopes of volcanic mountains.

On andesite flows—such as that which came from Sakura-jima, Japan, in 1914—the lava surface varies from typical aa to block lava (Fig. 50). Dacite flows on the ancient "Mount Mazama", at Crater Lake, Oregon, have the surface "covered with a thick pumiceous layer."[7] Obsidian (rhyolite glass) coulees are characteristically very rough and spinous, as minor relief is developed on them by overfolding of the stiff glass, such as is shown in Fig. 66, and also by the squeezing up of spines of lava through cracks in a hardened carapace and the piling up of a coarse, rubbly, block-field surface (Fig. 64).

Cumulo-domes

Very commonly lava of andesitic to rhyolitic composition (also alkaline lava, including trachyte and some phonolite) is extruded in a condition so viscid that it will not spread and flow but is merely squeezed up and accumulates so as to form more or less temporary projecting cylindrical plugs in craters and irregularly dome-like elevations over them. In most cases such portions of these lava masses as escape explosive disintegration shortly after their extrusion and while they are still hot are buried in debris and incorporated in growing volcanic mountains. A few of the larger members of this family, however, assume an independent existence

[6] A. Harker, The Geological Structure of the Sgurr of Eigg, *Quart. Jour. Geol. Soc.*, 62, pp. 40-69, 1906.

[7] J. E. Allen, Structures in the Dacitic Flows at Crater Lake, Oregon, *Jour. Geol.*, 44, pp. 737-744, 1936.

C. A. Cotton, photo

Fig. 66. Flow bands in a vertical attitude in the folded obsidian selvage of a comendite (soda rhyolite) lava flow, Mayor Island, New Zealand.

as landforms. These are hills, or even mountains, of accumulation, which may have a rather smoothly domed form but in other cases are less symmetrical piles also called by courtesy "domes".

Such forms were classed as *cumulovolcans* by Fouqué[8], who was one of the first to realise their nature. The designation "dome", which is often applied to them without qualification, and which has sufficed to distinguish them from "cones" of accumulation, has been amplified by Jaggar[9] to "cumulo-dome". This coinage, which follows Fouqué and at the same time recognises the dome shape which such masses may assume, serves also to separate these exuded bodies from the more perfect domes built of basalt flows, which are the volcanic domes *par excellence,* though sometimes erroneously referred to as cones. "Tholoid" is a convenient term in the sense in which it has been adopted by Escher (and used also by Wing Easton) for domes exuded in craters, that on Galunggung being taken as the type (Fig. 67).[10]

An attempt has been made by Daly[11] to mark the distinction between tholoids and basaltic domes of the Hawaiian type by describing the latter as "exogenous" and the former as "endogenous" domes; but Williams[12] has pointed out that such an attempt is without success, for the reason that cumulo-domes are in some cases not of endogenous growth, or may be so only in part, and "should only be referred to as such

[8] F. Fouqué, *Santorin et ses éruptions,* Paris, 1879.

[9] T. A. Jaggar, Seismometric Investigation of the Hawaiian Lava Column, *Bull. Seism. Soc. Am.,* 10, p. 164, 1920.

[10] B. G. Escher, L'éruption du Gounoung Galounggoung en juillet 1918, *Natuurk. Tijds. Ned.-Ind.,* 80, pp. 259-263, 1920; N. Wing Easton, Volcanic Science in Past and Present, *Science in the Netherlands East Indies,* pp. 80-100 (p. 94), 1930; the term "tholoid" was introduced by K. Schneider, *Die vulkanischer Erscheinungen der Erde,* p. 67, 1911. More or less exact synonyms are *Staukuppe* (A. Bergeat, Staukuppen, *Neues Jahrb. f. Min.,* Jub. Bd., pp. 184-329, 1927) and *Quellkuppe* (E. Reyer, *Theoretische Geologie,* p. 113, 1889).

[11] R. A. Daly, *Igneous Rocks and their Origin,* 1914; *Igneous Rocks and the Depths of the Earth,* p. 150, 1933.

[12] Howel Williams, The History and Character of Volcanic Domes, *Univ. Cal. Publ., Bull. Dep. Geol. Sci.,* 21 (5), p. 53, 1932. Williams does not include Hawaiian basalt domes ("shield volcanoes") among "volcanic domes" at all.

Fig. 67. The tholoid (cumulo-dome) which arose in the active crater of Galunggung, Java, in 1918. *L. van Vuuren, photo*

where it can be shown that they have grown chiefly or entirely by expansion from within, the magma having risen within a solid, fractured crust, rather than by normal effusion from a summit crater." This quotation from Williams supplies a precise definition of "endogenous" as applied to domes, and shows how the term must fall into disuse, like others which cannot be applied with confidence until after analysis has been made of the merits of their application to particular cases.

Essentially cumulo-domes are non-basaltic and are distinct from dome-shaped tumuli and small forms built of basaltic spatter. Some of these latter forms have been referred to as "mamelons", but the term "mamelon" may be reserved with advantage for the Grand Sarcoui type, or "trachyte dome", which ocurs in greatest perfection in Auvergne, and which, in the opinion of Williams (following Scrope)[13], owes its external form to exogenous growth. Flows may have been little more than sufficient in volume to cover the surface, however, and to give the mamelon Grand Sarcoui its "overturned-bowl"

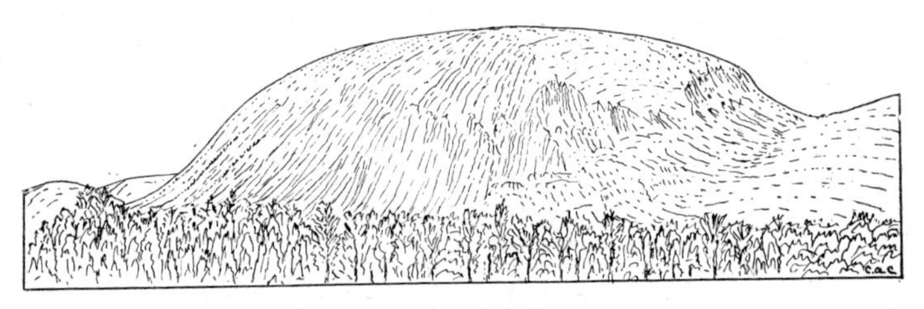

Fig. 68. The smooth, craterless Grand Sarcoui mamelon, Auvergne. (Drawn from a photograph by Tempest Anderson.)

form (Fig. 68). No trace of the vent from which such lava issued is to be found, moreover, on the surface of this dome as it survives to-day. It contrasts thus very strongly with the majority of the "puys" (volcanic hills) with which it is grouped in the Puy de Dôme (Auvergne) landscape, for these are typical basaltic scoria cones with open craters.

[13] "It may be recalled that Scrope recognised the dome [Grand Sarcoui, Auvergne] to have a concentric structure which he attributed to the piling up of successive flows of viscous lava from a central orifice" (H. Williams, *loc. cit.* [12], p. 115.)

Mayor Island, New Zealand, is an example of an effusive rhyolite volcano which has assumed a generally dome-shaped and steep-sided, though somewhat irregular, form. Its structure and mode of growth are revealed in the wall of an eccentric caldera (Fig. 167) and on an outer cliff (Fig. 69). At first sight the island appears to be maturely dissected, but closer inspection reveals that the relief of its surface is constructional.

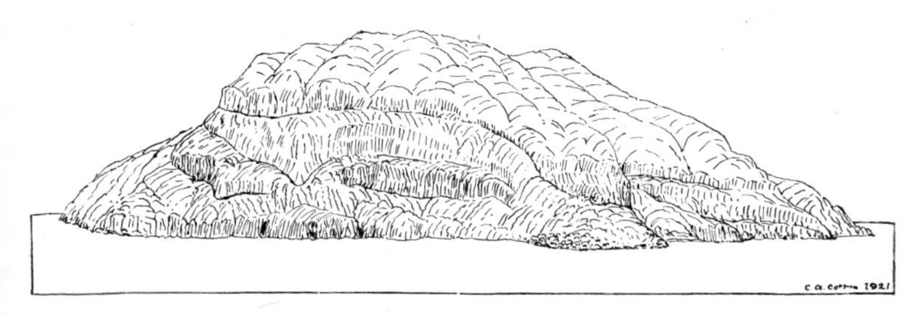

Fig. 69. Mayor Island extinct volcano, Bay of Plenty, New Zealand. This view shows the edges of four steep obsidian flows and the rugged constructional surface.

Viscid lava flows, following one another, have congealed each on the steep and rugged surface of its predecessor. Though viscid, this soda rhyolite seems to have had greater mobility than potassic rhyolites. It has been shown to be fusible at a comparatively low temperature.[14] Within the Mayor Island caldera a newer mound of obsidian has been built. Smoothly convex in profile as it appears from a distance, this is found to be rugged and spinous in detail.

Basaltic and Non-basaltic Lava Landforms in Contrast

A striking contrast is seen in some basaltic volcanic fields between normal basalt-lava forms and scattered small tholoids or mamelons built over the orifices through which comparatively small bodies of a more acid lava such as trachyte or perhaps phonolite have exuded, these having been derived apparently by a process of differentiation from the basalt

[14] P. Marshall, Acid Rocks of the Taupo-Rotorua Volcanic District, *Trans. Roy. Soc. N.Z.*, 64, pp. 323-366 (p. 354), 1935.

V. C. Browne, photo

Fig. 70. The "Chasm"—part of the Tarawera rift opened explosively in 1886—exposes in section the rhyolite of the Tarawera cumulo-dome (New Zealand). Beneath the superficial layer of basaltic scoria ejected in 1886 is another layer of white rhyolitic pumice; but some spinous asperities of the rhyolite cumulo-dome still project through these layers.

magma. As described by Daly,[15] such trachyte-built forms in a good state of preservation are dotted over the basalt field of Ascension Island, and there are remnants of similar forms composed of phonolite in dissected parts of St Helena Island. "Bulbous domes" of trachyte diversify a basaltic landscape also in western Maui (Hawaii),[16] and the fissures through which

[15] R. A. Daly, The Geology of Ascension Island, *Proc. Am. Ac. Arts and Sci.*, 60 (1), pp. 3-80, 1925; The Geology of Saint Helena Island, *ibid.*, 62 (2), pp. 31-92, 1927.

[16] H. T. Stearns, Geology and Ground-water Resources of the Island of Maui, Hawaii, *T.H. Div. Hydrog. Bull.*, 7, p. 148, 1942.

these exuded are occupied by dykes 8 to 25 feet thick. An example of a short, thick dyke expanding thus into a tholoid has been found by Speight[16a] on the Akaroa basalt dome, New Zealand.

Domes Chiefly of Endogenous Growth

Cumulo-domes, especially those which are squeezed up and out without any notable amount of overflow, are built by those magmas (non-basaltic in all cases) which are very viscous

Photo from Dr E. Marsden

Fig. 71. The steep side of a tholoid, Whakaipo Bay, Lake Taupo, New Zealand, showing columnar joints developed at right angles to a very steep cooling surface.

16a R. Speight, The Geology of Banks Peninsula: a Revision, Part 2, *Trans. Roy. Soc. N.Z.* (in press).

or even semi-rigid at volcanic temperatures. Such materials are plastic enough, however, to yield to a *vis a tergo,* and exude at the surface almost like cold metal from a press. They exude generally at temperatures lower than those at which their glass assumes a viscosity low enough—i.e. a liquidity sufficiently great—to be exploded by expanding and escaping gas (about 850° C. in the case of some rhyolite magmas at atmospheric pressure, but undoubtedly higher temperatures for some other glasses).

Typically jagged in outline, rounded or roughly level (though spinous in detail) on the top, and very steep-sided (Figs. 67, 70, 71), cumulo-domes of the commonest type (tholoids), which excludes the rarer smooth-surfaced mamelons, are formed by continuous injection of magma under a partly cooled and rigid carapace. Such injection lifts the carapace, which is cracked, however, and in many cases broken up into large blocks, so that lava may exude on its surface. Thus

> the growth . . . is . . . partly exogenous and partly endogenous, and it is likely that in most of the domes of which we have detailed knowledge this double process is continually in operation. . . . What is essential to the growth of such domes is a sufficiently low temperature and gas-content, and an effusion slow enough to enable solidification to proceed *pari passu* with uplift (Williams).

Cumulo-domes as Landforms

In contrast with the broad mamelon Grand Sarcoui (Fig. 68) the diameter of which is three times its height (it is 800 feet high) Puy de Dôme, in the same district of Auvergne, is a spike 1700 feet high and consists of a steep-sided rock core mantled by talus slopes (Fig. 72).

The domes of Auvergne are among the more stable and long-lived of such forms. Not many large landforms have been described as derived from ancient cumulo-domes. One such, however, is Tomichi Dome,[17] in south-western Colorado, the relief of which is about 2000 feet (Fig. 73). Another example, which, though of relatively small relief, is of great interest, as it is a buried or "fossil" volcano of middle Tertiary

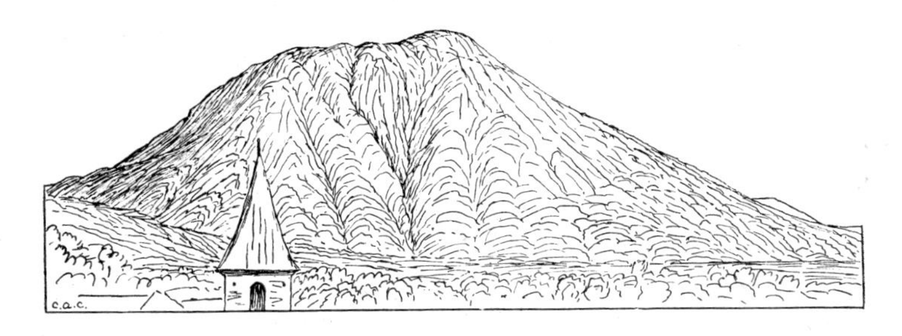

Fig. 72. The cumulo-dome (plug dome?) Puy de Dôme, as viewed from Laschamps, Auvergne. (From a photograph.)

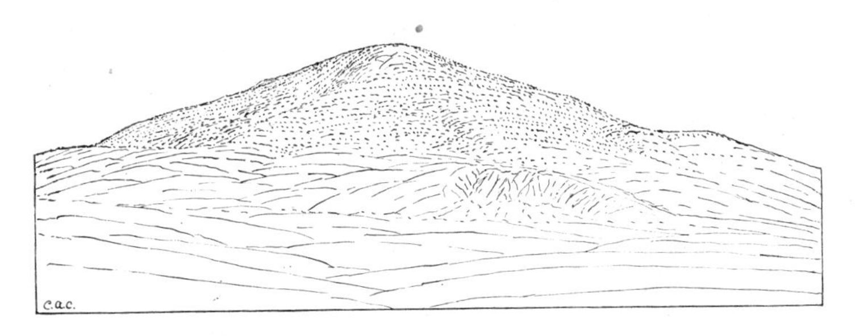

Fig. 73. Tomichi Dome, Colorado. (From a photograph.)

age exhumed by erosion since the dissection of the Ortiz pediment, has been described near Albuquerque, New Mexico.[17]

In the Rotorua district of New Zealand Grange[18] has described as "domes" rhyolite masses of two eruptive periods, though some of these are in part effusive, and there remains a slight doubt whether some of the older "rhyolites" thus described do not form part of the "ignimbrite" sheet (Chapter XII). Tarawera (Frontispiece and Figs. 36, 37, 74, 75), however, across which a new and deep volcanic rent was opened by the eruption of 1886, revealing the structure to some extent, is a cluster of large cumulo-domes which seem to be of endogenous growth.

[17] J. T. Stark and C. H. Behre Jr, Tomichi Dome Flow, *Bull. Geol. Soc. Am.*, 47, pp. 101-110, 1936; H. E. Wright, Cerro Colorado, An Isolated Non-basaltic Volcano in Central New Mexico, *Am. Jour. Sci.*, 241, pp. 43-56, 1943.

[18] L. I. Grange, *loc. cit.* (2), p. 72.

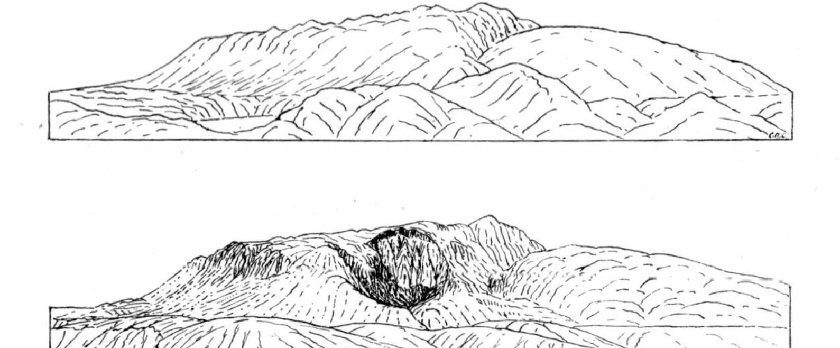

Fig. 74. Above: The Tarawera rhyolite dome cluster before the eruption of 1886, which opened a rent across it (compare Fig. 37). The profile shows the spinous surface characteristic of cumulo-domes, though the asperities had already been hidden to some extent by burial under a cover of rhyolite pumice. (View from south-west; from an old photograph.)

Below: The Tarawera mountain after the eruption of 1886. The surface has been further smoothed by a layer of scoria and bombs ejected during this eruption.

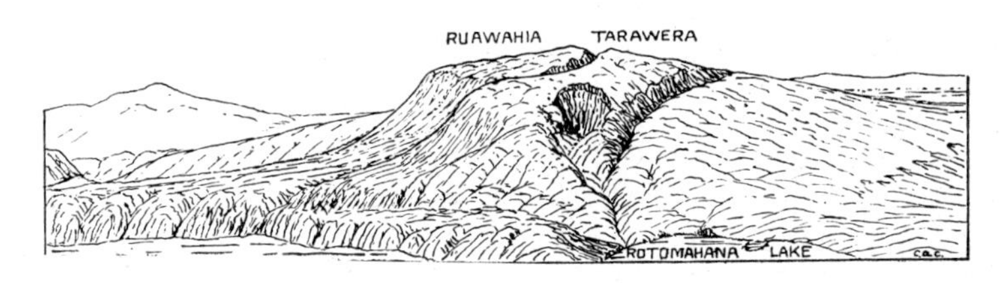

Fig. 75. Air view of the Tarawera dome-cluster from the south-west. This view shows separation of the component domes Ruawahia and Tarawera proper. (From a photograph.)

One of the largest of known cumulo-domes forms Lassen Peak, California, which is still an active volcano. The volume of the body of dacite lava rock composing the peak is estimated as three-fifths of a cubic mile.[19]

The great dome of Lassen Peak towers conspicuously above the cluster of minor domes on its south and north-west sides; elsewhere its slopes sweep down with diminishing angle. . . . Approximately,

[19] H. Williams, *loc. cit.* (12), pp. 69-72. See illustrations in N. E. A. Hinds, *Geomorphology,* photos pp. 273, 274, 1943.

the mountain has the form of a truncated pyramid owing to the enormous banks of talus which encompass it, . . . some of which have a fall of 3000 feet. . . . Although the steep-sided core of the mountain probably covers only about one square mile, the apron of talus extends over twice that area. Here and there, especially on the south and east flanks, high crags of solid dacite rise through the talus, but these are of small proportions compared with the volume of loose detritus. In this respect, as in many others, the dome of Lassen Peak is strikingly reminiscent of the famous Puy de Dôme of the Auvergne. . . . It must be emphasised that the great talus banks encircling Lassen are almost if not entirely due to the fracturing of the dome during its actual emplacement.[20]

Fig. 76. Cumulo-dome, or large tholoid, of the Soufrière of Guadeloupe, West Indies. (From a photograph by M. le Boucher.)

An example of a large cumulo-dome which has been built within a wide caldera or "Somma ring" is the Soufrière of Guadeloupe, in the Lesser Antilles. It is a volcano that is practically dormant, though it has shown slight symptoms of activity for some centuries. The dome stands with a relief of 1000 feet above the caldera floor (Fig. 76). A great many of the smaller domes, or tholoids, are thus contained within the walls of craters. Those of andesite and dacite lava are formed usually in summit or lateral craters of volcanic mountains (Figs. 67, 77, 81, 104); those of rhyolite lava, or obsidian, have made their appearance within rings of explosively ejected pumice which have been built just prior to the rise of the tholoid (Figs. 78, 79). Examples of the latter are found at Mono Craters[21] and at Salton Sea,[22] California.

[20] Howel Williams, Geology of the Lassen Volcanic National Park, California, *Univ. Cal. Publ., Bull. Dep. Geol. Sci.*, 21 (8), p. 316, 1932.

[21] W. C. Putnam, *loc. cit.* [3].

[22] V. C. Kelley and J. L. Soske, Origin of the Salton Volcanic Domes, *Jour Geol.*, pp. 496-509, 1936.

V. C. Browne, photo

Fig. 77. A tholoid in the crater of the extinct volcano Mt Egmont, New Zealand.

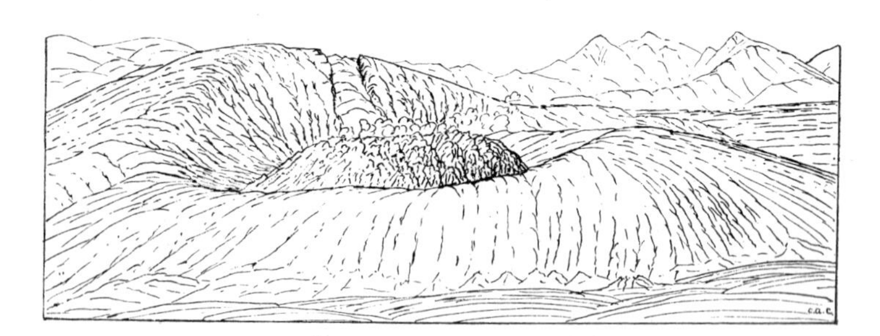

Fig. 78. The Novarupta tholoid, of glassy acidic lava, formed in 1912 near Katmai volcano, Alaska. Before the emergence of the tholoid a ring of ash had been built around the centre of eruption. (From a photograph.)

W. C. Putnam, photo

Fig. 79. A tholoid of obsidian developed within a crater ring of pumice in one of the (extinct) Mono Craters, California. (Courtesy of the *Geographical Review*, published by the American Geographical Society of New York.)

Tholoids which Contribute to the Bulk of Andesite Volcanoes

Numerous tholoids have emerged within the last few decades, and some have been kept under observation during their growth. Small as most of them are, and unlikely to survive as independent relief forms, they have yielded much information regarding the mode of growth of non-basaltic volcanoes. By crumbling during growth and by explosive disintegration while still hot—or perhaps when reheated by the volcanic furnace—such tholoids yield in the aggregate an immense amount of debris, both as block rubble and as fine ash. In the case of dacitic and andesitic volcanoes in particular it is now realised that much of the steep volcanic cone is built, in many cases, of the debris derived from numerous tholoids which have exuded successively at the summit. There is no reason to suppose, as one might be tempted to do where such a dome occupies a crater on the summit of an extinct volcano (Fig. 77), that the rise of a tholoid necessarily marks an expiring phase of volcanicity.[23] Hundreds—even thousands—of such domes, and the debris derived from them, may be built into the body of a great andesitic cone.[24]

Dome Flows

Convex lava forms intermediate in character between cumulo-domes and coulees of lava are present on the rim of the caldera that contains Crater Lake, Oregon—i.e. on the flanks of the former high volcanic cone "Mount Mazama". These consist of dacite lava which has piled up over lateral vents from which it exuded, attaining a thickness of about

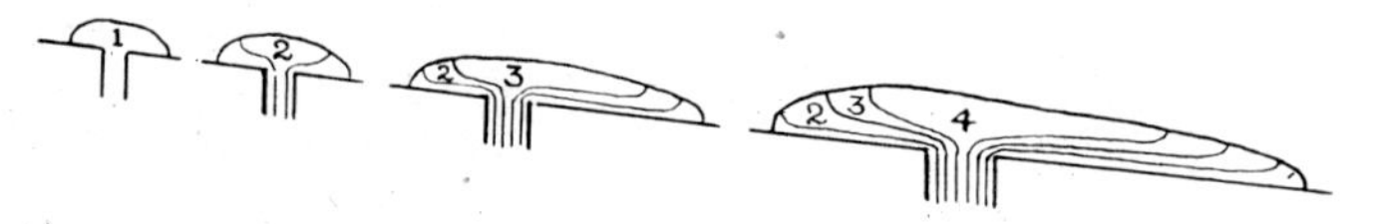

Fig. 80. Growth stages and displacement surfaces in convex dacite flows at Crater Lake, Oregon. (After J. E. Allen.)

1000 feet, spreading out all around, however, and not only flowing down but banking up also against the easy slopes of the great cone.[25] Such "dome flows" have contributed to later stages of the growth of this composite mountain. In the lava of the dome flows the observed "flow lines", or surfaces of shear and displacement, are arranged (Fig. 80) in a manner that suggests a spreading of the up-welling lava very much like that deduced theoretically by Reyer[26] in explanation of

[23] "The beginning of dome growth . . . marks the end of a more active period; indeed when lava stands in the principal vents too inert to participate in explosions, we may say: It is the beginning of the end. By this I do not mean the end of *all* activity. . . . Powerful explosions no longer propel their ejecta to great distances; . . . the *explosions* are now of lesser intensity than formerly; the debris thrown out now tends to *roll* down the slopes; and each new outbreak finally ends in the rise of a dome" (F. A. Perret, *The Eruption of Mt Pelée 1929-1932,* p. 112, Carnegie Inst., Washington, 1935).

[24] E. G. Zies, Surface Manifestations of Volcanic Activity, *Trans. Am. Geophys. Union,* 19, pp. 10-23, 1938.

[25] J. E. Allen, *loc. cit.* (7).

[26] *Loc. cit.* (10).

cumulo-domes—a theory that has not been found to be of general application, however (p. 176).

Tholoid Growth under Observation

The domes, or tholoids, successively piled above the floor of a crater on the summit of Mont Pelée in the two series of eruptions which began in 1902 and 1929 (Fig. 81), though

Fig. 81. Tholoid of dacite lava within a crater on Mont Pelée, Martinique, which was formed during the series of eruptions 1929-1932. Explosions on its flanks furnished the materials for innumerable *nuées ardentes*. To the left of the centre is a remnant of the large tholoid left over from the eruptions of the 1902 series, surmounted by a small spine. (Drawn from a photograph.)

similar in a general way to other new tholoids in Japan and Java, were especially notable in two respects: (1) in their highly explosive nature (related perhaps to rather high temperature of the lava), and (2) in the great size of the spines thrust up through chinks in the carapace. Their explosiveness led to progressive disintegration, with the release of a long succession of *nuées ardentes* (Chapters V and XII).

The tholoid shown in Fig. 104, which began to emerge from a lateral crater on the Santa Maria volcano, Guatemala, in 1923, is far down the flank of the mountain. Its dimensions and aspect in 1939 have been described as follows:

> This new edifice, called Santiaguito, has now, seventeen years after its first appearance, an altitude of about 300 metres above the crater floor and a diameter of roughly 700 metres. The bulk of this structure has been slowly but steadily increasing by virtue of the

extrusion of a hot (about 800° C. at the point of emergence), viscous, porous, and mostly crystalline igneous material. Up to the present time (1939) there has not been any well-defined flow of lava; but, instead, the freezing of the viscous rock-forming material has permitted the extrusion of ridges, spines, and hot fragmental material. The ridges and spines are often broken down by the pressure exerted against them from below, and as a result the outer surface of Santiaguito consists largely of loosely piled-up rocks, which are often at their angle of repose (about 32°). Intensive fumarole activity exists in various parts of the dome-like structure.[27]

Emission of a *nuée ardente* with a volume estimated at three million cubic metres from this tholoid in 1929 (Chapter XII) indicates that, like others of its kind, it consists of potentially explosive lava.

Tholoids within active craters, especially when they have assumed or have fallen into the condition of mounds of rubble, have been mistaken in some cases for inner, or "nested", cones. They are themselves craterless, however, though new craters may be blasted through them by explosions. The actual lava core generally develops with precipitous sides, and domes of largely endogenous growth are flanked, therefore, in practically all cases by masses of talus which slope at the angle of repose for coarse angular rubble (Fig. 81). Most of such talus has accumulated during the growth of the dome. Coulees of rhyolite are similarly steep-walled and talus-aproned (Fig. 82). Even a growing and progressively disintegrating and crumbling tholoid may, as Perret[28] has observed in the inferno of the active crater of Mont Pelée, develop the Hogarth "line of beauty", convex on the crumbling edge above and passing at a point of inversion into the smooth concave line of a talus slope below, a profile characteristic of all high-piled constructional volcanic forms.

To find cumulo-domes which are to some extent free of talus and thus exhibit more or less faithfully the steepness of the lava flanks one must turn to those which have risen from the sea and have been cleaned up progressively by marine

[27] L. H. Adams, Annual Report of the Director of the Geophysical Laboratory, *Year Book Carnegie Inst. Wash.*, 38, pp. 33-35, 1939.

[28] F. A. Perret, *loc. cit.* (23), Fig. 67.

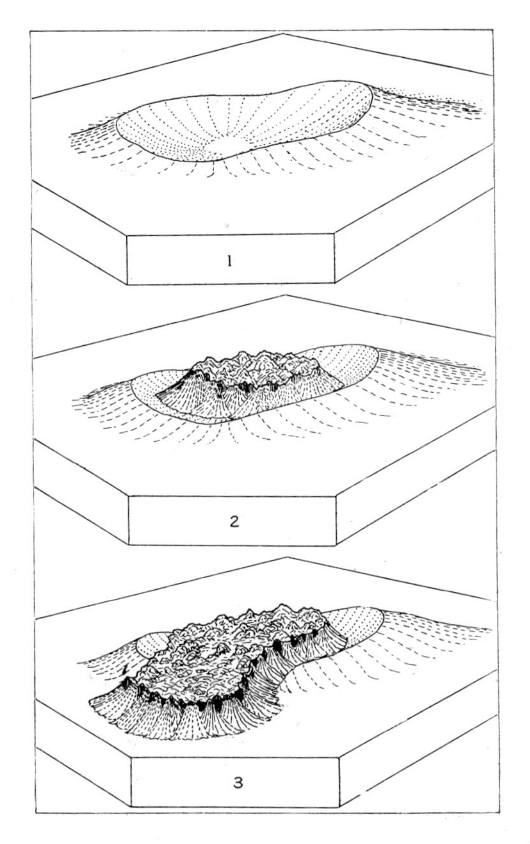

Fig. 82. Development of an obsidian dome (2) within a crater ring (1) in the manner characteristic of the Mono Craters, California (compare Fig. 79). At stage (3) the lava has pushed forward as a coulee, which has breached the tuff ring. Compare Fig. 64. (After W. C. Putnam, *Geographical Review*.

erosion. Such lava masses arise from the sea in the Bogoslof island group (Aleutians), but even there, though they are exposed to the attack of ocean waves, growth has generally been too rapid for contemporaneous removal of talus to take place. Jaggar[29] describes "rocky crags of the core and summit

[29] T. A. Jaggar, Evolution of Bogoslof Volcano, *The Volcano Letter,* 322, pp. 1-3, 1931.

of the mass, with slide-rock about its foot-slopes fed by the tumble from the rising cliffs above." Some of the newer of these have been destroyed by explosions within a few months of their first appearance; but some steeply cliffed pinnacle remnants of older lava masses survive (Chapter XV).

Some tholoids give the impression that they are mere heaps of huge blocks,[30] but this is illusory. Some are nearly flat and horizontal on the top, and old crater-floor scoria or gravel found on them, as has been the case at Keloet (Java) and at Usu (Japan), may indicate that they have risen almost like pistons. The question has been raised also whether the crust on the surface of a growing tholoid at Santorini volcano (Eastern Mediterranean) consists of loose blocks or whether it is continuous (except for cracks), as the term "carapace" used for such a crust by Lacroix and others suggests.[31] Most domes, however, though subject to deformation, fissuring, and faulting (Fig. 83), seem to have such a carapace. This is true especially of those on which—as in the case of the tholoid on Mont Pelée—magma from within is squeezed out through chinks; for a roof of blocks would be merely rearranged or re-piled by pressure from within instead of being pierced by spines. Apparently such a carapace is traversed by cracks, though these are discontinuous and do not divide the carapace into a mosaic. "Coronets" of encircling steam jets observed on the Fouqué Kameni island-forming tholoid at Santorini and on that at Novarupta, Alaska, have been interpreted as indicating the presence of encircling fissures—possibly groups of concentric encircling fissures — along which subsidence of a carapace might take place. At night this encircling crevice at Santorini was observed to glow, which was taken as an indication that it was deep enough to communicate with hot non-rigid lava within the carapace.[32]

[30] F. A. Perret (writing in 1930) described the early stages of growth of the new dome at Mont Pelée as the result of outflow of very viscous lava which "falls in incandescent blocks around its point of emission and there forms a dome" (*loc. cit.* (23) Appendix II, p. 124).

[31] H. S. Washington, Santorini Eruption of 1925, *Bull. Geol. Soc. Am.*, 37, pp. 349-384, 1926.

[32] H. S. Washington, *loc. cit.* (31), p. 369.

Fig. 83. O-usu, the eastern dome of Usu, Japan, showing a fault scarp on a carapace. (After Howel Williams.)

Plug Domes and Pitons

A variant from cumulo-domes of spreading form (tholoids) is the cylindrical or plug-like form produced by piston-like protrusion. This is the "plug dome",[33] or, as a landform, the "piton".[34] As an illustration of this mode of extrusion that which occurred when the twin domes O-usu and Ko-usu were uplifted in the crater of Usu volcano (Japan) has been taken. This has been contrasted by Tanakadate[35] (Fig. 84, B) with the spreading kind (Fig. 84, A) as exemplified by the tholoid of Tarumai, in which displacement lines fan out upward. These "displacement lines" and also "growth surfaces" are indicated in Fig. 84. The growth surfaces, which are successive profiles of the carapace during growth, are convex in both types, for even in the case of the piston-like plug dome (B) uplift is greatest at the centre and there is some differential movement along displacement lines, which in this case are vertical. The upper surface is less rugged than is usually the condition in case A, for much less stretching of the carapace has taken place. The plug domes of Usu

[33] Howel Williams, *loc. cit.* (12), p. 53. N. Wing Easton (Volcanic Science in Past and Present, *Science in the Netherlands East Indies,* pp. 80-100, 1930) has suggested that this form be called a "stop."

[34] F. A. Perret, The Volcano-Seismic Crisis at Montserrat 1933-1937, p. 10, 1939: "The grandest of non-explosive eruptive products, the *piton*."

[35] H. Tanakadate, Two Types of Volcanic Dome in Japan, *Proc. IV Pac. Sci. Cong.,* 2, pp. 695-704, 1930.

volcano have carried up thick layers of lake-deposited clay and gravel, baked by heat from the lava and scratched by movement. The top of the O-usu dome, though otherwise smooth, is broken by a fault scarp (Fig. 83), which may be explained as the surface expression of one of the vertical displacement surfaces on which movement has been localised.

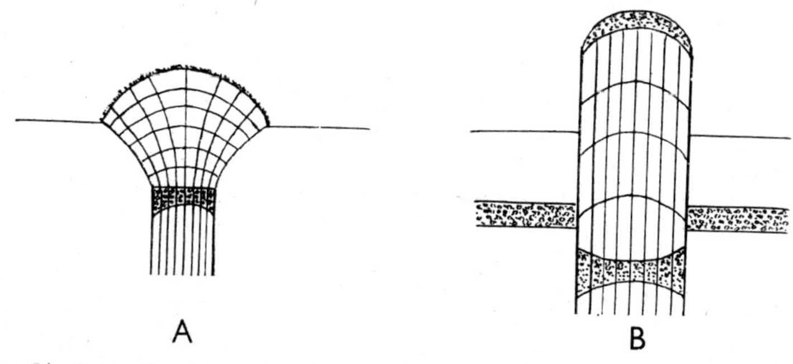

Fig. 84. Contrasting types of endogenous lava extrusions.
A: Tholoid of Tarumai, Japan.
B: Plug dome of O-usu, Japan.
Debris formerly lying on a crater floor has been carried up on a carapace in each case. (After Tanakadate.)

Such faulting may be attributed to the upthrust of a block of the carapace rather than to gravity subsidence.

The upward fanning out of displacement lines (Fig. 84, A), which is a structural peculiarity observed in the rock of some extrusions, is not thoroughly understood unless

> in early stages of growth the domes were marked by distinct [central] vents, which gradually became clogged as activity continued.

On the other hand, many tholoids

> show little internal structure. Growing larger by expansion from within, they are intensely fissured and brecciated, and the flow-planes, while rudely concentric with the surface, are obscure and much distorted. During protrusion irregular dykes are continually intruded between the solid parts and serve as feeders to surface flows, but on cooling these dykes resemble the wall rocks so closely as to be indistinguishable.[36]

[36] Quotations from Howel Williams, *loc. cit.* (12), pp. 144-145.

As has been noted by Williams,[37] the exceptionally large dome (p. 166) which forms a great part of Lassen Peak, California, appears to be roughly cylindrical—i.e. a plug dome as here defined, the peak being a piton as defined by Perret. Though its sides are deeply buried under a thick apron of talus blocks, at one point on the southern flank the solid lava rock

Professor Howel Williams, photo

Fig. 85. Smooth cliffs interpreted as moulded during vertical extrusion of a plug dome to form the piton Lassen Peak, California. (This pinnacle of rock is below the summit on the southern slope of the piton.)

emerges, exposing "beautifully smooth cliffs", 500 feet high, "apparently polished and striated during upheaval by the attrition of the solid skin of the mass against the circular walls of the vent, a process similar to that which smoothed the face of the Pelée spine." Vertical upthrust of a much smaller body of hot lava, a core only of the main plug dome, took place at Lassen Peak during the eruption period 1914-15, producing "a miniature copy of the rise of Lassen Peak itself into the pre-Lassen crater."[38]

The Pitons of Carbet, in Martinique, are craterless volcanic peaks which were explained by Lacroix[39] as monolithic cores

[37] Howel Williams, *loc. cit.* (12), p. 72.

[38] A. L. Day and E. T. Allen, *The Volcanic Activity and Hot Springs of Lassen Peak*, p. 75, Carnegie Inst., Washington, 1925; Howel Williams, *loc. cit.* (20) p. 320.

[39] A. Lacroix, *La Montagne Pelée et ses éruptions*, p. 158, Paris, 1904.

of cumulo-domes, originally of greater breadth, stripped by erosion. These together with similar peaks on the island of Montserrat (peaks of the Soufrière Hills) are regarded by Perret,[40] however, as "monolithic vertical extrusions", each of which "may represent an eruptive centre of the first order" —in other words, as plug domes, the mode of extrusion of which he compares with that of the great spine on Mont Pelée (described on a later page).

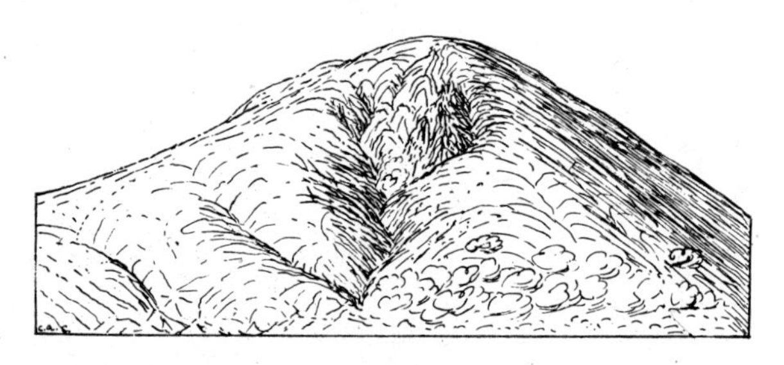

Fig. 86. Pointed, craterless summit, or piton, of the main peak of Acatenango, Guatemala, 3960 metres, exhibiting slight activity in 1925. (From a photograph by Sapper.)

Some much higher, in some cases isolated, steep and sharp cones, of which Merapi, in Java, is the most active example (while another example is Acatenango, in Guatemala) (Figs. 86, 110, 123) are built, apparently, of the debris of successively extruded plug domes (or tholoids) which disintegrate explosively. During the temporary presence of such a feature it forms the summit of the mountain, which is then craterless. Such a condition has been reported in many volcanoes both active and extinct, including Ararat (Fig. 123) and a number of South American peaks.[41]

Spines

Though it appears that viscous new—or even re-heated —lava may exude as a more or less cylindrical plug dome, it is much less probable that a quite rigid old pipe-filling such as may conceivably be left over from a previous eruption is ever thus hoisted "like the cork out of a bottle". Such,

[40] F. A. Perret, *loc. cit.* [34], p. 12.
[41] A. Stübel, *Die Vulkanberge von Ecuador,* pp. 165, 399-400, 1897.

however, has been a popular explanation of the truly remarkable short-lived pinnacle, *"aiguille"*, or "spine" which rose to a great height above the summit of Mont Pelée, Martinique, in 1903. This theory was maintained by Heilprin,[42] and his lead has been followed by a number of textbook writers who have failed to realise the nature of the dome, or tholoid, from which the spine arose, but have regarded it as a debris-built cone. An alternative and generally accepted explanation of the spine of Pelée—supported by many reliable observations made during the growth of this giant monolith[43] as well as of a great many lesser spines—is that "spines" in general result from the extrusion of semi-rigid new lava through cracks in the carapaces of cumulo-domes.

Small spines may exude through any local crack, but the larger spines may perhaps arise (as Perret[44] has thought) over fissures or gas-fluxed pipes of a more permanent nature at a greater depth. He has described aligned spines on the Pelée dome, "like a row of trees on a hillside", as related to "a hidden fissure with gas-bored chimneys at intervals".

One side, but one only, of the great spine of 1903 was smoothed and slickensided into grooved lava (Chapter IX), being moulded by contact with the curved wall of the crevice through which the semi-rigid material emerged. Though deceptively appearing cylindrical when viewed from the east (Fig. 87), the spine was seen from other points of view to have curled over, developing a sharp crest, and the western side was extremely jagged (Fig. 88). Observations made by Perret[45] on spines exuded from Mont Pelée in 1931 show that this asymmetrical form is characteristic.

What actually comes into view is only a part of the spine, perhaps half of it; it is like a horn split vertically, the half of which is extruded and matured as a spine. The core of the structure may be liquid, and from the base of the flat vertical face lava often exudes (PERRET).

[42] A. Heilprin, *The Tower of Pelée,* 1904; Tower of Pelée, *Rep. VIII Internat. Geog. Cong.,* p. 446, 1905.

[43] See especially A. Lacroix, *loc. cit.* (39), pp. 643-644.

[44] F. A. Perret, *loc. cit.* (23), p. 112.

[45] F. A. Perret, *loc. cit.* (23), p. 66.

The maximum relief attained by the great spine was nearly 1000 feet, and its most rapid rate of emergence was forty feet per day. But for progressive disintegration on the ragged side

A. Lacroix, photo

Fig. 87. The spine of Mont Pelée from the east. This aspect of the spine gave it a deceptive appearance of being cylindrical. Transverse ribs mark pauses in the extrusion of the material, which was only semi-rigid as it emerged.

it might have attained—so it has been estimated—a height of 2500 feet above its base on the carapace of the Pelée dome. In a few months after its extrusion ceased the spine had broken down into a heap of rubble.

The rise of many smaller spines has been observed. One rose, for example, to a height of 200 feet above the domed

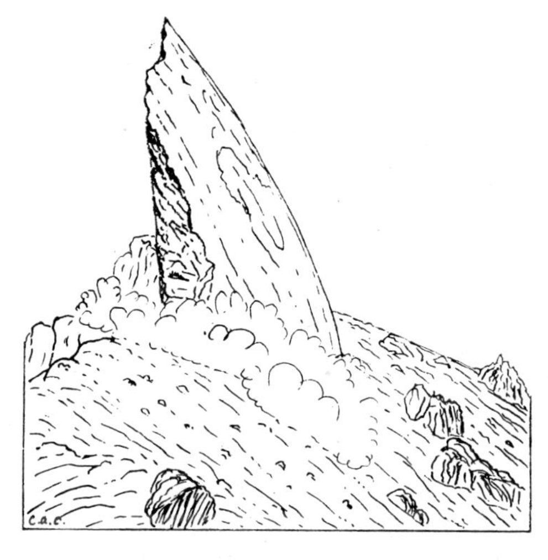

Fig. 88. The spine of Mont Pelée, 8th March, 1903, from the south, showing the eastern side convex and slickensided, while the western side was rough and ragged. (Drawn from a photograph.)

Fig. 89. The Peak of Fernando de Noronha, composed of phonolite, a plug dome or perhaps a spine that has proved resistant to erosion. (After Branner.)

summit of the tholoid Santiaguito (Santa Maria, Guatemala). After a few months this spine was, like many others, destroyed by explosion.

Some ancient spines may have survived as landforms, and such an origin has been suggested for the Peak (Fig. 89), a well-known landmark composed of phonolite lava rock, on the island Fernando de Noronha (South Atlantic)[46] It may be a plug dome, or piton, or it may be, on the other hand merely the resistant remnant of a broader cumulo-dome reduced to its pinnacle form by erosion. It is surrounded by an apron of talus.

Necks related to Tholoids

Many necks are features exposed only by deep differential erosion which lays bare the filling of an ancient volcanic pipe (Chapter XVIII); but some—the Hopi type of Williams[47] —have been developed by a very much smaller amount of erosion, sufficient only to remove a ring or cone of tuff that has been present around a cumulo-dome, some portion of which still survives as a butte or small mesa. Such features, examples of which occur in the Hopi country of the western United States, have been accentuated, rather than actually exposed, by erosion, as shown in Fig. 204, A.

[46] J. C. Branner, Is the Peak of Fernando Noronha a Volcanic Plug like that of Mont Pelée?, *Am. Jour. Sci.*, 10, pp. 442-444, 1909.

[47] Howel Williams, Pliocene Volcanoes of the Navajo-Hopi Country, *Bull. Geol. Soc. Am.*, 47, pp. 111-171, 1936.

CHAPTER XII

Ash Showers and Nuées ardentes

THE CHIEF PRODUCT OF BOTH VULCANIAN AND PELEAN ERUPTIONS is ash. It is impossible to avoid using the popular designation "ash" for what Russell[1] has preferred to call "volcanic sand and dust," even though, as Russell remarks, "the term 'ash' conveys the idea that they are the residue left by combustion," and is for that reason to some extent objectionable.

VOLCANIC ASH

Both Vulcanian and Pelean eruptions are "magmatic" in the sense that the motive power that causes them—the emission and expansion of gases, accompanied to an extent it is impossible to estimate by exothermic and explosive reactions in the gas mixtures—comes from within the magma, or from hot lava newly derived from magma; and the by-products of the explosions of magma are various grades of magmatic debris, including pumice lumps and lapilli, pumiceous ash, and vast quantities of the fine (sandy to dusty) grades of ash. Most ash originates in this way; but in addition much dust and sand, some material of the coarseness of lapilli, and also some coarser fragments, including angular blocks of large size, are produced by the fragmentation and comminution of old rocks—i.e. previously solid and cold materials—which are ejected by explosions from craters after collapse and landsliding of their walls.

From many descriptions of volcanoes and volcanic phenomena it might be inferred, but quite erroneously, that most if not all pyroclastic, or ejected, materials have been derived in this way from debris of older rocks. The fact that some Plinian and also minor explosive eruptions have not been accompanied by effusion of any liquid lava has been some-

[1] I. C. Russell, *Volcanoes of North America,* p. 75, 1897.

times interpreted as an indication that these have been simply explosions due to the outburst either of slowly accumulated gas[2] or of suddenly generated steam without the presence of eruptible magma even in the depths of the volcanic throats or "hearths" concerned. If such were the case, all the rock debris scattered by these explosions would consist of broken and comminuted rocks derived from previously built volcanic edifices and from the foundations of pre-volcanic strata on which these have stood.

Undoubtedly the comminution of older rock by explosions has been greatly overvalued as a source of ash. During the eruption of Vesuvius in 1906 the acute observer Perret was so impressed by the evidence there afforded of the production of abundant ash from new magma in or below the throat of the volcano—a process which had not previously been given the attention it deserved—that he felt it necessary to expose "the absurdity of the idea that volcanic ash is always formed by mutual attrition of lava fragments." Perret's article on "the direct conversion of lava into ash"[3] is of value not actually as the record of a new discovery, for, as Perret[4] himself has subsequently shown, his discovery of the magmatic origin of ash was not new, but as an indictment of the prevailing tendency to ignore what is undoubtedly the main source of ash and, in the case of some volcanoes, the source of much coarse debris as well.

[2] According to Perret's "gas-liquid differentiation" hypothesis such accumulation of gas may take place even under a body of magma which is eventually ejected by it in a "major explosive eruption" (F. A. Perret, *The Volcano-Seismic Crisis at Montserrat 1933-1937,* p. 75, Carnegie Inst., Washington, 1939).

[3] F. A. Perret, Volcanic Vortex Rings and the Direct Conversion of Lava into Ash, *Am. Jour. Sci.,* 34, p. 405, 1912.

[4] F. A. Perret, *The Vesuvius Eruption of 1906,* p. 86, 1924. As Perret here points out, the recognition of the derivation of ash from new lava is either explicitly stated or is implied in explanations of volcanic processes given by many earlier volcanologists, notably Scacchi, Johnston-Lavis, and Dana. A description of the ash-forming process—one which has not been mentioned by Perret—is contained in the following quotation from Chamberlin and Salisbury: "If . . . the lavas are surcharged with gases, and if these are restrained from free escape by the viscosity of the lavas, the gases gather in large vesicles in the lava in the throat of the volcano, and on coming to the surface explode, hurling the enveloping lava upwards and outwards, often to great distances. The violence of projection reduces a portion of the lava to a finely divided state constituting the 'ash' and 'smoke' of the volcano. Other portions less divided are inflated by the gases disseminated through them, and form 'pumice and scoria.'" (*Geology,* I, p. 617, 1905.)

Volcanic "ash" in the strictest sense is fine material, of the grain size of sand and dust. Some authors even exclude "volcanic sand" and refer to the dust grade only as ash. Much of this volcanic dust is so fine that when carried up by the belches of volcanic gases from paroxysmal Vulcanian or Plinian eruptions into high layers of the atmosphere it may remain long enough in suspension to travel around the world with air currents. Brilliant displays of sunset colours commonly result from the presence of such dust in the atmosphere, which has had its origin in eruptions in distant regions.[5]

Material of fine or ash grade graduates through pumiceous pellets, or lapilli, into coarse pumice. These are all derived from new lava (magma); but they may be mixed in any proportion with fine to coarse debris of older rocks of volcanic or non-volcanic origin which have been shattered and even to some extent comminuted by explosions. The whole of such a mixture is referred to as ash by many writers, and, though the usage is perhaps not strictly correct, it seems that this lead must be followed for want of a better inclusive term. On the other hand, the use of the term "cinders" seems objectionable if applied, as it has sometimes been, to all coarse grades of debris of mixed or non-magmatic origin as well as to scoria and pumice.

Pumiceous Inflation of Magma

As has been described already in Chapter V, the viscid (intermediate to acid) magmas tend, when at temperatures high enough to give them some degree of fluidity, to froth when relieved of pressure in the same way as does basaltic magma. Failing, however, to become sufficiently liquid to allow the gases liberated from solution to escape freely by

[5] In New Zealand in 1932, for example, there was evidence of the presence in the air of dust which must have travelled two-thirds of the way around the world from South America, carried by the westerly air currents and taking four to seven weeks on the journey. Effects observed included a long series of brilliant sunsets over a period of several weeks; notably a pink afterglow persisted for more than an hour after sunset. Before the sun set the dust was visible like a haze of diffused smoke which appeared to lie at quite a low level. Sometimes the sheet of haze was "visible to an elevation of 15° and resembled smoke from bush fires as seen in Australia." (E. Kidson and R. G. Simmers, Volcanic Ash Phenomena . . . observed in New Zealand, *N.Z. Jour. Sci. and Tech.*, 14, p. 319, 1933.)

effervescence, they are inflated to make a foam which if solidified becomes pumice; and much of this foam explodes, producing finer grades of ash. Jaggar has described the process of magmatic explosion as follows:

> The dominant lava of an explosive volcano unrelieved by abundant pyromagmatic effervescence at high . . . temperature . . . is . . . . slowly honeycombing, heating [owing to exothermic gas reactions], and expanding below the surface with accumulated gas. Disruption finally relieves pressure, effervescence is rapid and paroxysmal, and the gas reactions are explosive.[6]

Adiabatic expansion of escaping gas will have a cooling effect, but, as chemical reactions between the gases are such as to produce heat, this blast-furnace effect may balance the cooling due to expansion, or may even be dominant. Thus the escaping mixture of gas and ash will not be quenched, but may even become hotter as it is emitted.

Volcanic Gases

The gases which inflate and explode magma so as to produce and eject pumice and ash are very generally considered to be of the same nature as those which have been collected when escaping from the pyromagma of Kilauea (Chapter III). One experienced volcanologist, however, has expressed his conviction that the gaseous mixture emitted during great volcanic explosions differs but little from atmospheric air in composition.[7] Poisonous compounds of chlorine and sulphur

[6] T. A. Jaggar, Seismometric Investigation of the Hawaiian Lava Column, *Bull. Seism. Soc. Am.*, 10, pp. 155-275 (p. 167), 1920.

[7] F. A. Perret states: "On several occasions . . . the writer has been completely enveloped for fifteen or more minutes at a time in the cloud of gas and ash proceeding directly from the crater of a volcano during a paroxysmal eruption. In every case there was no noxious gas—no HCl, SO_2, H_2S, SO_2—in perceptible amounts, although these are present in distressing quantities during phases of minor activity. The conclusion is inevitable that the paroxysmal gases—i.e. the gases which produce a great eruption and which have been the cause of the formation of volcanoes—consist mainly of the same ingredients as atmospheric air. . . . I have found only a slight feeling of oppression, which may be due to the high temperature or possibly to a slight deficiency in the proportion of oxygen during the subterranean travel of the gases." (Report on the Recent Great Eruption of the Volcano "Stromboli," *Smithsonian Report for 1912,* pp. 285-289, 1913.)

Respirability of volcanic gases was reported also in the case of the Bandaisan eruption in Japan; but it was regarded by Sekiya and Kikuchi as evidence that the explosion was not of magmatic origin. (The Eruption of Bandaisan, *Jour. Coll. Sci. Imp. Univ. Tokyo,* III, 2, 1889.)

are, indeed, less abundant in samples of gases obtained from magma than they are in the gases emitted by many fumaroles and solfataras; but it is permissible to suppose that the clouds which, despite their dense, smoke-like appearance, attributable to the presence of both condensed water and volcanic dust, have proved innocuous to observers enveloped in them have contained the magmatic gases only in a state of extreme dilution. If not, it is difficult to understand how their temperatures have been reduced; the undiluted gases must be lethal because of their high temperature.

VESUVIUS: THE VULCANIAN ASH PHASE

The eruption of Vesuvius in 1906 provided a spectacular example of Vulcanian gas-and-ash emission (Fig. 10). When voluminous outflow of pyromagma had relieved the pressure on the lava column under the volcano in the first phase of the eruption, an enormous belch of gas followed as the second phase recognised by Perret (Chapter VI). It has been estimated by I. Friedländer that during this phase the steam emitted was the equivalent of 1·75 cub. km. of liquid water, but was accompanied by only ·25 cub. km. of ash. During this "gas" phase of the eruption, "in comparison with the stupendous volume of the escaping vapour the solid matter was quantitatively almost negligible;" but in the third, or "ash", phase of the eruption, which soon followed and continued thereafter for many days, the gas, outrushing now less rapidly, was very heavily laden with ash, much of which was of fine grade.[8]

VULCANIAN SHOWER DEPOSITS OR SUBAERIAL TUFFS

Basaltic magma of the more explosive kind—that, for example, which rose in the craters, "chasms", or nearly continuous rift (Figs. 36, 37) opened by the eruption of

[8] F. A. Perret, *loc. cit.* (4). According to Perret's "gas-liquid differentiation" hypothesis (*loc. cit.* (2)), though the gas emitted during the "ash phase" was newly derived from the magma, that of the "gas phase" had separated prior to the eruption. The idea of such separation of a large body of gas is contrary to the views of most modern volcanologists, however. Thus, B. G. Escher states: "It cannot be asserted in general that in the magma chamber before the eruption there has been a separation between gas and magma" (On the Formation of Calderas, *Proc. IV Pac. Sci. Cong.*, 2, pp. 571-589, see p. 584, 1930).

Tarawera (New Zealand) in 1886 (Chapter VIII)—produces vast quantities of vesicular or pumiceous lapilli, of which a pure deposit covers hundreds of square miles of country north-eastward of Tarawera.[9] Sorted during passage through the atmosphere, the Tarawera lapilli are in any particular locality of very even size—coarser near to and finer far from the source. At a distance of about thirty miles north-eastward the lapilli grade into fine ash (Fig. 90, Tw). Only in the immediate vicinity of the source do lapilli give place to scoria, bombs, and also ejected blocks which are derived from the wall rocks of the rift. The maximum thickness of this accumulation of coarse debris, near the rift, is 250 feet. On the rim there is some agglutinated spatter which has produced "solid lava . . . formed by the running together of molten erupted fragments,"[10] and here, except for the admixture of blocks of non-basaltic rocks torn from the walls, the deposit resembles the products of basaltic fire-fountaining. It appears almost that the emission of lapilli and scoria from the Tarawera rift, which did not conform at all closely to the Vulcanian type of eruption, may be classed as an example of fire-fountaining on a scale perhaps unprecedented among modern eruptions. In Iceland, in 1783, however, the Laki rift, twenty miles long, emitted not only the flood of basalt that has often been described but also a vast quantity of scoria and lapilli, together with much dust, which was spread far abroad and exceeded in volume the whole of the lapilli and dust produced by the Tarawera eruption.[11]

The description given by Perret of lapilli ejected from Vesuvius in 1906, and undoubtedly derived from new magma in the volcanic throat, fits the Tarawera lapilli very well.

[9] It is now generally recognised that lava rose in the Tarawera rift (Chapter VIII). In spite of the abundance of ejected material quite obviously derived from new lava recognition of the magmatic character of the eruption was long delayed. In 1899, however, B. Friedländer remarked: "After a personal inspection of Tarawera and Ruawahia I never doubted but that there had been molten rock." (Some Notes on the Volcanoes of the Taupo District, *Trans. N.Z. Inst.*, 31, pp. 498-510, 1899).

[10] L. I. Grange, The Geology of the Rotorua-Taupo Subdivision, *N.Z. Geol. Surv. Bull.*, 37, p. 83, 1937.

[11] K. Sapper, *Vulkankunde*, p. 129, 1927.

At Vesuvius, however, lapilli were formed only in small quantity as compared with finer ash. Because of the bulk —nearly a cubic mile—of the Tarawera lapilli shower deposit and its purity, its material might be considered typical of basaltic lapilli. Such lapilli are "exceedingly light in weight because of their great vesiculation, and are really vitreous, but of earthy appearance. They have apparently been blown directly from the effervescing lava in free contact with the atmosphere."[12]

Andesitic magma produces much ash of sandy to finer grades, which is carried far abroad by winds, and along with this lapilli, rock fragments and blocks, and some bombs of the breadcrust variety are ejected, accumulating in the vicinity of the source. Andesitic ash has been produced abundantly, for example, by the Tongariro-Ngauruhoe volcano group, in New Zealand, in very recent times. Major eruptions are prehistoric, but light sprinklings of fine ash are derived from mild eruptions of Ngauruhoe from time to time. The bulk of a widespread shower deposit that contains also some debris of coarser grade came, no doubt, from this ash-building cone, perhaps with some contributions as well from craters on the broad summit of Tongariro. This "shower" mantles to a depth of several feet the previously dissected flank slopes of the Tongariro mountain, where it is mixed with some lapilli and angular lava fragments. The layer covers a large area, but thins to a few inches at twenty miles from the source (Fig. 90, N).

Dacitic and *rhyolitic* magmas produce much inflated pumice, rhyolitic pumice being especially abundant; and some volcanoes in which these magmas rise scatter breadcrust bombs. One of the thickest and most widespread of known accumulations of rhyolitic pumice and fairly coarse pumiceous ash is that comprising "shower"[13] deposits which mantle the

[12] F. A. Perret, *loc. cit.* (4), p. 88.

[13] These widespread layers of superficial unconsolidated tuff have been termed and mapped as "showers" by New Zealand pedologists (L. I. Grange, A Classification of Soils of Rotorua County, *N.Z. Jour. Sci. and Tech.*, 11, pp. 219-228, 1929; Volcanic-ash Showers, *ibid.*, 12, pp. 228-240, 1931; *loc. cit.* (10). A synonym is "pumice fall" (S. Kôzu, The Great Activity of Komagatake in 1929, *Tchermaks M.u.P. Mitt.*, 45, pp. 133-174, 1934).

surface thickly over thousands of square miles in the centre of the North Island of New Zealand, surrounding Lake Taupo. Over a considerable fraction of the area mapped in Fig. 90 as covered by the Taupo shower the material consists largely of pumice lumps. Other less extensive shower deposits of pumice which also are of very recent origin cover the land

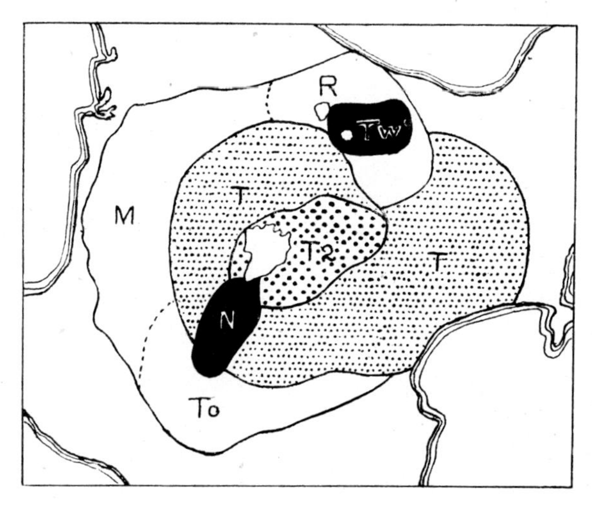

Fig. 90. Ash showers of the North Island of New Zealand. The district covered by ash from various sources is enclosed by a heavy line. Taupo shower, stippled: T, more than 6 inches thick; T2, more than 2 feet thick. Tw, Tarawera lapilli shower. N, Ngauruhoe andesitic ash. M, R, and To, more ancient showers. (After L. I. Grange.)

surface farther to the north-east, and somewhat older showers, which have escaped burial under Taupo ash, still form the surface in western districts.

The Taupo shower not only ranks among the most recent showers of rhyolitic pumice, but its thickness and extent testify also to the occurrence in this district of one of the most violent paroxysmal eruptions of comparatively recent (though prehistoric) times. The deposit is mapped as covering an area of 8800 square miles. Abundant pumiceous ash has been carried by the prevailing wind to the east coast (Fig. 90), where it is six to fifteen inches thick; while around Lake Taupo the pumice layer is more than two feet thick over an area of upwards of 1000 square miles. Throughout this area there is a layer of large pumice lumps at the base, and this passes up into a mixture of smaller lumps with ash of coarse sandy grade. Central North Island rivers built extensive aggraded plains of this material.

It has been estimated that the eruption which scattered the pumice of the Taupo shower ejected four or five cubic miles of material; and from the charred condition of timber buried under it it has been inferred that the temperature of the newly fallen pumice must have been above 250° C. over a wide area.[14] Various investigators have regarded Lake Taupo, or rather some crater within its wide perimeter, as the source of the Taupo-shower pumice.

Cones and Showers

Only one of the New Zealand ash showers mentioned in the preceding section has had as its source a vigorously growing ash cone (Ngauruhoe). Volcanic ash and coarser debris are built into cones when ejected from a central orifice by explosions or fountaining of such moderate intensity that the coarser lumps at least have trajectories sufficiently low to bring them down immediately around the source. Thus the lesser, or cone-building, eruptions (Chapter XIII) pile a considerable proportion of the ejected material into mounds or mountains, though commonly they produce also much fine ash which is carried with the accompanying gas clouds to a great height and, being spread abroad by winds, falls as landscape-mantling showers. The distinction most easily made, indeed, between true explosive-ash production and the fire-fountaining and pyro-explosion of basalts is not so much that the latter processes produce an absolutely large volume of coarse ejected material as that the proportion of ash of finer grades produced along with this is small; it may be almost infinitesimal as compared with the production of ash by the more explosive magmas.

In New Zealand the most prolific (Vulcanian or perhaps Plinian) shower-making eruptions have been unaccompanied

[14] L. I. Grange, *loc. cit.* (13) (1931), p. 230.

by cone-building. These include the outbursts which supplied the Taupo pumice shower and the smaller Rotorua shower (Fig. 90, R). This latter came, wholly or in part, from the explosion-made pit now occupied by the small blue lake Tikitapu (Fig. 91). Other similar showers have provided the materials for successive beds of tuff now buried under the more recent showers; and the basaltic lapilli ejected in 1886 from Tarawera are spread as the most recent "coneless",

C. A. Cotton, photo

Fig. 91. Lake Tikitapu, the source of emission of a pumice shower.

or paroxysmal, shower, which is comparable with those of Iceland.[15]

Where abundant ash and fragments are projected two causes may contribute to their accumulation as a coneless shower.

(1) The material may be ejected so violently that a very high proportion of it falls far from the source. This must be the explanation of the wide distribution of the Taupo pumice.

[15] A. Geikie remarks that, despite the scattering of ash over wide areas, "round the Icelandic explosion craters the rim of fragmentary material is very little higher than the adjacent ground. Great though the amount of ejected stones and dust must be, it seems to be scattered with such force that only a small part of it falls back around the orifice." (*Textbook of Geology*, 4th ed., p. 343, 1903.)

(2) The proportion of material of coarser grade, such as will fly with a short trajectory and fall in the vicinity of the source, may be very small. This has been the case at Tarawera, where the prevailingly abundant light, vesicular lapilli were carried far away by a strong wind.

Unlike those pumice showers, as described above, which have been belched from and distributed widely around mere holes in the ground, many paroxysmal ash eruptions have come from volcanoes previously built up as large cones of dacitic and andesitic materials during long periods of moderate Vulcanian or non-Vulcanian activity. Not only have such volcanic mountains been decapitated and either eviscerated or undermined (Chapter XVI) as the result of great blasts of gases of magmatic origin fountaining through their craters, but vast quantities of magma-derived ash have been ejected along with the columns of steam and other gases and have been distributed far and wide. It is indeed by ejection of magmatic material in this manner that the volcanoes have been undermined.

Many great outbursts of this kind have occurred in historic times, and even within the last two centuries, in the East Indies, Japan, and Central America. Perhaps the greatest of these eruptions that has been observed was that of Tamboro, East Indies, in 1815, which ejected and spread abroad 150 cubic kilometres of debris,[16] destroying, it is said, a hundred thousand human lives. The eruption of Krakatau, a volcanic island in the Strait of Sunda, which took place in 1883, ejected eighteen cubic kilometres of material,[17] covered the adjacent sea with floating pumice which travelled far over the Indian Ocean, and produced so much fine dust that it provided the world with spectacular sunset effects for several years.

One of the most recent and at the same time one of the greatest of ash-distributing eruptions was that which occurred at the volcano Katmai, Alaska, in 1912, when a vast quantity of rhyolitic pumice and ash was ejected. The bulk of this was about five cubic miles, which, though but a fraction of

[16] R. D. M. Verbeek, *Krakatau*, pp. 140-141, 1886.
[17] *Loc. cit.* (16).

that attributed to the explosion of Tamboro, is equal to the whole of the pumice and ash of the Taupo shower. The quantity was sufficient to bury deeply several thousand square miles of the surrounding region, and at Kodiak, a hundred miles from its source, the weight of ash which accumulated on roofs was sufficient to cause houses to collapse. It has been pointed out, indeed, that if such an eruption were to occur in a populated country instead of an Arctic waste it might cost millions of lives.

H. T. Ferrar, photo

Fig. 92. Successive ash beds of shower origin near Mairoa, in the west-central district of the North Island of New Zealand. The land surface is constructional, having been built up by the ash showers. The lowest outcrop consists of underlying (prevolcanic) sandstone.

Soil-forming Ash Showers

As already described, a large part of the North Island of New Zealand has experienced burial under widespread ash showers not once only but many times. Thus over about a third of the island all soils and subsoils are of shower origin except on some steep slopes from which the shower deposits have been washed. In the western (windward) part of this region of shower soils, where the beds of ash are relatively thin and overlie a non-volcanic terrain, their thickness amounts to an aggregate of about twenty feet[18] (Fig. 92).

Among these beds two distinct types have been recognised, the origin of which can be explained as the results of the two types of Vulcanian eruption previously mentioned—the cone-building and the paroxysmal. The cone-building process is progressive and comprises a long series of eruptions of moderate intensity; it is accompanied by the regional accumulation of "ash beds of intermittent origin" (so termed by Taylor[19]) which are formed gradually. "Each coating of ash is generally thin, the existing vegetation is often but little affected; and the soil-forming processes are continuous, although they are modified to some extent by the addition of unweathered material to the surface." Thus the beds are not only fine throughout, but are also weathered for their full depth. Such a deposit may be described as volcanic loess.[20]

In contrast an "ash bed of paroxysmal origin" is less uniformly fine and the ash is sorted into layers, the coarser material below and finer above. As a result of a single paroxysmal eruption "a relatively thick layer of ash is deposited, which completely covers the former soil and destroys most of the existing vegetation." Later "the bed becomes weathered on the top, while the lower parts are still comparatively fresh."[21] Where abundant fine material has filled and levelled up hollows, wind may roll pellets of the volcanic mud, which

[18] N. H. Taylor, Soil Processes in Volcanic Ash Beds, *N.Z. Jour. Sci. & Tech.*, 14, pp. 193-202, 1933.

[19] N. H. Taylor, *loc. cit.* (18), p. 195.

[20] Loess deposits, including one bed eight to ten feet thick which has been buried under thin lava flows, have been described in Hawaii by H. S. Palmer (Loess at Ka Lae, Hawaii, *The Volcano Letter,* 350, pp. 1-3, 1931).

[21] Quotations from N. H. Taylor, *loc. cit.* (18).

sometimes then accumulate as layers of spherical "pisolites", or "chalazoidites."

Ash-shower Modification of Landforms

Ash that is spread as showers accumulates generally in layers parallel to or concentric with the curvature of the surface on which it falls, but blurs the outlines of the minor relief by smoothing over some of the asperities of a dissected surface (Fig. 93). A common relation between older and newer (unconformable) layers is that the former are approxi-

H. T. Ferrar, photo

Fig. 93. Older ash beds covered by later shower materials which conform to an eroded surface, Mairoa, New Zealand.

mately horizontal, while the latter follow the irregularities of a surface either dissected by erosion (Figs. 93, 94) or pitted by a crater- or caldera-making process (Fig. 95).

Another effect produced by very permeable shower deposits, such as those of rhyolite pumice, is to cause the disappearance of surface streams, to produce thus an extensive development of dry valleys, and so to introduce much delay into the process of dissection which may have been active and progressing rapidly before the shower fell. There are many examples of

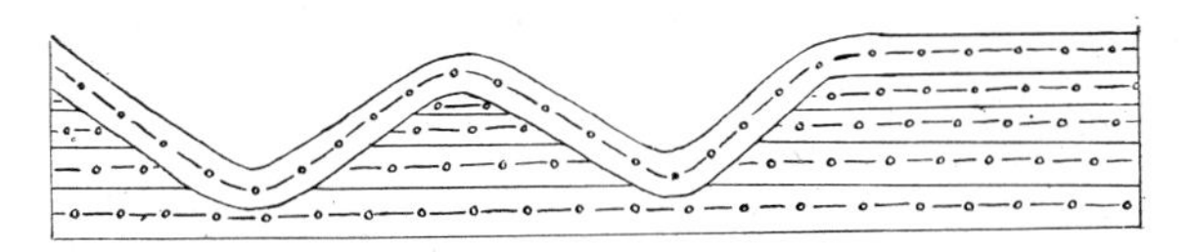

Fig. 94. A common relation of older, dissected to newer, blanketing ash- or pumice-shower deposits.

C. A. Cotton, photo

Fig. 95. Cliff section showing unconformable relation of inclined (younger) beds of pumice lumps and lapilli, deposited on the wall of a caldera, to horizontal (older) beds of similar material, Mayor Island, New Zealand. The section has been exposed by marine erosion.

erosion thus checked in the pumice-buried portion of the North Island of New Zealand, especially within the area covered by the Taupo shower (Fig. 90). Deep, winding ravines, in particular, which trench lake terraces at Taupo have been thus affected, so that their development has practically ceased.

Ash which, on the other hand, contains a high proportion of fine material or some interstitial clay, of whatever origin, is relatively impermeable, and thus a land surface covered by it rapidly develops innumerable consequent rills, especially where the initial surface is inclined. These will dissect a

C. A. Cotton, photo

Fig. 96. Badland dissection of volcanic mud of shower origin, Rotomahana, New Zealand.

thick shower deposit into a fine-textured pattern of ridges, spurs, and narrow ravines so as to produce a badland landscape. That developed immediately after the shower of 1886 on the blanketed hills around Lake Rotomahana (New Zealand) affords an example (Fig. 96). The initial surface of this accumulation consisted of lake silt and the debris of decomposed (kaolinised) rhyolite ejected by a locally phreatic facies of the Tarawera eruption. The Tarawera rift (Figs. 36, 37), which elsewhere emitted basaltic lapilli traversed in this vicinity the basin of the former Lake Rotomahana, and the eruption here emitted only lake mud and the debris of rock material already deeply rotted by volcanic steam and acid gases.

SANDFLOWS AND PUMICE FLOWS OF *Nuée ardente* ORIGIN

Falling from the upper air ash may be intensely cold, and in most cases pumice fragments and glass shreds of widespread shower origin are sufficiently cool to be quite solid before they reach the ground, though it appears that when such ash is in great abundance, as in the case of the Taupo shower, the material is still hot enough to char wood when it falls even far from the source. Similar particles and fragments that are not so far travelled and have not suffered cooling during a long and high flight may be so hot that they are still plastic and become agglutinated when they fall, so as to become a firm—in some cases even compact and almost lava-like—rock which has been called "ignimbrite".[22]

It is doubtful whether such agglutination of particles derived from Vulcanian ash showers ever takes place; and the material which forms ignimbrites seems generally to have been of Pelean origin. It may be necessary, however, to appeal to Vulcanian ejection to account for the distribution of a veneer of obsidian-textured ignimbrite in south-eastern Idaho, which is reported to be spread in a layer only about twenty feet thick over a land surface with very strong relief.[23]

Accumulation of glass shreds so hot as to be still in a viscous, plastic condition seems to be the result generally of the emission of *nuées ardentes* by one or other variant of the Pelean mode of ejection (Chapter I) or by an up-welling of similar material from fissures (Katmaian type). The most voluminous, or first-order, *nuées ardentes,* such as have frothed over either from craters or from fissures, have been required to make extensive and thick sheets of firmly agglutinated

[22] P. Marshall, Notes on Some Volcanic Rocks of the North Island of New Zealand, *N.Z. Jour. Sci. & Tech.*, 13, pp. 198-200, 1932; Acid Rocks of the Taupo-Rotorua Volcanic District, *Trans. Roy. Soc. N.Z.*, 64, pp. 323-366, 1935.

[23] G. R. Mansfield and C. S. Ross, Welded Rhyolitic Tuffs in South-eastern Idaho, *Trans. Am. Geophys. Union,* 16, pp. 308-321, 1935.

rock. The formation and emission of such an "emulsion" or intimate mixture of incandescent particles (still emitting gas, as observers in Alaska and Martinique have clearly recognised) and the hot gases derived from them, into which a viscous magma may resolve itself, has been likened by Lacroix to the boiling-over of milk. Though such eruptions have proved very destructive to human life in Martinique, Java, and elsewhere, some volcanologists regard them as an indication of decadence in the volcanoes emitting them, for it has been assumed that they occur because of the presence in the volcanic throat of magma at a temperature too low to flash into a gas-and-ash mixture such as will be projected forcibly upward in Vulcanian fashion.[24] Second-order *nuées ardentes* are generated by the explosive disintegration of portions of cumulo-domes. These, however, produce smaller volumes of gas and ash, though this mixture may be generated at a very high temperature. While the aggregate of material produced in this way by successive *nuées ardentes* is quite commonly very large, the quantity produced at one time is rarely of sufficient volume, apparently, to retain its high temperature long enough to form agglutinated rock.

Nuées ardentes of the First Order

A *nuée ardente* of the first order which frothed over from the crater of Mont Pelée, Martinique, in 1902 was sufficiently voluminous to spread widely and to flow with great velocity as a glowing avalanche down the slope of the mountain, completely wrecking and burning out the town of St Pierre on its way to the sea. The temperature of this glowing avalanche as it passed through the town was below the melting point of copper (1058° C.) but sufficiently high to melt bottle glass (650°-700° C.).

A similar, though probably less voluminous, glowing avalanche or *nuée ardente* of the first order which coursed down from the summit of Mont Pelée by way of the valley

[24] F. A. Perret, *The Eruption of Mt Pelée 1929-1932,* p. 112, Carnegie Inst., Washington, 1935. N. Wing Easton (Volcanic Science in Past and Present, *Science in the Netherlands East Indies,* pp. 80-100, 1930) also has remarked: "An active volcano in its old age is much more dangerous to the surrounding population than in its youth."

of Rivière Blanche, along the track followed later by a multitude of smaller *nuées ardentes,* was witnessed from the sea by Anderson and Flett two months after the destruction of St Pierre. As described by these observers:

In an incredibly short time a red-hot avalanche swept down to the sea. . . . It was dull red, with a billowy surface reminding one of a snow avalanche. In it there were larger stones which stood out as streaks of bright red, tumbling down and emitting showers of sparks. In a few minutes it was over. . . . It was difficult to say how long an interval elapsed between the time when the great glare shone on the summit and the incandescent avalanche reached the sea. Possibly it occupied a couple of minutes: it could not have been much more. Undoubtedly the velocity was terrific.

A. Lacroix, photo

Fig. 97. A *nuée ardente* rushing towards the sea from Mont Pelée, Martinique. From it arises a cauliflower cloud of steam or "fume."

The sight of this (compare Fig. 97) conveyed a clear impression of the nature of such eruptions:

As soon as the throat of the volcano is thoroughly cleared [by preliminary explosions], and the climax of the eruption is reached, a mass of incandescent lava rises and wells over the lip of the crater in the form of an avalanche of red-hot dust. It is a lava blown to pieces by the expansion of the gases it contains. It rushes down the slopes of the hill, carrying with it a terrific blast, which mows down everything in its path. The mixture of dust and gas behaves in many ways like a fluid.[25]

During the last and probably the greatest eruption of the once high volcano "Mazama", when its summit collapsed to form the caldera Crater Lake (Oregon), an immense volume of pumiceous material frothed out as *nuées ardentes* in quick succession. "The total volume blown out of Mount Mazama during this culminating cycle of activity is between 10 and 12 cubic miles. All this escaped within a short space of time."[26] Some of the pumiceous tuff which flowed down flanking valleys as glowing avalanches has become agglutinated, forming ignimbrite rock, a characteristic feature of which is columnar jointing.[27]

Emission of a great volume of *nuée ardente* material took place in Alaska in 1912 immediately before the paroxysmal eruption of the adjacent Katmai volcano, though not from the crater of the volcano. The phenomena were not under observation, but it appears from discoveries subsequently made that *nuées ardentes* emerged simultaneously from many fissures which opened in a valley, and, being trapped in the valley, the flood of very mobile, incandescent, partly gaseous "emulsion" spread out flat so as to become when it settled down a plain of sandy tuff 53 square miles in extent. The broad floor of the valley has thus been covered to a depth of 100 feet or more and has become the Valley of Ten Thousand Smokes, so called because of the great number of fumaroles—jets of steam and acid gases—which issued from the tuff for years

[25] T. Anderson and J. S. Flett, Preliminary Report on the Recent Eruption of the Soufrière in St Vincent and of a Visit to Mont Pelée in Martinique, *Proc. Roy. Soc. London*, 70, pp. 423-445 (p. 444), 1902.

[26] Howel Williams, Calderas and their Origin, *Univ. Cal. Publ., Bull. Dep. Geol. Sci.*, 25 (6), pp. 272-273, 1941.

[27] "I remember that in 1933 . . . I felt I was looking, in one of these valleys, at a cooled representative of what had been a glowing avalanche, because it showed well-developed columnar jointing. A glowing avalanche is, after all, a close relation to a lava flow." (E. B. Bailey, Review of *Crater Lake: the Story of its Origin*, by Howel Williams, *Nature*, 150, p. 3, 1942.)

after its emplacement (Fig. 98).[28] Though fumaroles may be expected to issue for a time from such material as a result merely of the escape of trapped gases,[29] in this case the fumarolic activity continued for ten years, while a linear arrange-

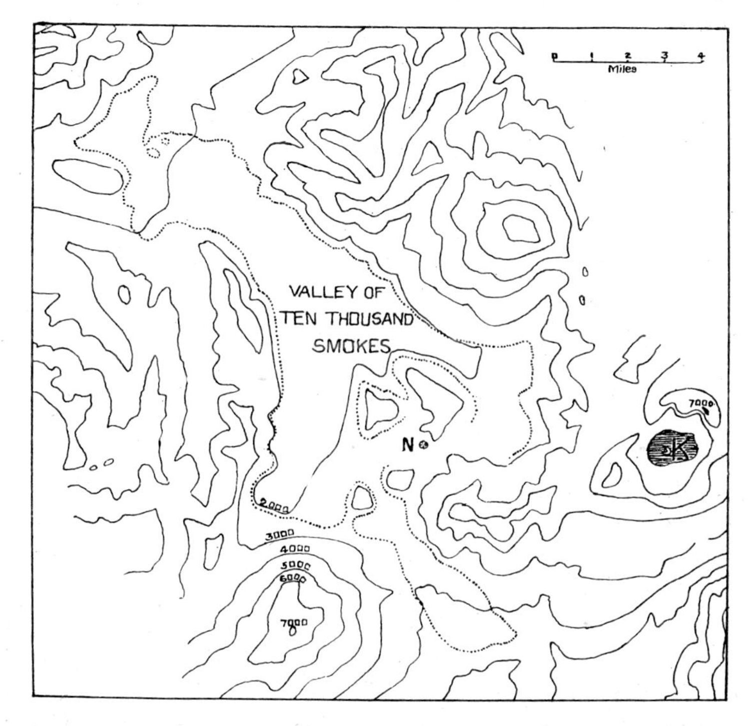

Fig. 98. The Valley of Ten Thousand Smokes, Alaska. A dotted line encloses the plain formed by the surface of the "sandflow" of 1912. K, Katmai volcano; N, Novarupta new volcano. (After C. N. Fenner.)

[28] This explanation of the "sandflow" of Ten Thousand Smokes given by C. N. Fenner (The Katmai Region, Alaska, and the Great Eruption of 1912, *Jour. Geol.*, 28, pp. 569-606, 1920) and R. F. Griggs (Our Greatest National Monument, *Nat. Geog. Mag.*, 40, pp. 219-292, 1921) has been further elaborated by Fenner (The Origin and Mode of Emplacement of the Great Tuff Deposit of the Valley of Ten Thousand Smokes, *Nat. Geog. Soc. Katmai Series,* 1, 1923), who has been able to give good reasons for rejecting another explanation offered by B. G. Escher (On the Hot Lahar of the Valley of Ten Thousand Smokes, *Proc. K. Akad. Wetens Amsterdam,* 24, pp. 283-293, 1922); according to Escher the phenomenon was a "hot lahar," or mudflow, from the Katmai crater lake.

[29] Such activity continued for six weeks at many points on the surface of deposits termed "pumice flows" (of *nuée ardente* origin) which were emitted from the crater of Komagatake (Japan) in 1929. (S. Kôzu, *loc. cit.* [13].) Fumaroles of this kind are sometimes described as "without roots."

ment of the vents seems also to indicate that the gases emanated from fissures beneath the tuff—the fissures from which, in all probability, the "sand" of the tuff was vomited as *nuées ardentes.*

Though there are some indurated patches of considerable extent[30] in it, the sand composing the Ten Thousand Smokes tuff is generally loose at the surface and also where exposed to a considerable depth in stream-cut ravines. The temperature of most of the material of the sandflow when it came to rest must, therefore, although very high, have been below that at which shreds of glass would cohere. The depth also was insufficient to compress the lower layers in such a way as to induce "welding". Temperature observations on "pumice flows" of *nuée ardente* origin in Japan, which have been recorded by Kôzu,[31] have indicated an average of 350° C. at a depth of forty centimetres eight to eleven days after the eruption, and the deposit has taken two years to cool off completely. Such a high internal temperature is in strong contrast with that of a contemporaneous "pumice fall", or shower deposit, which had cooled off in a few hours.

Plateau-building Ignimbrite Sheets

In a thick sandflow or *nuée ardente* deposit some or all of the material may become agglutinated, though probably in most cases portions of it remain incoherent volcanic sand or tuff. The term "ignimbrite"[32] is applicable to all rocks formed by such agglutination. These grade from indurated but open-textured and porous tuff to firmly compacted rocks which have been mistaken for flow rhyolites. This is the condition assumed by deeply buried material which has been compressed under the weight of the upper layers of the deposit. For this process the expression "welding" is perhaps admissible; but

[30] C. N. Fenner, Earth Movements Accompanying the Katmai Eruption, *Jour. Geol.*, 33 pp. 116-139, 193-223, see Fig. 6, 1925. Though at one time willing to admit the possibility of agglutination due to "cohesion of the fragments of hot glass", Fenner has since rejected this possibility, preferring to regard the induration of portions of the tuff as a result of very rapid "pneumatolytic action accomplished by the magmatic gases" (Tuffs and Other Volcanic Deposits of Katmai and Yellowstone Park, *Trans. Am. Geophys. Union,* 18, pp. 236-239, 1937).

[31] S. Kôzu, *loc. cit.* (13).

[32] *Loc. cit.* (22). "Welded tuff" is a synonym.

the welded product no longer presents any resemblance to a "tuff".[33] It rings under the hammer, and may closely resemble a flow rhyolite, or even obsidian, though generally distinguishable under the microscope from rocks of lava-flow origin.

It is in the main the initial temperature (which will be maintained for some considerable time by continued emanation of hot gases) that determines whether an incoherent tuff

Dr P. Marshall, photo

Fig. 99. Gorge of the Mangakino (tributary of the Waikato River) through the ignimbrite plateau, North Island of New Zealand.

or an ignimbrite shall result from the accumulation of *nuée ardente* sand in great volume;[34] but the physical properties of the parent magma, in so far as these depend on chemical composition, will determine in some cases whether the glass shreds are or are not plastic enough to cohere at the temperature that prevails in a newly accumulated volcanic sand. Tests

[33] In the Bishop "tuff" in eastern California C. M. Gilbert has found complete "welding" only at depths of three hundred feet and more in the deposit, while upper layers, though agglutinated, remain porous and somewhat incoherent. "At the base vitric constituents are compressed, distorted, and aligned in the horizontal plane." (Welded Tuff in Eastern California, *Bull. Geol. Soc. Am.*, 49, pp. 1829-1862, 1938.)

[34] "Its great volume, its high temperature and slow cooling, and the gases [rising] through it for a considerable time after its emplacement." (C. M. Gilbert, *loc. cit.* (33).)

made by Marshall[35] indicate that the temperature at which the *nuées ardentes* issued to form the New Zealand ignimbrite was in the vicinity of 1000° C. He found that obsidians which may be assumed to be derived from the same magma as the ignimbrite did not actually fuse at much below 1200° C., but swelled to form pumice at 1000° C.

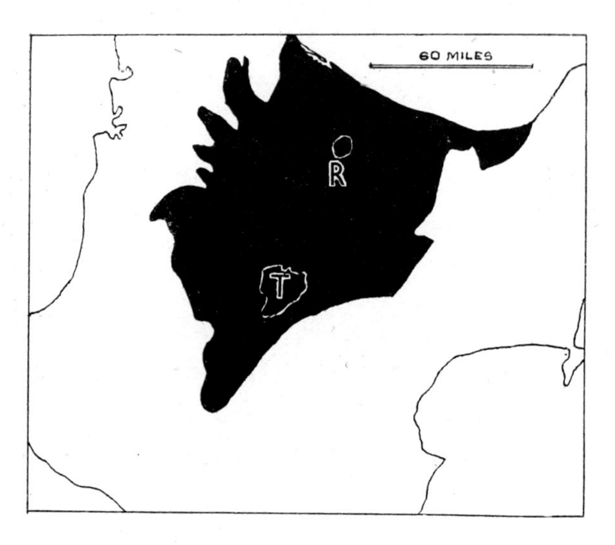

Fig. 100. The ignimbrite sheet in the North Island of New Zealand, including areas covered by younger volcanic deposits, lakes, and lacustrine and alluvial accumulations (boundaries approximate). R, Rotorua; T, Taupo.

The volume of the ignimbrite sheet in New Zealand (Figs. 99-101) is not far short of 2000 cubic miles, and the whole of this material may have been emitted almost instantaneously, though perhaps from many vents, for in some sections exposed to depths as great as 400 feet the mass is monolithic (except for shrinkage jointing) as Marshall and others have observed. Columnar jointing is common and may be on a small or a large scale (Fig. 101). This ignimbrite formation was known to geologists until recently as the "rhyolite" plateau and was believed to consist of a lava flood or flows; and it

[35] P. Marshall, *loc. cit.* (22) (1935). Other and more fusible obsidian subjected to tests at the same time was of more sodic (comendite) composition.

Fig. 101. Escarpment of the New Zealand ignimbrite plateau at Pairere, near Hinuera. *Dr P. Marshall, photo*

has been suggested that other plateaux consisting of great horizontally extended bodies of "rhyolite" are also in reality of ignimbritic construction.[36] Such sheets, even if of obsidian-like texture, will not generally have pumiceous surfaces, nor, in the case of stony-textured ignimbrites, will the rock masses have obsidian selvages.

The supposed rhyolite of the Yellowstone Park is about 3000 square miles in extent. According to Daly[37] it is "decidedly massive and perhaps monolithic," and in the Madison Canyon section there is not "in the rhyolite a single break that would indicate more than one outflow of the rhyolite in an exposure 400 metres high." Brouwer[38] also reports that "large portions of the broad volcanic plateau of the central Yellowstone Park consist of horizontal or slightly dipping lava flows, which often show columnar structure." This supposed rhyolite mass fills to a great depth a "depression in the pre-rhyolite topography". In contrast with this sheet, true coulees, or tongues, of glassy flow rhyolite in Yellowstone Park, which Brouwer has described, are small and steeply inclined, clinging to slopes as steep as 45° and even 50°.

The ignimbrite sheet in south-eastern California known as the Bishop "tuff" is 400 square miles in extent and its average thickness is 500 feet. It is characterised "by an absence of bedding, by columnar structure, and by a surface which is remarkably even despite the irregularities of the buried topography."[39]

Another vast deposit of rhyolitic tuff which has assumed an ignimbritic condition at least in some parts, with the characteristic columnar jointing, surrounds Lake Toba, in northern Sumatra,[40] When emitted, apparently as a *nuée ardente* of the first order or Katmaian type in a single eruption, that which preceded and caused the first and major subsidence concerned

[36] P. Marshall, *loc. cit.* (22), pp. 355-356.

[37] R. A. Daly, *Igneous Rocks and the Depths of the Earth,* p. 143, 1933.

[38] H. A. Brouwer, On the Structure of the Rhyolites in Yellowstone Park, *Jour. Geol.,* 44, pp. 940-949, 1936.

[39] C. M. Gilbert, *loc. cit.* (33).

[40] R. W. van Bemmelen, The Origin of Lake Toba (North Sumatra), *Proc. IV Pac. Sci. Cong.,* 2, pp. 115-124, 1930; The Volcano-tectonic Origin of Lake Toba (North Sumatra), *De Mijningenieur,* 6, pp. 126-140, Batavia, 1939.

in forming the Lake Toba volcano-tectonic depression, this material, though it spread over a region of considerable relief, assumed a level plateau surface. Its volume was of the order of 2000 cubic kilometres. The lowest layer of the tuff shows columnar jointing and has been mistaken for flow rhyolite. It appears therefore to be typical ignimbrite. A fair degree of compactness throughout, which may be perhaps ascribed to agglutination, is indicated by the fact that the outlet of Lake Toba, by way of a re-excavated channel in a tuff-filled valley, has been lowered only gradually to the extent of 250 metres. In fact the lowering is still in progress.

The Feeding Channels of Ignimbrite Eruption

In the case of at least some plateau-making ignimbrite eruptions supply of the material seems to have taken place entirely through fissures and not by way of pipe-like orifices of the kind generally associated with Vulcanian eruption. The "dykes", consisting generally of typical ignimbrite, which will remain plugging such fissures after an eruption have not necessarily any topographic expression. Though it may be true that some prominent necks mark the sites of former ignimbrite eruptions, no features that can be so interpreted are associated with the New Zealand ignimbrite plateau. The only landform the appearance of which might suggest such an origin is a very conspicuous isolated pinnacle, Powhateroa, at Atiamuri, which has been developed by erosion in the valley of the Waikato River and shows some resemblance in form to pipe-filling necks and to pitons. Though it has been described as intrusive, Powhateroa appears, in reality, to have no "root", but to be merely an outlier of the ignimbrite plateau, for it is a high, narrow butte surrounded by a talus-fringed, precipitous escarpment formed by a thick vertically jointed cap-rock,[40a] which is probably a thoroughly agglutinated patch of ignimbrite. This butte has originated as one of a number of spur-ends more or less cut off by insequent dissection (others of which are as yet less completely isolated) along the margin of the plateau where it borders the deeply eroded valley of the Waikato.

[40a] *Loc. cit.* (10), Plate XI.

Fig. 102. A thick dyke of ignimbrite which plugs a feeding fissure at Ongarue, North Island of New Zealand. Polygonal cross-jointing indicates shrinkage during cooling after solidification by agglutination has taken place. *Dr P. Marshall, photo*

As in the case of the tuff in the Valley of Ten Thousand Smokes and of the Bishop "tuff" of California, the vents from which the ignimbrite-forming material of New Zealand has issued are probably numerous fissures, most of which are hidden under the ignimbrite plateau. One such vent, however, has been discovered by Dr P. Marshall near Ongarue, where it traverses underlying strata at the frayed western edge of the plateau (Fig. 102). It is no longer open, but is plugged, as might be expected, with agglutinated material identical with that of the plateau-forming ignimbrite. This is traversed by shrinkage cracks which divide it into small polygonal "columns"—not vertical, however, but transverse. It is highly probable that *nuées ardentes* from many fissures joined forces, overlapping and intermingling with one another, though as is the case at Katmai and in California, the New Zealand ignimbrite formation in some localities seems to have been deposited continuously, showing no traces of intervals between eruptions. Along the western border of the plateau, however, Marwick has noted the presence of what appear to be successively consolidated sheets, at the margins of which escarpments are developed.[40b]

Nuées ardentes OF THE SECOND ORDER

Very numerous glowing avalanches have had their origin on the sides of the tholoids which have exuded during the eruptions of 1902-1903 and 1929-1932 on the summit of Mont Pelée, Martinique (Chapter XI).[41] These are *nuées ardentes* of the second order, such as have occurred frequently in Pelean phases of eruption in historic times, notably at volcanoes in Japan, Java, and Central America. An intimate mixture of incandescent ash and expanding gas similar to that which spreads out widely as Katmaian sandflows and ignimbrite sheets is formed and ejected by explosion of disintegrating

[40b] "The ignimbrite of the Te Kuiti Subdivision cannot be relegated to a single eruption such as deposited the Bishop Tuff. . . . Sections near Takapiko and in the Ongatiti Valley show soft unwelded beds between columnar-jointed welded ones" (J. Marwick, The Geology of the Te Kuiti Subdivision, *N.Z. Geol. Surv. Bull.*, 41, in press).

[41] A. Lacroix, *La Montagne Pelée et ses éruptions*, 1904; *La Montagne Pelée après ses éruptions*, 1908; F. A. Perret, *The Eruption of Mt Pelée 1929-1392*, Carnegie Inst., Washington, 1935.

magma. The emission of *nuées ardentes* from some volcanoes (as exemplified in the present century by Mont Pelée and Merapi in particular) has been repeated at frequent intervals over long periods, and is thus of prime importance as a cone-building process.

The extreme mobility of the gas-and-ash mixtures causes them to course down slopes at terrific speed as glowing avalanches. In some, but not all, cases, however, the force of the explosion which gives rise to a *nuée ardente* propels it initially down the slope, giving it perhaps even greater velocity than it would attain as an avalanche. From observations made by Stehn[42] on *nuées ardentes* descending from the summit of the volcano Merapi, in Java, it may be inferred that these outbursts are generated on all sides of a tholoid or plug dome—not only on the side facing a gap in the rim of a surrounding crater (through which they escape). Most of them are evacuated by gravity from the ring crater through such a gap, and rush thence down the slope of the mountain as "glowing avalanches". The behaviour of these is so similar to that of the *nuées ardentes* impelled down the slopes by the initial explosions (in a manner described by Lacroix) that it would be undesirable to maintain a hard and fast distinction in nomenclature between them.

The crater-ring gaps through which *nuées ardentes* escape have perhaps been formed initially where the ring has been broken down by lateral pressure exerted by a spreading tholoid ("pressure valleys"[43]). Such breaches are enlarged, however, by the eroding activity of block avalanches derived from the tholoid. Through a gap in the crater rim surrounding the tholoid of Pelée vast numbers of glowing avalanches have been seen to issue. Debris derived from them has filled the valley of the former Rivière Blanche to the brim as a great dry talus fan (Fig. 103), which spills over into adjacent valleys and which reaches the sea as a delta. The fan is composed to some considerable extent of blocks and smaller lava frag-

[42] C. E. Stehn, Beobachtungen an Glutwolken . . . Merapi . . . 1933-35, *Handel VII Ned. Ind. Nat. Cong.*, pp. 647-656, 1936.

[43] N. Wing Easton, *loc. cit.* (24), p. 97.

Fig. 103. Avalanche debris derived from *nuées ardentes* fills the valley of Rivière Blanche (Martinique). (From a photograph.)

ments which true ash avalanches have swept along. Much of the fine ash has passed on and has built a large submarine delta which reaches out into ocean depths.

A *nuée ardente* similar in origin to these and of considerable magnitude which was ejected from the flank of the Santiaguito tholoid at Santa Maria volcano, Guatemala, in 1929, was

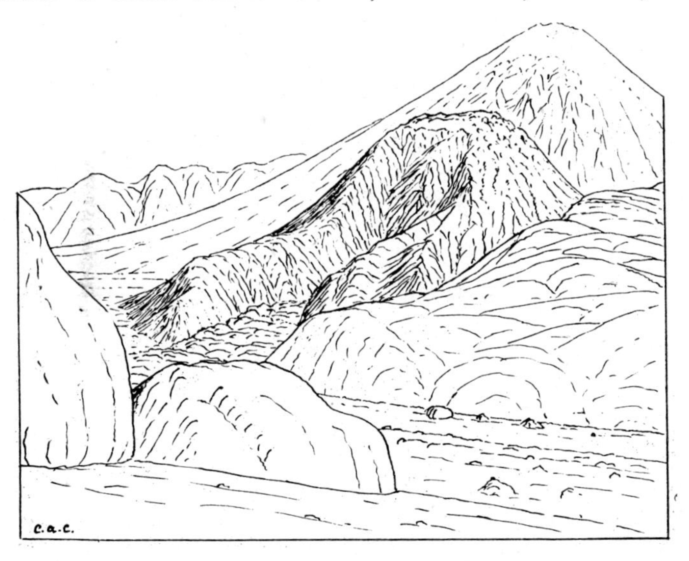

Fig. 104. The tholoid at Santa Maria volcano, Guatemala, showing the scar left after the emission of a powerful *nuée ardente* from it in 1929. Boulders in the foreground are part of the coarser residue of the glowing avalanche formed by the *nuée ardente*. (From a photograph, Sapper and Termer, *Zeits. Vulk.*, 13 (2), 1930).

observed at close quarters to be a red-hot sandflow, the glowing appearance of which caused it to be mistaken at the time for molten lava. This glowing torrent travelled far from the source, following the courses of rivers, so that Jaggar has described it as combining the features of a lahar (volcanic mudflow) and *nuée ardente*.[44] It filled to overflowing a canyon 100 metres deep and 80 metres wide. After its passage

W. C. Putnam, photo

Fig. 105. An explosion pit on a rhyolite cumulo-dome, Mono Craters, California. (Courtesy of the *Geographical Review*, American Geographical Society of New York.)

the residue of coarser fragments left behind it contained many enormous blocks (Fig. 104). The derivation of this *nuée ardente* from the Santiaguito tholoid, which arose a few years earlier (Chapter XI), is indicated by the opening of a large explosion cavity in the side of this dome (Fig. 104) at a spot which had appeared red-hot for several days previously.

[44] T. A. Jaggar, *The Volcano Letter*, 356, pp. 1-4, 1931.

There is a similar pit on one of the obsidian domes at Mono Craters, California,[45] from which it appears that a charge of ash and rock fragments derived from the obsidian while it was still very hot has been projected vertically or obliquely upward (Fig. 105).

Though such discharges may be directed upward, more generally they emerge laterally. They are generated in pockets of the partly congealed lava mass that contain magma either remaining at a high temperature or re-heated. The outburst of gas and ash results, according to Perret,[46] from disintegration by frothing of the magma of such a pocket, which is inherently explosive. This explanation of the origin of Pelean *nuées ardentes* differs somewhat from that previously offered by Lacroix,[47] which is that the explosions ejecting them result from sudden release of gas (derived from newly risen magma) under the carapace of the dome so as to pulverise a portion of it. There is agreement, however, that the cause of the explosions is release either in the interior or on the surface of the dome of gas which has previously been in solution.

The ash particles and fragments are still giving off gas, the continued emission of which under pressure maintains separation between them, cushioning them and giving the whole mixture its emulsion-like character and great mobility, by virtue of which it flows "swiftly, however slight the incline."[48] The gas cushion eliminates friction between particles and fragments, and the rush of the glowing avalanche is silent.[49]

Though the debris resulting from auto-explosions such as are described by Perret is largely ash similar to that of *nuées ardentes* of the first order, such as make ignimbrite, these Pelean glowing avalanches carry along with them cargoes of blocks and fragments of unexploded rock. Such material (Figs. 103, 104), as well as the debris of ladus, or block flows (Chapter X), accumulates as volcanic breccias.[50]

[45] W. C. Putnam, The Mono Craters, California, *Geog. Rev.*, 38, pp. 68-82, 1938.
[46] F. A. Perret, *loc. cit.* (41). [47] A. Lacroix, *loc. cit.* (41) (1904), p. 358.
[48] F. A. Perret, *loc. cit.* (41), p. 84.
[49] "The *nuée ardente* is *soundless*. This marvellous characteristic, the cushioning effect of compressed gases, reveals as nothing else does the absence of attrition between the solid fragments of the avalanche." (F. A. Perret, *loc. cit.* (41), p. 95.)
[50] A Lacroix, Contribution à l'étude des brèches et des conglomérates, *Bull. Soc. Géol. Fr.*, 6, pp. 635-685, 1906.

CHAPTER XIII

Ash-built and Stratified Cones

COARSER DEBRIS EJECTED BY THE MILDER VULCANIAN ERUPTIONS accumulates as cones, and the complex Pelean process which consists of the exudation or extrusion of tholoids and plug domes followed by crumbling and disintegration of these as *nuées ardentes* builds also conical structures.

CENTRAL VENTS

It is obvious that a localised, or "central", vent is a necessity for cone-building. Analogous forms—ridges of debris—aligned on fissures are rare,[1] though paroxysmal, shower-producing eruptions from fissures have occurred in Iceland and in New Zealand (Chapter XII).

RELATION OF CENTRAL VENTS TO FISSURES

The relation of central vents to underlying fissures or fissure swarms (rift zones) in the crust is rarely questioned, but the fissures are in nearly all cases hypothetical. "The volcanologist would point out very simply that in an alignment of eruptive centres, whether straight or curved, a fissure is deduced—a feeding fissure common, at no matter what depth, to the various vents."[2]

Some volcanoes, in particular those of East Africa and some in western North America (Fig. 31), are close to great faults the fissures of which, it is reasonable to suppose, allow escape of magma; and sub-oceanic volcanic lines such as have been

[1] Hekla, in Iceland, which is an example of such a ridge, "is made up of successive sheets of lava and tuff which . . . have not been formed into a cone but into an oblong ridge, fissured in the direction of its length and bearing a row of craters along the fissure" (A. Geikie, *Textbook of Geology,* 4th ed., p. 343, 1903). F. von Wolff states the dimensions of Hekla as twenty-seven kilometres long and two to five kilometres wide (*Der Vulkanismus,* 2 (2), p. 889, 1931).

[2] F. A. Perret, *The Volcano-seismic Crisis at Montserrat 1933-1937,* p. 1, Carnegie Inst., Washington, 1939.

inferred in the Central Pacific region[3]—notably that mapped by Dana[4] through the Hawaiian Islands—may also mark the courses of tension faults along which the magma of a basaltic substratum may find egress.

In the case of continental central vents (with which must be included those of volcanoes on the island festoons fringing continents) actual fissures do not appear at the surface, and nothing is known of the depth at which postulated fissures may communicate with underlying bodies of magma such as may have already developed in or have been injected into the crust.

The process of development of a central vent from a fissure is also hypothetical. Conduits have been opened perhaps where rift zones intersect.

A volcanic centre of such permanence [as Vesuvius] is generally regarded as the intersection of two fissures that cross at considerable angles. The plausibility of this needs no comment; but, aside from the questionable postulate of such cross fissures under all the great volcanoes, and in view of the known behaviour of lateral and eccentric vents originating by simple fissure [Perret[5] is] inclined to doubt that an original cross fissure in the rifted strata is always present. If the volcano originates over a single fracture, the edifice that is built up by the erupted material will first be elongated in the direction of the fissure. This ridge would subsequently (under continued or renewed upthrust from below) tend to break across, forming a second fissure that would localise the vent at its intersection with the first and, by its effusions, tend to render the volcanic structure equal in length and breadth, giving a ground plan that will be more or less quadrilateral. Later, the upper edifice may be superficially fractured in all azimuths, with consequent approach to a circular shape.

It has been suggested by Daly[6] that central lava conduits have been formed at certain points where portions of fissures

[3] J. D. Dana, *Manual of Geology,* 4th ed., p. 36, 1894; Coleman Phillips, The Volcanoes of the Pacific, *Trans. N.Z. Inst.*, 31, pp. 510-551, 1899, *ibid.*, 32, pp. 188-212, 1900; P. Marshall, Oceania, *Handbuch reg. Geol.*, 7 (2), pp. 3-4, 1912.

[4] J. D. Dana, *Am. Jour. Sci.*, 36, map on p. 170, 1888.

[5] F. A. Perrett, *The Vesuvius Eruption of 1906*, p. 13, Carnegie Inst., Washington, 1924.

[6] R. A. Daly, The Nature of Volcanic Action, *Proc. Am. Ac. Arts & Sci.*, 47, pp. 47-122 (p. 69), 1911; *Igneous Rocks and the Depths of the Earth*, p. 360, 1933.

have been enlarged by gas-fluxing. The fissures themselves without such enlargement, though wide enough to allow of the escape of hot gases, may be too narrow to allow the passage of viscous magma at, or perhaps below, volcanic temperatures. Such matters are highly speculative; but some indication of the trends of rift zones under volcanic fields is given by occasional strict alignment of several volcanoes, and in the arrangement of larger numbers of volcanoes in

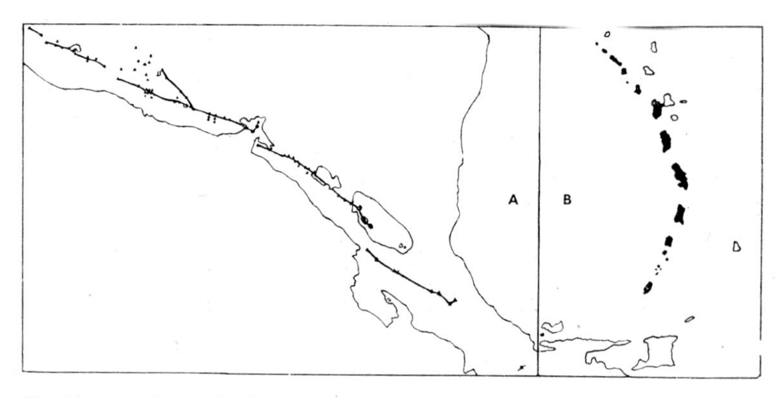

Fig. 106. A: Linear distribution of volcanoes in Central America. (After Sapper.) B: The Antillean arc of recent volcanic and seismic activity. (After Perret.)

belts of regional extent. The extensive zone of volcanoes (active, dormant, and extinct) close to the western sea-margin of Central America[7] and the arc of the Lesser Antilles[8] are conspicuous examples of such belts or "volcanic zones" (Fig. 106); and an even more striking alignment is found in some eighty volcanoes distributed along the Java belt, or Sunda arc, which comprises the islands Sumatra, Java, Bali, Lombok, Sumbawa, and Flores. Arrangement in belts is characteristic also of the volcanoes of Japan and of the Aleutian Islands and, indeed, of all the volcanic regions around the Pacific Ocean. It is questionable whether these volcanic zones are better regarded as marking the crests of geanticlines or as belts of fissuring parallel to such crests.

[7] K. Sapper, Die vulkanische Tätigkeit in Mittelamerika im 20 Jahrhundert, *Zeits. Vulk.*, 9 pp. 156-203, 231-271, 1925-6; P. C. Putnam, Magma Mass Underlying Central America, *Jour. Geol.*, 34, pp. 807-823, 1926.

[8] F. A. Perret, *loc. cit.* (2), pp. 1-2.

In New Zealand a number of active and dormant or recently extinct volcanoes are in a north-easterly-trending line from Ruapehu to White Island: this is the "Taupo zone" of Hochstetter.[9] Within this zone the several vents showing signs of lingering and recent activity on the summit of Tongariro mountain are strictly in a line, which coincides with the trend of the Taupo zone itself, as has been pointed out by Speight[10] (Fig. 107).

Fig. 107. Alignment of vents in the Tongariro-Ruapehu group, New Zealand. View south-westward along the line of Blue Lake crater, Red Crater (breached), and Ngauruhoe, with Ruapehu in the distance. (From a photograph by Professor R. Speight.)

Alignment of closely-spaced small vents suggests the presence of a fissure at a shallow depth, and this is indicated more particularly by the tendency of the locus of activity in some volcanoes to migrate in a constant direction throughout a series of eruptions, so as to trace the line of a hypothetical fissure.[11] An example of such migration has been described at Talakmau (Sumatra), where not only are the adult volcanoes Nilam, Pasaman, and Talakmau in line, but signs of a further extension of an underlying crack towards the

[9] F. von Hochstetter, Geologie von Neuseeland, *Reise d. Novara,* Geol. Teil, I Bd., pp. 92-95, 1864.

[10] R. Speight, in L. Cockayne, Botanical Survey of the Tongariro National Park, pp. 7-12, *N.Z. Parl. Paper C. 11,* 1908.

[11] N. Wing Easton, Volcanic Science in Past and Present, *Science in the Netherlands East Indies,* pp. 80-100 (p. 99), 1930.

surface are seen in the progressive migration of the centre of activity on the summit of Talakmau.[12] Evidence of similar linear migration is seen on Monte Rotaro, Ischia, and on Galunggung, Java[13] (Fig. 129). This migration of the active centre on Galunggung was the cause, according to Escher, of the initiation of a great lahar (volcanic mudflow) which will be described later in this chapter. Migration from the first position, I, central in the cone, was followed by the formation of a crater lake in the lateral crater, II; and a later eruption in a new crater, III, destroyed the rim of the crater of stage II, allowing sudden outbreak of the water of the lake.

In some cases similar migration, however, perhaps with transference of all activity to a lateral ("parasitic") crater, may take place merely along a radial crack developed in the original cone without guidance by any deep-seated fissure—and, therefore, in any azimuth. Such is the explanation suggested, for example, for the migration of activity on the summit of the Soufrière of St Vincent and for the opening of a lateral crater far down on the flank of the Santa Maria cone, in Guatemala.[14]

Serious doubt has been thrown by Geikie[15] on the theory that *all* volcanoes are related to fissures in the crust or even to zones of weakness. His objection has been based on the absence of evidence of fissuring, or, at any rate, of faulting, among the rocks intersected by volcanic vents where these are exposed by erosion. This is the case notably in central Scotland, where it seems as though the orifices had been punched through overlying rocks by explosions (in a manner that has been reproduced experimentally on a small scale by Daubrée[16]) after a body of magma had already risen, by intrusion, to within easy striking distance of the surface.

[12] M. Neumann van Padang, Shifting Craters of the Talakmau Volcano, Sumatra, *Jour. Geomorph.*, 3, pp. 218-226, 1940.

[13] A. Rittmann, *Vulkane und ihre Tätigkeit*, Fig. 6, p. 78, 1936; B. G. Escher, L'éboulement préhistorique de Tasikmalaja et le volcan Galounggoung, Java, *Leidsche Geol. Meded.*, 1, pp. 8-21, 1925.

[14] T. A. Jaggar, The Crater of Soufrière Volcano, *The Volcano Letter*, 359, pp. 1-4, 1931.

[15] A. Geikie, *loc. cit.* (1), p. 279.

[16] A. Daubrée, *Bull. Géol. Soc. Fr.*, 19, p. 317, 1891.

Such vents, opened by explosions, have been termed "diatremes", whether situated on fissures or not.

Daly has postulated a process of upward protrusion of magma which will allow it to approach the surface sufficiently closely to initiate volcanic activity, the vents being finally opened either by gas-fluxing or by explosion: "Round bosses or small stocks", he writes, "are characteristic cupola ornaments of batholiths. . . . Some bosses closely simulate volcanic necks." The "thin roof" of a cupola of magma "may be punctured by blowpiping fusion or by pure explosion."[17]

Ash Cones

In the non-paroxysmal type of Vulcanian activity only ash of the finest grade—that which forms part of the high "cauliflower" cloud (Fig. 10) characteristic of such activity—is carried far away by winds. Coarser "ash", including debris which grades through lapilli into pumice or scoria, is built into "ash cones" (Figs. 108, 109). This material when consolidated is termed "tuff". Thus "tuff cone" and "ash cone" are synonymous terms. "Although it is well recognised that some of the beds in these cones contain sufficient coarse material to make them agglomerates, the main bulk of the cone is of the size of ash."[18] Mingled with the ash in varying proportions there may be breadcrust and other bombs and also angular fragments of rock, which are derived from "bench" lava and lava crusts formed in the vents during quiet intervals between explosions, as well as from older lavas and prevolcanic rocks forming the perilith around the feeding pipe or fissure at some depth beneath the volcano.

Growth of an ash cone begins with the building of a palisade-like ring of ejected material around a crater which is wide in proportion, and the ring increases in height as layer after layer of new material is added to it; it thus becomes a mound and later a cone. The summit may remain rather broadly convex; but, if the trajectory for coarse projected frag-

[17] R. A. Daly, *loc. cit.* (6) (1933), p. 361 and Fig. 119.

[18] H. T. Stearns, Geology and Ground-water Resources of the Island of Oahu, Hawaii, *T. H. Div. Hydrog. Bull.*, 1, p. 17, 1935.

ments is consistently short and the supply of such projectiles is maintained, a sharp cone with only a small summit truncation is formed—that of Fujiyama (Fig. 108), for example—such as is rather characteristic of andesitic volcanoes except in the case of those that have blown their heads off or suffered central collapse. The higher such a cone becomes the sharper does the summit appear, though the diameter of the central pipe through which lava and ash ascend is unlikely to change appreciably.

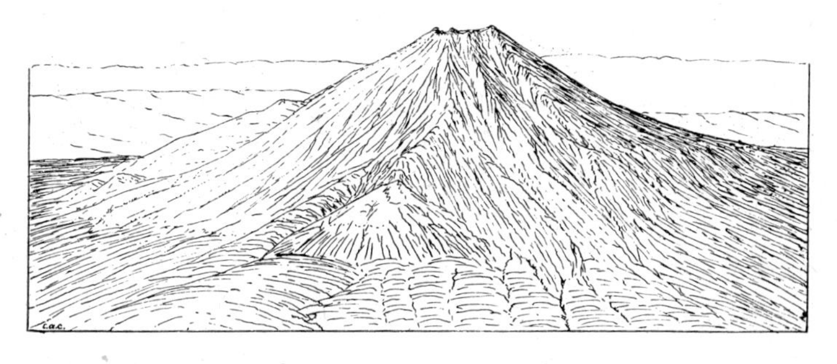

Fig. 108. Fujiyama volcano, Japan, a smooth cone with concave profile. (From a photograph.)

On the truncated summit there is typically a hollow crater, bowl-shaped or precipitous-walled (Chapters XIV, XVI). During a phase of quiet ash-cone building the inner slopes of the crater, like the outer slopes of the cone, may stand at the angle of repose of the pumiceous debris and ash (Figs. 141, B; 158, A), though the inner slopes are more commonly steepened as a result of explosions and of internal landsliding (Fig. 157, B). On many summits such craters have been more or less obliterated by the protrusion of tholoids or plug domes within them (Chapter XI). Many volcanoes which have been reported craterless are normal cones, built largely of ash during Vulcanian activity, with the craters thus obscured or plugged.

Incorporated with the ash and coarser ejected debris of Vulcanian origin on a cone there may be much rubble derived from progressive disintegration of cumulo-domes that have emerged at the summit from time to time. In perhaps exceptional cases cones may be built entirely of such material, as in the case of the talus fringe around the plug dome of Lassen Peak (Chapter XI) and of the *nuée ardente* debris building the flanks of Merapi, Java, described later in this chapter.

Young Ash Cones

The rate of growth of ash cones may be very rapid, hundreds of feet being added to the height in a single eruption lasting only a few days; but such eruptions may be infrequent,

Turner, photo

Fig. 109. The new cone Vulcan, at Rabaul, New Britain, built in 1937.

so that thousands of years may be required for the production of a mountain. A number of examples are known of new cones of rapid growth, in which activity has died out or become dormant. Such cones may or may not eventually grow into great volcanoes. A cone of trachytic ash was built, for example, on the shore of the Bay of Naples in 1538. This cone, which has been called Monte Nuovo, grew to a height of over 400 feet in a few days, but has shown no signs of renewed activity since.

One of the youngest of new cones is that recently named Vulcan,[19] in Blanche Bay, Rabaul, New Britain. On its site, which is one of several active points within the caldera or "Somma" ring formed by the ruins of an ancient large volcano, there had previously been mild activity, but it was not until the eruption of 1937 that cone-building began. Then pumice

Rafael Garcia, photo: National Geographic Magazine.

Fig. 109A. A growing cone of pumice and ash, Paricutin, Mexico, 1943. Night photograph showing "trails" which trace lines of ascent and descent of large glowing rocks.

and pumiceous ash were ejected abundantly, and much of the ejected material fell immediately around the orifice so as to build the new cone (Fig. 109). Finer ash was at the same time distributed widely and fell as a shower over the surrounding district. Explosions followed one another in quick succession, resulting in an intermittent fountaining or belching

[19] There is another and larger active volcano in the same region which has been previously given the name Vulcan—an island north of New Guinea, in Lat. 4° S., Long. 145° E.

of gases laden with ash. A smoke-like column rose to a height of about five miles, and this supplied the fine ash which was wind-borne. Around the orifice a ring of the coarser debris grew up rapidly into a cone, which attained a height of 600 feet during the first day of the eruption but within three or four days had ceased to grow, being then 742 feet high.[20]

In contrast with such cases of short-lived activity, Jorullo, a volcano in Mexico which was new in 1759, was in active eruption for seven months. Its neighbour Paricutin (Fig. 109A), the first eruption of which took place in February, 1943, grew to a height of 500 feet in the first week, continued very active, and was 1100 feet high after ten weeks and 1500 feet high a year later, having meanwhile emitted ten lava flows from points at the base of its debris cone. These, in common with the lava of Jorullo, are of rather basic composition, but of a viscous, andesitic habit.

Large Ash Cones

Nearly all large volcanic cones are "composite", or stratified, for some overflow of lava generally takes place before an ash cone attains the stature of a mountain. It is commonly assumed that all cones more than a few hundred feet high are of composite construction even if they show no visible lava. In many of the largest volcanoes, however, the proportion of lava flows is so small as to be almost negligible, and ash buries and smooths over the small surface irregularities made by the occasional extrusion of coulees.[21] In form, at least, such volcanoes are ash cones. The proportion of flow lavas to fragmentary materials in the volcanoes of the East Indies (Java belt) has been computed as only 1 : 370.[22] Some of this fragmentary material is of *nuée ardente* origin, but it takes part in the building of volcanic mountains, becoming mixed and interbedded with Vulcanian ash.

[20] N. H. Fisher, Geology and Vulcanology of Blanche Bay and the Surrounding Area, New Britain, *Terr. New Guinea Geol. Bull.*, 1, Canberra, 1939.

[21] G. F. Becker, The Geometrical Form of Volcanic Cones, *Am. Jour. Sci.*, 30, pp. 283-293, 1885.

[22] K. Sapper, *Zeits. Vulk.*, 11 (3), 1928.

The volcano Agua, in Guatemala, has the appearance of being constructed entirely of ash.[23] It is a perfect cone, 11,000 feet high, the "grace and symmetry" of which "indicate that it was built up by long continued but comparatively mild explosive eruptions." Another very symmetrical cone, Mayon, in Luzon, has occasionally emitted lava flows, but the ash, which buries these, governs the form.

Fig. 110. The peak of Merapi volcano, Java, showing a small lava tongue (coulee) which has flowed from a summit crater. (From a photograph.)

Composite or Stratified Cones

High and pointed cones occasionally pour out a little lava from summit craters (Fig. 110); larger flows, however, generally break through the weak ash-built cone structure and emerge on the flanks of a cone, though the immediate source of such lava is probably accumulated pyromagma in the central pipe. An outbreak of somewhat liquid andesite in considerable volume may take place also if a flow from the summit is sufficiently mobile to erode and break down ("breach") the crater wall (Fig. 107). Most summit flows from the more acid non-basaltic volcanoes are of such small dimensions, however, that they solidify as narrow, steep-sided tongues (coulees)

[23] I. C. Russell, *Volcanoes of North America,* p. 171, 1897.

on the steep slopes of a cone. Where the stratified layers of which a cone is built are exposed in section the number of such small flows is commonly found to be large. Such exposure of the volcano structure is often seen on the inner walls of craters (Figs 111, 112) that have been enlarged to

V. C. Browne, photo

Fig. 111. Outer and inner ("nested") craters of Ngauruhoe volcanic cone, New Zealand, showing edges of ash and lava layers in the wall of the outer crater.

funnel shape, either by explosions or by internal landsliding, so as to expose the edges of outwardly-dipping layers, among which are seen lava flows as well as tuff (ash) beds. Commonly cross-sections of dykes are exposed also in this manner, showing that lava from the central pipe has entered cracks in the cone. Sills also may have been injected between earlier-deposited layers. One such massive intrusive layer exposed in the wall of the deep crater of the Soufrière of St Vincent (as photographed by Tempest Anderson in 1907) is described by Jaggar[24] as several hundred feet thick and columnar.

[24] T. A. Jaggar, *loc. cit.* (14).

Injected dykes are particularly abundant in the high section of the Krakatau volcano which was exposed as a great cliff on the island Rakata by the Plinian explosive eruption of 1883 in the Strait of Sunda. They are seen in many other "Somma rings", for example in the scarp descending to Crater Lake, Oregon (Fig. 198A), and abundantly in the cliffs of the Somma remnant of the ancient Vesuvius, which was eviscerated or engulfed by the Plinian eruption of A.D. 79 (Fig. 112).

Fig. 112. Stratified lava and tuff layers, intersected by dykes, in the wall of Monte Somma, Vesuvius. (Drawn from a photograph by Dr A. L. Day.)

Sections afforded by inner slopes of the crater of Vesuvius as they were exposed partly by explosion and partly by internal land-sliding in 1906, when the throat of the volcano was cleared and the crater was enlarged in the shape of a funnel (Chapter XVI), showed a remarkably similar structure in the modern cone.

Individual flows of lava on composite cones are narrow tongues (Fig. 109), but where lava flows are very numerous they may collectively form continuous or nearly continuous sheets, the individual flows overlapping and interfingering in section to give the "pile-of-flat-fish" appearance noted by Jaggar.[25] Externally a sheath so formed may encase and yet preserve the constructional form of an ash cone; and this

[25] T. A. Jaggar, *loc. cit.* (14).

perhaps is the explanation of a few very symmetrical steep cones which appear to be built of lava—for example, the dacite cone Edgecumbe in New Zealand (Fig. 113) and the rhyolite cone East Butte, Idaho (Fig. 114).

C. A. Cotton, photo

Fig. 113. The lava cone Mt Edgecumbe, New Zealand.

Fig. 114. The rhyolite cone East Butte, Idaho. (Drawn from a photograph.)

More massive flows of viscous lava, with a tendency to develop a broadly convex form, which commonly break out on the flanks but may also breach crater walls, convert cones into the more irregular forms assumed by some lava piles (Fig. 63).

Multiple Volcanoes

Some large andesitic volcanoes are complex piles of materials accumulated as the consequences of eruptions that have taken place either simultaneously or (more commonly) in succession at a group of vents. Where the locus of activity migrates, "parasitic" cones may grow on the sides of others that have already attained the stature of mountains (Fig. 115). If the earlier-built volcano ceases to erupt, the parasitic cone which

F. G. Radcliffe, photo

Fig. 115. Mt Egmont and Fantham Peak, New Zealand. The latter is sometimes regarded as a parasitic cone.

Fig. 116. The broad summit of the multiple volcano Tongariro, New Zealand, overtopped by the growing cone Ngauruhoe. View from south-east.

continues to grow beside it may overtop it and perhaps eventually bury it. Commonly, however, two or more vents remain intermittently active, as has been the case on Tongariro, New Zealand (Figs. 116-118). Ruapehu (Tongariro's neighbour) also has been in the past the scene of multiple activity. It is seen in the background in Fig. 118.

Both these mountains are of hypersthene-andesite construction. Each has a very broad summit, and they have been described as truncated by explosion.[26] As the volcanoes are multiple, however, it is by no means certain that they have levelled their summits by blowing their heads off. Benedict Friedländer, an experienced observer, who examined them in 1896, has made no such suggestion. He remarks, however: "Tongariro is not a single mountain, but a highly complicated volcanic system." He has found a similar complexity in Ruapehu, though the features of the summit plateau are in that case obscured to a great extent by a covering of glacier ice. An extensive hollow in the plateau is described as a "vast oblong level plain", which appears to be a pair of large craters, and quite distinct from these is the smaller crater, containing a tepid lake, in which some activity lingers.[27] A small portion of the summit area of multiple craters appears in Fig. 119.

If removal of the summits has taken place, the theory of explosion, though it may present difficulties, seems preferable to an alternative hypothesis of collapse in the case of these mountains. There is not now a single caldera on either mountain, nor a definitely continuous "Somma ring". On the Tongariro summit, though there has been recent activity, the lavas of the older mountain are covered only in part by newer eruptive materials, and the plateau is pitted by maar-like "crater lakes" (Figs. 107, 118). These occupy an area which as a whole is of the dimensions of a caldera.[28] Though not deep these hollows are steep-walled[29] (Figs. 117, 120).

[26] R. Speight, *loc. cit.* (10).

[27] B. Friedländer, Some Notes on the Volcanoes of the Taupo District, *Trans. N.Z. Inst.*, 31, pp. 498-510, 1899.

[28] The area of the summit plateau on Tongariro is eight or ten square miles, and it is half as large again if a "Somma ring" which extends partly around the Ngauruhoe cone is included with it.

[29] "All these craters . . . have flat floors surrounded by cliffs which show clearly the edges of the lava flows of the old mountain. They do not seem to have poured out molten rock, but a considerable amount of semi-fused scoriaceous material . . . may have been thrown out of them." (R. Speight, *loc. cit.* (10), p. 10.)

L. Cockayne, photo

Fig. 117. Craters on the broad summit of the Tongariro multiple volcano, New Zealand.

V. C. Browne, photo

Fig. 118. View looking south-south-west across the summit of Tongariro volcano, New Zealand, a field of multiple volcanic activity. The cone Ngauruhoe is on the flank of the broad-topped Tongariro mountain. Ruapehu, another multiple volcano is in the distance.

V. C. Browne, photo

Fig. 119. The north-western slope of Ruapehu multiple volcano, New Zealand.

On both the mountains newer and more symmetrical cones overlap to some extent the partly destroyed older stumps or pedestals. A small ash or scoria cone, Red Crater, on the summit plateau of Tongariro has recently been breached by the escape of a considerable flow of fluid lava (Figs. 50, 107), and the young (and still growing) ash-and-lava cone Ngauruhoe, which stands within a "Somma ring" on the flank of Tongariro (Figs. 116, 120), must be regarded as part of the large multiple volcano.

Fig. 120. View north-eastward from Ruapehu (N.Z.), showing the stump, or wreck, and composite caldera of Tongariro beyond the smooth cone of Ngauruhoe, which stands within a "Somma ring."

The smaller dacite volcano Tauhara, near the northern shore of Lake Taupo, New Zealand, shows also some complexity, as it appears to have grown in two stages (Fig. 121).

C. A. Cotton, photo

Fig. 121. The Tauhara extinct volcano from the north-west, Taupo, New Zealand. (Relief, 2000 feet.)

At the north-west side is a craterless dome, but a cratered cone, apparently of later growth, forms the summit.[30]

[30] J. Marwick and H. Fyfe, in L. I. Grange, The Geology of the Rotorua-Taupo Subdivision, *N.Z. Geol. Surv. Bull.*, 37, p. 24, 1937.

Concave Volcanic Profiles

Concavity of external slopes as they appear in profile is characteristic of volcanic cones whether built of ash alone or of ash together with lava flows (Figs. 108, 122); it appears also on the lower slopes of basalt domes (Chapter VII). In the case of cones that incorporate much lava as small coulees

Fig. 122. El Misti, Peru, a composite cone showing concave profiles. The cone has suffered some dissection and is fringed by fans of water-deposited debris. (Relief, 11,500 feet.)

from the summit crater the upper, lava-covered slopes are steep, and in the case of some andesitic volcanoes rather mobile flows have spread out somewhat widely also from the base, simulating a basaltic habit. Commonly, however, the concave profile is not thus determined by lava slopes, but sweeps smoothly down instead by way of an extensive ash surface to a surrounding shower-mantled plain.

There are various ways in which a concave profile may be developed on cones, several or all of which may contribute in a particular case. A common cause that naturally suggests itself is the gradual thinning of the projected and air-borne ash deposit with increasing distance from the source. The correctness of this theory has been tested experimentally by

Linck[31] and Kuenen,[32] who have made model cones by vertical projection of sand grains of mixed sizes, reproducing in this way (Figs. 142, 158) a concavity of profile such as is developed on outer slopes of scoria mounds and small ash cones rapidly built of vertically ejected fragments—e.g. Vulcan (Fig. 109).

On the assumption of vertical, or Vulcanian, projection of material consisting of fragments of mixed sizes mathematicians have devised equations for various curves which the concave profile may assume.[33] In the growth of large volcanoes, however, the process is complicated in various ways, and for them such theoretical curves have little significance.

An equally general cause contributing to concavity of profile is transfer of debris by gravity from upper steep slopes near the crater rim towards the periphery of the cone. Rubble from disintegrating tholoids and lava tongues rolls down talus slopes,[34] and after a cone is subject to dissection much ash is transported by water streams, the water-borne debris being built into an apron of fans around the base (Fig. 122). During episodes of active growth of the cone landsliding is a process of some importance also, slumped material being spread abroad as short mudflows.

In addition there are three processes which are strictly volcanic in that they are in operation only during eruptions. These are: (1) Mobile *nuées ardentes*; (2) hot, dry avalanching of newly-fallen ash in which there is a considerable proportion of fine grade; and (3) lahars (volcanic mudflows).

[31] G. Linck, Ueber die äussere Form und den inneren Bau der Vulkane, *Neues Jahrb.*, Jub.-Bd., pp. 91-114, 1907.

[32] P. H. Kuenen, Experiments on the Formation of Volcanic Cones, *Leidsche Geol. Meded.*, 6 (2), pp. 99-118, 1934.

[33] J. Milne, On the Form of Volcanoes, *Geol. Mag.*, 5, pp. 337-345, 1878; *ibid.*, 6, pp. 506-514, 1879; G. F. Becker, The Geometrical Form of Volcanic Cones, *Am. Jour. Sci.*, 30, pp. 283-293, 1885; G. Linck, *loc. cit.* (31); see also F. von Wolff, *Der Vulkanismus*, I, pp. 476-480, 1914.

[34] It is on record that on several occasions enormous red-hot blocks, which have been derived along with ladus from disintegrating tholoids, or merely break off from the viscid lava masses as these thrust outward unsupported from the steep summit of Merapi (Java), have rolled to immense distances, starting fires in the surrounding country.

Transfer of Debris by Glowing Avalanches

The symmetrical concave slopes of Merapi, in Java (Fig. 123) are added to almost continuously by glowing avalanches of *nuée ardente* origin which either break away from the sides of disintegrating lava plugs at the sharp summit or emerge from under these.[35] It would appear, indeed, that some such sharply-pointed cones—Ararat (Fig. 123) may be

Fig. 123. Above: Profile of Merapi volcano, Java (from a photograph). Below: Ararat (with steepness of slopes exaggerated), after a field sketch by A. K. Lobeck.

another example—are built almost entirely in this way. In December of 1930 vast quantities of lava debris descended thus with great velocity through a narrow debris-eroded ravine on the upper slope of Merapi and then spread out very widely below, distributing as dry fans so as to overspread many villages and much cultivated land and to cause great loss of life. It is recorded that in one month, January 1935, Merapi emitted ninety-one *nuées ardentes*. The cycle of activity of the volcano includes a period of relatively quiet growth of a tholoid or plug dome at the summit; then gas explosions burst through the dome; *nuées ardentes* are emitted and the dome is progressively destroyed; and the cycle closes with the emission of a little flowing lava which solidifies on the slope as a coulee[36] (Fig. 110). During the phase of plug-dome-protrusion the cone becomes craterless and sharply pointed.

[35] M. Neumann van Padang, De uitbarsting van den Merapi . . . 1930-31, *Vulk. en Seism. Meded.*, 12, Batavia, 1933.

[36] M. A. Hartmann, articles on Merapi eruptions summarised by B. G. Escher, Rapport sur les phénomènes volcanologiques dans l'Archipel Indien . . . , *Bull. Volc.*, 1, pp. 127-177, 1937.

Hot Ash Avalanches

The slope of repose on a growing ash cone built by Vulcanian emission of ash is not assumed in all cases as a result of mere rolling down of fragments such as takes place on a scoria mound. An avalanching process has been observed which incidentally excavates on the slopes of accumulation a pattern of radial gashes which may easily be mistaken for consequent ravines cut by water streams. The gullying which develops also on the inner slopes of craters is due in nearly every case to a similar process, the avalanche debris being there for the most part re-engulfed. On outer slopes the avalanche debris is distributed rather widely around the periphery of the cone, for avalanches of very hot, dry ash are extremely mobile and come to rest in forms resembling the dry fans in which the coarser debris of *nuées ardentes* accumulates (Fig. 103). The mobility of the material is not so great, however, as that of glowing ash of *nuée ardente* origin, which is still emitting gas.

Fig. 124. The cone of Vesuvius in 1906, showing the parasol ribbing pattern resulting from hot avalanching on exterior slopes. (From a photograph by Mercalli, taken after the cessation of the eruption.)

The avalanching of newly fallen, hot, dry ash has been studied on Vesuvius, and a pattern of gashes separating sharp-edged radial spurs similar to that developed on Vesuvius in 1906 (Figs. 124, 125) is the "parasol pattern."[37] It must be distinguished from a somewhat similar pattern resembling badland dissection (Fig. 96) in which radial ravines have been

[37] *Structure en ombrelle* of A. Lacroix, *La Montagne Pelée après ses éruptions, etc.*, p. 98, Acad. Sci., Paris, 1908; A. Geikie, *loc. cit.* (1), p. 332.

eroded by water streams—e.g. on the ash cones Bromo and Batok, in Java.[38]

An account given by Perret[39] of the hot avalanches observed on Vesuvius during the third, or "ash", phase of the eruption of 1906 is as follows:

> The . . . material . . . although it was still hot had come to rest on the outer slopes of the cone. Its motion, therefore, was simply gravitational and not due to explosive propulsion. . . . The sculpture of the cone gives evidence of the number of avalanches, . . . especially when it is remembered that each groove was the carrier of many slides.

Fig. 125. Vesuvius in 1906 from another point of view, showing the parasol pattern.

The ash comes to rest in fans of concave radial profile and in less regular piles of blocks around the base of the cone. The newly deposited, very hot ash is like quicksand while it still contains some hot gas trapped among the solid particles, but it subsides eventually into "a compact condition, . . . having lost with its heat the peculiar qualities which gave it its power of extreme mobility."

Mudflows; Lahars

Among the strictly volcanic processes which widely distribute debris both coarse and fine is the volcanic mudflow —termed in Java "lahar". Small mudflows of a quite ordin-

[38] See N. E. A. Hinds, *Geomorphology,* photo p. 247, 1943.
[39] F. A. Perret, *loc. cit.* (5), pp. 90-01.

ary — i.e. non-volcanic — kind occur commonly and help to distribute and re-distribute the apron of volcanic ash deposited by various agencies around a cone after the material has cooled and has become waterlogged, perhaps by the torrential rains that accompany and follow some eruptions. In Java these are termed "rain lahars". Rain due to condensation of vapour of volcanic origin is not a necessary accompaniment of eruptions, as has sometimes been supposed, but rain falling in the ordinary course of events, especially tropical rain, may be quite sufficient to cause extensive sliding and redistribution of unconsolidated ash by mudflows.[40] Major lahars, on the other hand, are phenomena of a larger order.

At Santa Maria (Guatemala) in 1929, and on various occasions at Merapi, in Java,[41] lahars have resulted from a mingling of *nuées ardentes* with river waters; and many lahars are fed at the source by the water of crater lakes ejected or escaping suddenly through breached walls — an event recorded as occurring with disastrous results at the outbreak of many eruptions.[42] Thus a mudflow which evacuated the muddy contents of an ancient crater on Mont Pelée (Martinique) devastated the lower valley of the Rivière Blanche, completely burying a village, in advance of the still more disastrous *nuée ardente* phase of the eruption of 1902.[43] In the same year the contents of a large crater lake on St Vincent (Antilles) were ejected so as to generate extensive mudflows which rushed down various radial valleys to the sea. In this case also the evacuation of the lake occurred as a preliminary to the emission from the same crater of *nuées ardentes* of the first order.[44]

The active volcanoes of Java have given rise to numerous great lahars, many of them due to the rapid evacuation of crater lakes. Escher[45] distinguishes two kinds of these.

[40] Howel Williams, Volcanology, *Geology 1888-1938*, pp. 365-390, Geol. Soc. Am., 1941.

[41] Howel Williams, *loc. cit.* (40).

[42] See, for example, A. Geikie, *loc. cit.* (1), pp. 270-271.

[43] A. Lacroix, *La Montagne Pelée et ses éruptions*, pp. 171-176, 1904.

[44] T. Anderson and J. S. Flett, Preliminary Report on a Recent Eruption of the Soufrière, *Proc. Roy. Soc.*, 70, pp. 423-445, 1902.

[45] B. G. Escher, *loc. cit.* (13).

Those of the first kind, "hot lahars", result from uprise of cumulo-domes in the lakes or from eruptions which blow up through them, as in the case of the St Vincent example mentioned above, while the others, termed simply "lahars", "cold lahars", or "lahars *de rupture*" are caused by the sudden escape of crater-lake waters liberated as a result of an eruption or otherwise. Mixing with ash the torrent of water and mud from a crater lake—either expelled explosively as a "hot lahar" or escaping through a breached wall—forms a vast muddy stream, and this sweeps along also coarser debris, including in some cases great quantities of large boulders of lava rock. The whole stream travels far and may spread out and deposit its load of debris on level land at the base of the volcano.

Other destructive mudflows, which may be classed also as lahars, have been ascribed to the melting of snow and glacier ice by volcanic heat. This has occurred in Iceland, and also during the eruption of Cotopaxi (Ecuador) in 1877.[46] Such melting is attributable in some cases not so much to the heating of an underlying rock surface (which may be negligible) but to the heat from lava flows, from showers of hot ash, and from *nuées ardentes* which may overspread the snow or ice surface.

Destructive Crater-lake Lahars

A crater lake on the summit of Keloet (Klut) volcano (Fig. 126), in Java, broke out in 1875 as a result of an abnormal rise of the water level; and there was another outbreak in 1919, this time the result of an eruption. On each occasion lahars were started which travelled far from the source and were very destructive. In the disaster of 1919 more than 5000 lives were lost; and in order to guard against similar lahars in the future the lake has been drained to a low level by means of a tunnel through the mountain. A similar disaster was caused by the outbreak of a crater lake on the extinct volcano Agua (Guatemala) in 1541. "The crater presents all the characteristics of a former lake

[46] A. Geikie, *loc. cit.* (1), p. 312; F. von Wolff, *loc. cit.* (33), p. 399.

basin, and upon the side of the mountain an immense ravine can be clearly seen, departing from a place where the rim of the crater is broken, and extending in the direction of Ciudad Vieja," a town on the site of a former village which was overwhelmed by the lahar.[47]

Fig. 126. Crater lake on the Keloet (Klut) volcano, Java, the level of which has been lowered artificially in order to prevent its outbreak, which might cause a disastrous lahar. (From a photograph by N. J. M. Taverne.)

Rain Lahars; Mud Lavas

Mud flows in rain-saturated ash deposits—"secondary" mudflows of Sapper[48]—may be included among the cold lahars. Perret, who studied them at Vesuvius after the ash falls of 1906 refers to them as "mud lavas". These lahars of the second order spread far abroad the Vesuvian ash which had already travelled far down the slopes of the cone as hot avalanches. They "carried boulders and blocks of all sizes and constituted . . . a peril second only to that

[47] I. C. Russell, *loc. cit.* (23), p. 170.
[48] K. Sapper, *Vulkankunde,* 1927.

of the lava flows." Stiffer flows came to rest in thick convex tongues, but in the case of those containing much water the surface became finally "that of a thin fluid"—i.e. spread out almost horizontally.[49] Ancient mud lavas are an important element in the structure of many volcanic regions, as Perret has recognised, for example, in the island of Montserrat (Antilles). As is probably the case quite commonly, however, the materials of the mud lavas of Montserrat had their origin in *nuée ardente* deposits.[50]

The earliest recorded disaster attributable to a lahar is the burial of Herculaneum under mud lava during the eruption of Vesuvius in A.D. 79. In this case the ancient mud deposit has itself been covered by lava.

Volcanic Conglomerates

Mudflows, or lahars both hot and cold, are the chief makers of volcanic conglomerates.[51] Characteristics of a volcanic conglomerate of minor lahar origin in Oahu, Hawaii, are summarised as follows by Stearns.[52]

> The . . . deposit . . . has . . . poorer bedding than . . . tuff cones. Large blocks do not cause bomb sags in the bedding such as are so characteristic of subaerial tuff deposits. The bedding has dips parallel to the valley floor [or surface on which it is laid down]; hence it lacks the high dips found in tuff cones. Considerable round and sub-angular stream gravel and boulders are found mixed with the ejecta, and in places the deposit shows stream sorting. . . . On mapping it proves to be an alluvial deposit filling a valley to a certain depth.

Williams remarks:

> The lack of bedding and the chaotic assemblage of large and small blocks in a gravelly or sandy matrix give the [mudflow] deposits

[49] F. A. Perrett, *loc. cit.* (5), p. 102.

[50] F. A. Perret, *loc. cit.* (2), p. 16.

[51] A. Lacroix has attributed the origin of volcanic conglomerates to mudflows and that of volcanic breccias to dry flows of the nature of hot avalanches and *nuées ardentes* (Contribution à l'étude des brèches et des conglomérates, *Bull. Soc. Géol. Fr.*, 6, pp. 635-685, 1906).

[52] H. T. Stearns, *loc. cit.* (18), pp. 19-20.

a strong similarity to morainic detritus, but much of the material is considerably water-worn.[53]

Fenner[54] has described the transportation of boulders, which were originally blocks of volcanic origin, for long distances at Katmai (Alaska) as "great boulder flows or landslides." Outbreak of a lake which had been dammed earlier by a landslide—not a crater lake in this case—on the lower slope of Katmai mountain resulted in a cold lahar which carried a mass of material consisting largely of boulders through a gorge and then spread it out widely. Here "the pounding and grinding undergone by the boulders seem to have been equivalent in their rounding effect to those produced on the ordinary boulders of river channels through hundreds of years of stream action" (FENNER). In this way one may find the explanation of contrasts between volcanic breccias and conglomerates, when it is borne in mind that, as quoted already in the preceding chapter, a *nuée ardente* (which is the most likely depositor of coarse breccia) "is *soundless*. This . . . reveals as nothing else does the absence of attrition between the solid fragments of the avalanche" (PERRET). On the analogy of the Katmai lahar-borne conglomerates Fenner explains enormous deposits of volcanic conglomerate in the Absaroka Mountains, Yellowstone Park, which consist mainly of very large but well-rounded boulders. A fossil lahar has been recognised by Scrivenor[54a] also in a boulder conglomerate of the Pahang volcanic series of Malaya.

In New Zealand an immense bouldery accumulation of volcanic conglomerate underlies the plains and low country westward and southward of the volcano Ruapehu (Fig. 127). "It is generally unstratified, but in a few places layers of boulders are interbedded with beds of clay. The boulders range from small blocks up to masses six and eight feet in diameter."[55]

[53] Howel Williams, Geology of the Lassen Volcanic National Park, California, *Univ. Cal. Publ., Bull. Dep. Geol. Sci.,* 21 (8), p. 321, 1932.

[54] C. N. Fenner, Tuffs and other Volcanic Deposits of Katmai and Yellowstone Park, *Trans. Am. Geophys. Union,* 18, pp. 236-239, 1937.

[54a] J. B. Scrivenor, *The Geology of Malaya,* pp. 91-93, 1931; see also The Mudstreams (Lahars) of Gunong Keloet, in Java, *Geol. Mag.,* 66, pp. 433-434, 1929.

[55] J. Park, On the Glacial Till in the Hautapu Valley, *Trans. N.Z. Inst.,* 42, pp. 575-580, 1910.

A tongue of the conglomerate extends far south-eastward along the Hautapu valley. Here

for the first twenty-five miles nearest Ruapehu the deposit consists almost entirely of clays or sandy material containing blocks of andesite sparsely throughout it. About a mile and a quarter north of Turangaarere andesite boulders are numerous and small in size; but going down the Hautapu—that is, farther from Ruapehu,

A. Hamilton, photo

Fig. 127. Volcanic conglomerate, probably of lahar origin, a mile south of Turangaarere, south-west of Mt Ruapehu, New Zealand. (Courtesy of Professor James Park; from *The Geology of New Zealand.*)

the source of the andesite—the blocks become more and more abundant, and also of greater size, until we reach the ridges between Mataroa and Taihape, where we find masses up to 30 tons in weight perched . . . at a point over thirty miles from their source at Ruapehu.[56]

The largest of these boulders that has been described is as far distant as forty-eight miles from Ruapehu. It is stranded on the side of the wide valley of the Rangitikei River below the

[56] J. Park, Glaciation of the North Island of New Zealand, *Trans. N.Z. Inst.*, 42, pp. 580-584, 1910.

junction of the Hautapu, which is a tributary, and rests 16 feet above the lowest terrace, which is 175 feet above the river. The block is 14·5 feet long, 6 feet wide, and 5·5 feet high, and has been estimated to weigh thirty-seven tons. This block is well rounded, but is also deeply scored longitudinally and diagonally, very probably as a result of the continued passage of a lahar over it after it had come to rest. Quite naturally the striated surface has been claimed as a proof of glacial transport.[57] It is unnecessary, however, to assume a glacial origin of these conglomerates to account for the transport of large boulders, as it is known from observation of present-day lahars in Java that they have ample mobility and transporting power for the purpose. Melting of glaciers by avalanches of hot material from an adjacent crater may, however, have contributed much of the water for Ruapehu lahars. The absence of angular blocks may also be explained without appealing to recent weathering in explanation of rounding, if the theory of glacial transport is rejected in favour of that of lahar origin.

Broad platforms of volcanic agglomerate, which contain enormous boulders in great abundance, have been described around the Bosavi and Favenc extinct volcanoes in southern New Guinea, and appear to be largely of lahar origin. The agglomerate apron around Favenc extends for 30 miles from the volcano; it is somewhat dissected.[57a]

Lahars in Java, as they gather debris, erode vertical-walled ravines on upper slopes of volcanoes, passing on, however, to leave these empty but for some perched blocks.[58] The destructive Klut (Keloet) lahar of 1919, consisting of thirty to forty million cubic metres of water and debris (including large

[57] J. Park, On the Occurrence of a Glaciated Erratic Block of Andesite in the Rangitikei Valley, *Trans. N.Z. Inst.*, 48, pp. 135-137, 1916.

According to observations (as yet unpublished) made by M. T. Te-Punga, the original lahar-deposited conglomerate in this locality was at a higher level, and large boulders derived from it have assumed the positions and attitudes in which they now lie as a result of being let down after the manner of sarsen stones during the later stages of rapid excavation of the Rangitikei valley in soft mudstone.

[57a] S. W. Carey, The Morphology of New Guinea, *Australian Geographer*, 3 (5), pp. 3-31 (p. 11), 1938.

[58] F. Junghuhn, *Java*, p. 241, 1851; G. L. L. Kemmerling, De uitbarsting van den G. Keloet . . . 1919, *Vulk. Meded.*, 1921; C. E. Stehn, Keloet, Excursion E2a, *IV Pacific Sci. Cong.*, Java, 1929.

boulders) passed in half an hour through a gorge only twenty to thirty metres wide and eighty to one hundred metres in depth. This, however, was a ravine which had been in existence since 1848. Lahars follow mainly channels already existing, filling them temporarily to the brim with rushing torrents but leaving them empty again and eventually depositing their loads of debris on low ground many miles beyond. Geikie[59] has noted the difference in behaviour between water streams and lahars.

Whereas in the former case a portion of the surface is swept away, in the latter, while sometimes considerable demolition of the surface takes place at first, the main result is the burying of the ground under a new tumultuous deposit by which the topography is greatly changed, not only as regards its temporary aspect, but in its more permanent features, such as the position and form of its water-courses.

Lahar Landforms

The eventual settlement and draining out of lahar-transported material explains the origin of very extensive fields of mounds and hummocky landscapes on the plains peripheral to large volcanoes. An idea of the characteristics of these may be gathered from descriptions of lahar-hillock landscapes of recent development.

Bandaisan

The landslide generated by volcanic explosion at Bandaisan (or Kobandai) in Japan in 1888 was perhaps not a lahar as strictly defined, but the material dislodged was thoroughly saturated with water and moved as a great mudflow. There is no question of the volcanic origin of the slide.[60] Twenty explosions occurred, all within about two minutes, and, according to a contemporary account, a great part of the mountain (Fig. 128), estimated at about a third of a cubic mile of rocky material, was overthrown—or projected laterally—so that it was split into mighty fragments, which were thrown down much after the manner of a landslip. Descending the mountain-side with ever-accelerating velocity the components of these avalanches were

[59] A. Geikie, *loc. cit.* (1), p. 312.

[60] S. Sekiya and Y. Kikuchi, The Eruption of Bandaisan, *Jour. Coll. Sci. Imp. Univ. Tokyo,* 3, pp. 91-172, 1890.

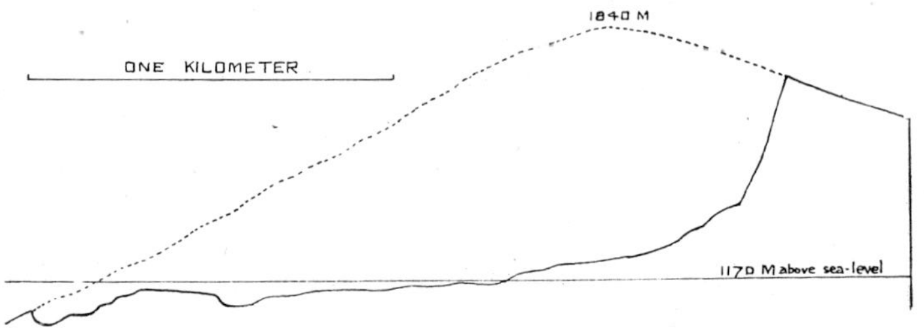

Fig. 128. Bandaisan, Japan. Above: The scar left by the volcanic landslide and the mound field in which the resulting lahar came to rest. Below: Profiles of the mountain before and after the occurrence. (Outlined after a sketch by Sekiya and Kikuchi.)

dashed against obstacles in their way and against each other, and were thus rapidly reduced to confused masses of earth and rocks.[61]

It appears probable that lateral explosions of moderate intensity near the base of the mountain opened a cavity in such a way as to leave the summit unsupported, thus precipitating a great landslide.

Jaggar,[62] who has described the mound field now existing where the debris of Bandaisan came to rest, finds that the material has sunk away from its highest level, which "corresponded with the tops of the mounds". Most of the mounds have hard bouldery cores.

The first rush of the landslide was a thick fan, while the huge rocks were grinding up; then the vast amount of water acquired by the finer material made this finer stuff rush much farther as a mud flood, carrying smaller boulders and boulder clots and spreading as a thinner layer. This outspreading of the lower fan drained down the deposit . . . to the final level, and the larger rock

[61] J. B. Stone, quoted by T. G. Bonney, *Volcanoes,* 2nd ed., p. 18, 1902.
[62] T. A. Jaggar, *The Volcano Letter,* 286, p. 3, 1930.

fragments in units and groups remained in relief as mounds which resisted farther progress.

Galunggung and Tasikmalaja

At the base of the volcano Galunggung, in Java, the debris of a gigantic lahar is spread out and forms the "ten thousand hillocks" of the plain of Tasikmalaja (Fig. 129). Accompany-

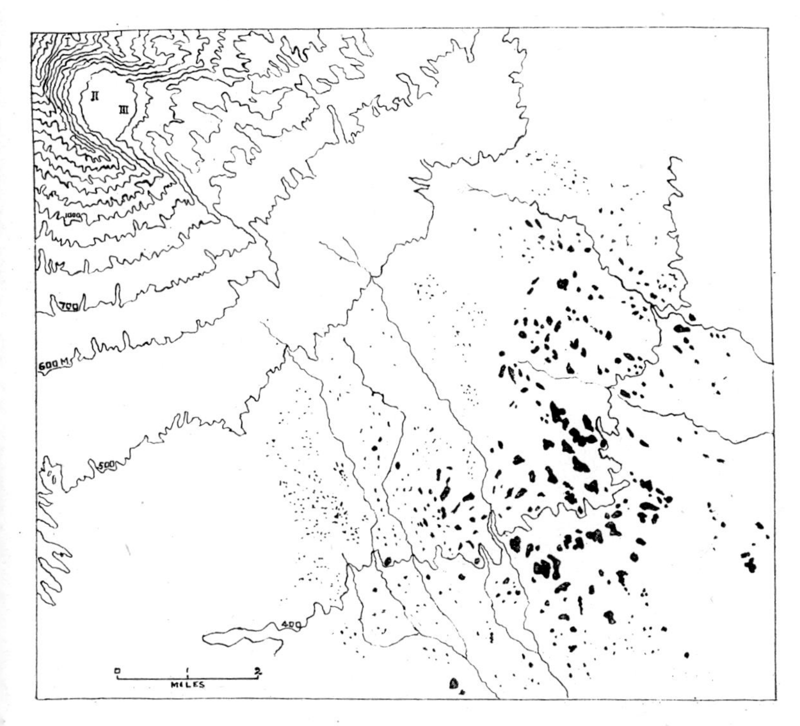

Fig. 129. The summit of Galunggung volcano, Java, and the mound field ("ten thousand hillocks") of Tasikmalaja. (After Escher.)

ing an eruption of this volcano in 1822 deep mud of lahar origin came to rest over a wide area extending many miles from the mountain. It invaded, it is said, 114 villages and caused the loss of 4011 lives.[63] A prehistoric lahar, however, which came to rest as the mound field of Tasikmalaja was

[63] B. G. Escher, L'éruption de Gounoung Galounggoung en juillet 1918, *Natuurk. Tijds. Ned.-Ind.,* 80, pp. 260-264, 1920.

on a vastly larger scale than this. Its source was, apparently, an amphitheatre-shaped scar near the summit of the mountain, in which is a crater now active.[64] The crater once contained a lake more or less similar to that on the Klut volcano to-day (Fig. 126); and an unstable divide between this lake and the head of a ravine already excavated by erosion on the side of the mountain seems to have been destroyed by an eruption (p. 220). The escaping waters of the lake are credited

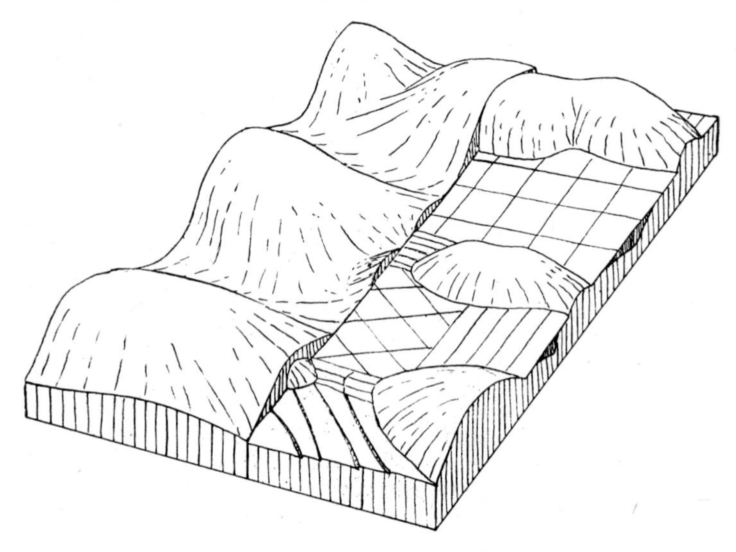

Fig. 130. Artificial modification of the mounds of Tasikmalaja to form rice fields. (After diagrams by Escher.)

with the enlargement of the ravine to form the great gash in the mountain shown in Fig. 129, and the resulting debris has been spread out over a plain 100 square miles in extent. This area, now accurately mapped, includes the "ten thousand hillocks," among which are extensive artificially levelled and terraced rice fields (Fig. 130).

Though a carefully made survey of the volume of material in the "ten thousand hillocks" shows it to be only one-twentieth part of that of the great gash in the mountain Galunggung,[65]

[64] B. G. Escher, *loc. cit.* (13).
[65] B. G. Escher, *loc. cit.* (13).

there seems little doubt that much of the debris of a great lahar underlies them. Some of the finer debris from it has very probably been carried farther on, moreover, and deposited eventually in the sea.

Observation of the signs of man's patient labour throughout centuries in levelling rice fields has led to the formulation of an alternative theory[66] that the hillocks themselves are of human construction; but the incredible labour involved in such a project would have produced results of comparatively little value. In view, moreover, of the close analogy in form and arrangement of this concourse of hillocks with other similar though less extensive mound fields in lahar landscapes, together with its obvious relation to the mountain in the background, the lahar explanation of its origin seems unassailable.

L. I. Grange, photo

Fig. 131. Hillocks of mudflow (lahar) origin in the crater of White Island volcano, New Zealand.

New Zealand Lahar Landforms

A miniature example of a mound-field landscape which was developed in 1914 by a mudflow on the small volcano White Island (New Zealand) is shown in Fig. 131. This serves as an introduction to larger mound fields which are undoubt-

[66] F. X. Schaffer, Die zehn tausend Hügel von Tasikmalaja, *Centralbl. f. Min.*, pp. 207-209, 1926.

edly also of lahar origin. The White Island mudflow carried with it a

> mass of rock fragments of all sizes. The larger . . . exceeded 10 feet in average dimension. . . . The coarse debris has been left in somewhat conical mounds dotted here and there . . . ; some are quite small, but others rise to a height of over 50 feet above the level of the sand-and-mud plain at their base.[67]

The origin of extensive mound fields near the bases of the New Zealand volcanic mountains Ruapehu and Egmont remained a mystery until the lahar explanation was invoked by Grange[68] to account for them. The Ruapehu example, in

C. A. Cotton, photo

Fig. 132. Part of a field of mounds of lahar origin in the Tongariro National Park, New Zealand. Quarrying for roadmaking material has shown these mounds to consist of angular andesite rubble and boulders.

the Tongariro National Park, is shown in Fig. 132, and an extensive landscape of mounds on the south-west side of the extinct volcano Egmont in Fig. 133.[69]

[67] J. A. Bartrum, White Island Volcano, *N.Z. Jour. Sci. & Tech.*, 8, pp. 261-266, 1926.

[68] L. I. Grange, Conical Hills on Egmont and Ruapehu, *N.Z Jour. Sci. & Tech.*, 12, pp. 376-391, 1931.

[69] These are the "conical hills" reported by P. G. Morgan and W. Gibson (Egmont Subdivision, *N.Z. Geol. Surv. Bull.*, 29, 1927), who regarded each of the hillocks as a separate volcano. It has been suggested, on the other hand, by L. Bossard (*N.Z. Jour. Sci. & Tech.*, 10, p. 124, 1928) that the hillocks have been built by gas fountaining or explosions on the surface of a lava flow.

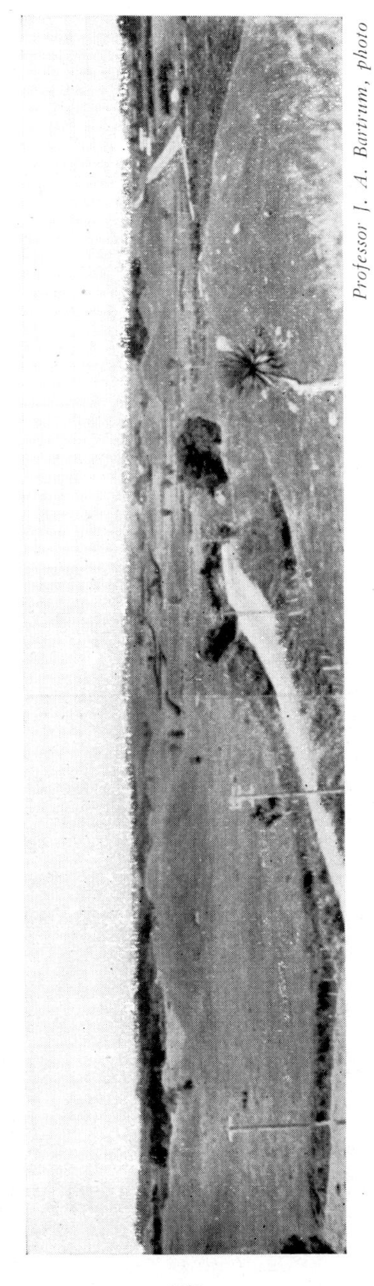

Professor J. A. Bartrum, photo

Fig. 133. A mound field near Opunake, New Zealand, believed to be the debris of lahars that have descended the slopes of Egmont.

CHAPTER XIV

Maars and Tuff Rings; Meteor Craters

FORMS THAT ARE RING-ENCLOSED CRATERS RATHER THAN elevations of the land surface—or may even be mere bowl-shaped hollows without raised rims—result generally from sporadic explosions, occurring either singly or in a brief series, followed by complete cessation of activity. Explosion craters are not all volcanic, but have community of origin and exhibit family resemblance whether they result from volcanic or non-volcanic explosions.

MAARS

A hollow excavated by an explosion provides a natural well if deep enough to intersect the water table. Such a hollow, when occupied by a lake, becomes a rather conspicuous landscape feature. It is termed a "maar," the name being taken from examples in the Eifel district of Germany (Fig. 134). While most maars are clearly volcanic craters there are a few isolated examples for which an origin by non-volcanic explosion has been suggested, as will be explained later in this chapter.

It would appear that each volcanic maar has been the result of a very brief episode of activity, following which the "embryonic"[1] or "abortive"[2] volcano has shown no further signs of life. Craters were thus produced by single or very brief series of explosions in New Zealand in 1886. An example of these is the Inferno Crater (Fig. 135) towards the south-west end of the Tarawera-Rotomahana rift (Fig. 36). Since that date these craters have exhibited fumarolic activity, and, seeing that they perforate a volcanic terrain, they cannot be regarded

[1] W. Branco, Schwabens 125 Vulkanembryonen, *Jahresh. Ver. f. vaterl. Naturk.*, pp. 505-997, Würtemberg, 1894.

[2] E. de Martonne, *Traité de géographie physique*, p. 737, 1935.

as quite dead. The pit occupied by the small New Zealand lake Tikitapu (Fig. 91) is similar in a general way to the basin of the larger Italian lake Nemi, near Rome (Fig. 136). These also are in volcanic terrains, and each pit may have been opened by a single explosive eruption. From each of them an immense amount of ash may have been ejected, which has

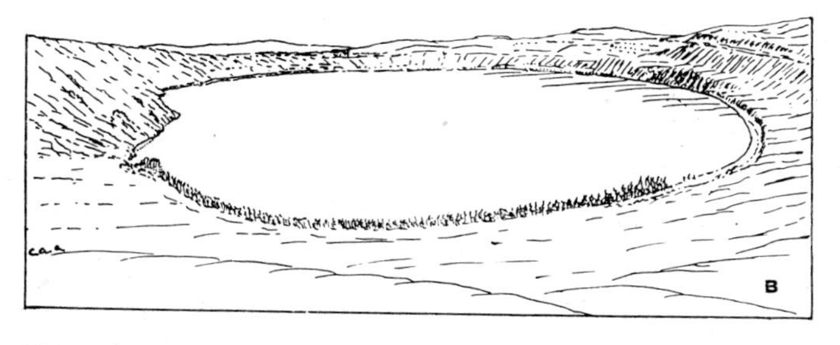

Fig. 134. Maars of the Eifel, each less than a quarter of a mile in diameter. These are members of a group of three, and are half a mile apart.
A: Gemündener Maar. B: Weinfelder Maar. (From photographs.)

been distributed far and wide. Unlike these latter explosion pits, the maars of the Eifel are funnels opened in a non-volcanic terrain and the explosions which opened them have expelled only small quantities of debris. These are maars in the strictest sense.

Included under the head of embryonic volcanoes, but better excluded from the category of maars, if that term is to be employed for landforms, are a number of debris-filled

C. A. Cotton, phot

Fig. 135. The Inferno Crater, opened by the explosive eruption of 1886, in New Zealand.

Fig. 136. Lago di Nemi, in the Alban Hills, Italy.

pipes or funnels of Miocene age perforating Jurassic strata in Swabia[3] and others of Permian age perforating Carboniferous strata in Fife, Scotland,[4] which have no longer any surface expression as landforms or which owe their present landscape form to differential erosion. All such structures, and most if

[3] W. Branco, *loc. cit.* (1).
[4] A. Geikie, Geology of East Fife, *Mem. Geol. Surv.*, 1902.

not all true maars, have been regarded by many geologists as the products of "phreatic" explosions—so termed by Suess[5]—such as may occur when "juvenile hydrogen" (to use Suess's words) or ascending magma[6] encounters water near the surface. The pipes made thus by explosions are diatremes (Chapter XIII) punched through overlying rocks generally without relation to or control by fissures. It is commonly assumed that grouped or clustered diatremes result from the presence of an injected sheet of magma at no great depth. Such phenomena are said to be characteristic of the foreland —northern border—of the European Alps.[7] (Some of the peculiar features of the Alpine foreland—such as the depression known as the Rieskessel—are not simply maars but are of more complex origin. They will be mentioned again later in this chapter as "cryptovolcanic" features.) Very numerous clustered diatremes in Arizona,[8] on the other hand, seem to be related to fault fissures which tap an extensive deep-seated body of basic but explosive magma.

The presence of sills or laccoliths has been postulated to account for the abundance of diatremes both in Swabia and in Fife, the igneous intrusive body immediately underlying the Jurassic formation in the former and the Carboniferous in the latter case. If derived thus from intrusive magma at no great depth instead of being connected directly with a deeper source of magma, each of the maars and small explosion pipes is in Daly's sense a "subordinate" as contrasted with a "principal" volcano.[9]

As an alternative to the phreatic hypothesis to account for the explosions which have made maars and other diatremes it has been suggested that the holes have been drilled by the explosive escape or release of gas which has accumulated in a

[5] E. Suess, *The Face of the Earth,* English ed., IV, p. 568, 1909.

[6] H. J. Johnston-Lavis, Geology of Mt Somma and Mt Vesuvius, *Quart. Jour. Geol. Soc.,* 40, pp. 35-119 (see p. 40), 1884.

[7] A. Rittmann, *Vulkane und ihre Tätigkeit,* 1936.

[8] J. T. Hack, Sedimentation and Volcanism in the Hopi Buttes, Ariz., *Bull. Geol. Soc. Am.,* 53, pp. 335-372, 1942.

[9] R. A. Daly, The Nature of Volcanic Action, *Proc. Am. Ac. Arts & Sci.,* 47 (3), p. 119, 1911; *Igneous Rocks and the Depths of the Earth,* p. 393, 1933. In his earlier work Daly denied to maars the status of "volcanoes," however, preferring to class them as "non-volcanic" (*loc. cit.,* p. 101).

shallow igneous intrusive layer during the progress of its crystallisation (Chapter V). In explanation of the more or less analogous explosive eruption of Kilauea, Hawaii, however, in 1924 (Chapter IV) the phreatic hypothesis (of volatilisation of ground water) has been preferred. Catastrophic explosions of a similar nature took place in southern Sumatra also in 1933 —the Pematang Bata eruption. This eruption, which occurred within a graben of "volcano-tectonic" origin (Chapter XVI), took place fifteen days after a severe earthquake and apparently as a consequence of it, steam pressure having gradually accumulated after water had come in contact with hot rocks or magma as an after effect of the earthquake.[10] Such eruptions, due to explosion of steam generated by heat from magma "at rest," are termed "secondary" by Wing Easton[11] in contrast with "primary" eruptions, "in which *moving* magma takes an active part."

Ubehebes

As examples of volcanoes that are little more than embryos —dry maars, or "ubehebes," as they might be called—the two Ubehebe Craters in Death Valley, California, may be cited. They have been described by von Engeln.[12] Beyond the rims of these craters (Fig. 137) are outward slopes on the surfaces of flat rings (or low cones) which have been built up chiefly (as is the case with true maars) of rock fragments derived from the immediately underlying terrain. In this case the rock fragments have been baked or scorched as well as broken by the heat accompanying the explosions that ejected them. Mingled and interbedded with these are magma-derived lapilli, and fine ash from the same source has been spread widely over the surrounding landscape. The relief of the ring around the smaller crater (Fig. 137) is 150 feet. A layer of lava exposed in this crater wall indicates that a small overflow

[10] C. E. Stehn, Die semivulkanischen Explosionen des Pematang Bata 1933, *Nat. Tijds. Ned. Ind.*, 94 (1), pp. 46-69, 1934.

[11] N. Wing Easton, Volcanic Science in Past and Present, *Science in the Netherlands East Indies,* pp. 80-100, 1930.

[12] O. D. von Engeln, The Ubehebe Craters and Explosion Breccias in Death Valley, California, *Jour. Geol.,* 40, pp. 726-734 1932; *Geomorphology,* p. 597, 1942.

has occurred prior to the final explosions by which the building of the ring was completed; the lava is basalt.

Professor O. D. von Engeln, photo

Fig. 137. The smaller of the two Ubehebe Craters, Death Valley, California.

Basaltic Tuff Rings

In the basalt fields of south-eastern Australia[13] and Auckland, New Zealand, there are a number of small tuff-ringed lakes which resemble maars (Figs. 138, 139). Those at Mount Gambier, South Australia, have been classed by Fenner as "collapsed craters." In the rings around all theses lakes the proportion of magma-derived material is much higher than it is on the rims of the maars of the Eifel. Beds of basaltic lapilli and ash forming the ring around the Auckland example, Lake Pupuke, are shown in Fig. 140. This tuff-making ash has come to rest with regular bedding, at low angles, parallel to

[13] C. Fenner, The Craters and Lakes of Mount Gambier, South Australia, *Trans. Roy. Soc. S. Aus.*, 45, pp. 169-205, 1921; E. S. Hills, *Physiography of Victoria*, p. 177, 1940.

Fig. 138. A cluster of tuff rings enclosing lakes in western Victoria (Australia). (From a photograph.)

Professor J. A. Bartrum, ph

Fig. 139. Pupuke Lake, Auckland, New Zealand, a maar-like lake enclosed by a ring of basaltic tuff.

outer slopes and in some cases also towards the centre of the ring.[14] (Fig. 141A). A laboratory model of a volcano made by Linck[15] by vertical ejection of sand from a pipe has reproduced the tuff-ring form remarkably well (Fig. 141B).

C. A. Cotton, photo

Fig. 140. Beds of lapilli and ash in the tuff ring surrounding Lake Pupuke, New Zealand. The small faults have resulted from settling of the tuff.

The bedding of such subaerial tuffs is extremely regular, almost the only irregularities being the "bomb sags" due to the falling of bombs or large ejected blocks, and the necessary thinning and inclination of beds where they pass over such deposited lumps and at places where an uneven floor is buried. Commonly beds of ash and lapilli of fairly uniform grade follow one another in alternation (Fig. 140).

[14] F. von Hochstetter, Geologie von Neuseeland, *Reise d. . . Novara,* Geol. Teil., I (1), p. 161, 1864.

[15] G. Linck, Ueber die äussere Form und den inneren Bau der Vulkane, *Neues Jahrb.,* Jub.-Bd., pp. 91-114, 1907.

One of the most perfect of the tuff rings of the Auckland (N.Z.) volcanic field, that at Crater Hill (Fig. 142), has been described by Firth.[16] The base of the whole structure is in this case a little below the level of a surrounding alluvial plain, which has apparently been aggraded subsequently to the

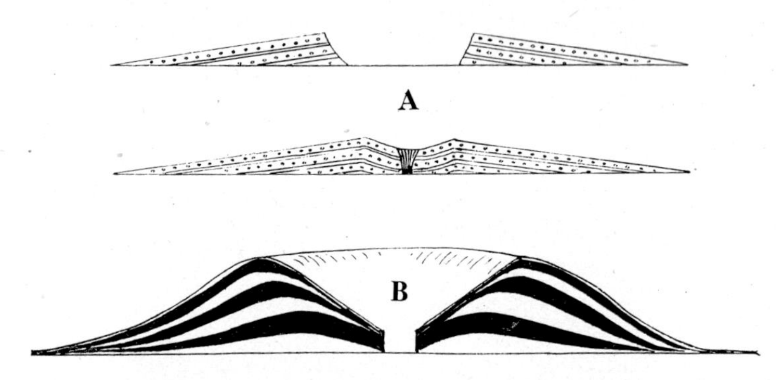

Fig. 141. A: Tuff-ring structures at Auckland, New Zealand. (After Hochstetter.) B: Tuff-ring form and structure produced artificially on a small scale. (After a photograph by Linck.)

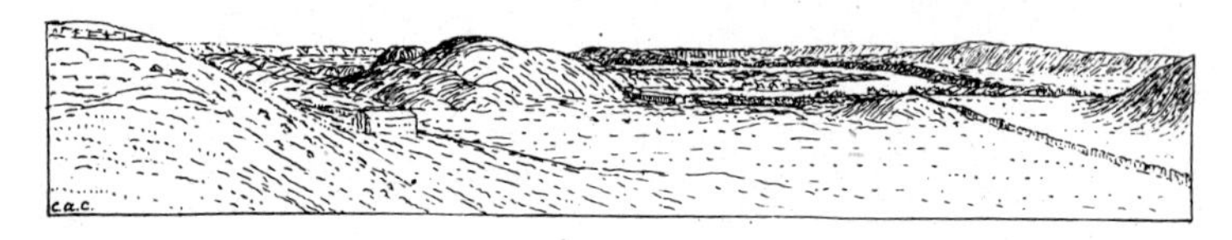

Fig. 142. Tuff ring, Crater Hill, Auckland, New Zealand. A marshy lake within it is fringed by a bench of basalt lava, and scoria mounds have been built in a final phase of activity. (From a photograph.)

volcanic outbreak. This brings the base very close to sea-level, while the tuff ring rises to 120 feet above sea-level. The ring is half a mile in diameter and its interior is floored with basalt. The basalt floor has been reduced to a bench, apparently during the brief period of activity, when it seems that a sink has been formed and that an inner pit crater has contained a lava lake. The sink is now a marsh. From the floor rises a cluster of small scoria mounds built during the final phase of activity.

[16] C. W. Firth, The Geology of the North-west Portion of Manukau County, Auckland, *Trans. N.Z. Inst.*, 61, pp. 85-137, 1930.

The tuffs of the enclosing ring (as described by Firth) are splendidly bedded in layers from ½ inch to 3 inches thick, which dip towards the crater on the inside of the rim, though radially outwards elsewhere, and consist dominantly of rounded basaltic lapilli with subordinate sedimentary material, though rare blocks of scoriaceous basalt are occasionally met with adjacent to the crater. There are some basalt-coated bombs of foreign rock.

It seems beyond question that the contrast between such wide tuff rings and the scoria mounds (Chapter X) of the Auckland basalt field depends entirely on the mode of injection of the material composing them. Some of the scoria mounds are built in great part of basalt lapilli differing but little from the material of the tuff rings. If the scoria mounds must be attributed, as seems probable, to fire-fountaining, the rings, which are of earlier origin than scoria mounds associated with them, have resulted from explosions which have scattered projectiles farther abroad.

The presence of fairly abundant fragments—large blocks in some cases—derived from underlying sedimentary formations is consistent with the view that such explosions were either phreatic or of the kind termed by Stearns[17] "phreatomagmatic." This explanation, which has been applied by Stearns to the well-known tuff rings of south-eastern Oahu—the most conspicuous of which as coastal landmarks are Diamond Head and Koko Crater (Figs. 143, 144)—postulates emission of basalt lava from a vent below sea-level, so as to make contact with water either in the open sea or in a cavernous coral reef. In the case of the Auckland tuff rings which are all at low levels, contact with water seems to have been made in saturated deltaic silts or in alluvium.[18] The "phreatomagmatic" process

[17] H. T. Stearns, Volcanism in the Mud Lake Area, Idaho, *Am. Jour. Sci.*, 11, p. 358, 1926; Geology and Ground-water Resources of the Island of Oahu, Hawaii, *T. H. Div. Hydrog. Bull.*, 1, p. 15, 1935.

[18] J. A. Bartrum, Geology of Area near Auckland City, *A. and N.Z. Assoc. Adv. Sci. Handbook for New Zealand,* pp. 17-19, 1936; Geography and Geology of Auckland and Environs, *Auckland,* pp. 7-13, A. and N.Z. Assoc. Adv. Sci., 1937. It has been shown by Bartrum that the Auckland isthmus, though very recently drowned by a positive movement, has been subject previously, like other regions, to eustatic fluctuation of sea-level. At the time when the tuff rings were built it appears that sea-level was at least as high as it is at present and that the sites of these eruptions may have been submerged.

Professor J. A. Bartrum, photo

Fig. 143. Diamond Head tuff ring, Oahu, Hawaii. View looking north.

Professor J. A. Bartrum, photo

Fig. 144. Koko Crater tuff ring, Oahu, Hawaii. Being in the trade-wind belt, the Diamond Head and Koko rings are highest on the south-west side owing to wind transport of projected material.

differs from "phreatic" explosion in that the fragmentation of lava, with the production of abundant lapilli and some finer ash, is ascribed to escape of gas from the magma when the pressure on it is relieved by the occurrence of steam explosions,[19] and these explosions (taking place under a shallow cover) are accountable for the projection of non-volcanic rock debris as well as the disintegrated basalt debris to a very considerable distance. A quick-firing series of such explosions may pile up much debris as a tuff ring in a very short time—in a few days, perhaps—after which there is no further activity, the supply of outflowing basalt magma having come to an end.

Broad-floored craters within tuff rings (Fig. 145) and steep inner walls in which the edges of outwardly-dipping beds of lapilli are exposed (Fig. 141A) have resulted either from engulfment of the walls of a deep funnel-shaped central cavity (Fig. 141B) during the brief period of ring-building or from a later slumping and erosion from the sides which have filled a deep initial crater.

As both inner and outer slopes of the ring ridges of Diamond Head and Koko Crater (Figs. 143-145) are steep, their crests are sharp. Some of the sharpening of the crest-line may be due to centripetal landsliding and to erosion which has supplied materials to level the crater floor; but few scars or traces of this are seen. The inner slopes are remarkably free from dissection, and, though some deep consequent gullies have been eroded on the outer slopes, parts of these also still preserve the initial form and slope at the angle of repose for lapilli. Thus it is seen that the rings, especially that of Diamond Head (Fig. 145), have been initially of wide radius; and such wide rings can be explained only by assuming that the lapilli of which they are built have been projected to a great altitude.

[19] "Material collapsing from the walls of the newly-formed crater, supplemented by rocks falling back into it, would tend to make a temporary plug, so that the water rushing into the crater would then cause another steam blast. The consequent sudden relief of pressure on the gases in the magma column probably caused the magma to fly apart violently." H. T. Stearns, *loc. cit.* (17) (1935), p. 16.

Vulcanian explosions almost invariably project the bulk of the material vertically upward,[20] though rare cases of obliquely directed projection are recorded, one being the

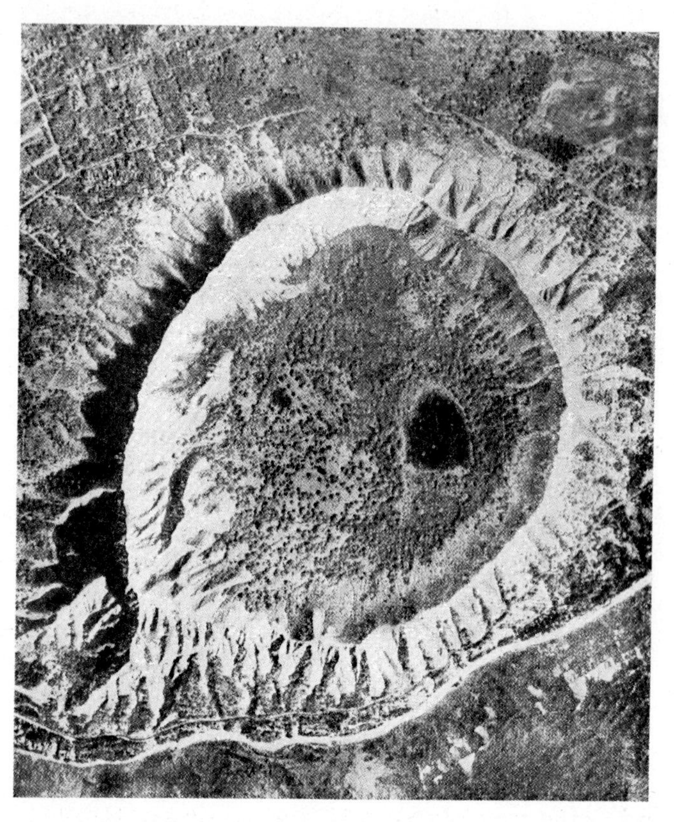

Fig. 145. The broad-floored crater within the Diamond Head tuff ring, which is of large diameter—compare Fig. 143. (After Stearns.)

remarkable north-eastward ejection of ash in great bulk from Vesuvius in 1906.[21] Thus crater rims and rings are normally built up symmetrically except in so far as material is carried to one side by a prevailing wind or, in the case of a brief out-

[20] J. D. Dana, *Am. Jour. Sci.*, 35, p. 220, 1888.

[21] Attributed by F. A. Perret (*The Vesuvius Eruption of* 1906, pp. 43, 63, 1924) to the shape of the conduit. "This inclined jet was the cause of great disaster to the cities of the plain lying in that direction [north-east]."

burst, by any strong wind blowing during the eruption. The well-known asymmetry of the ash and scoria cones of Ascension Island[22] and of the tuff rings of Oahu (Figs. 143, 144) is attributable to carriage of the material to the north-west in the former and south-west in the latter case by the trade winds. In a similar way the crater rim of Ngauruhoe (New Zealand) is highest on the east side (Fig. 117) probably because the prevailing wind is westerly.

Erosion may follow immediately on the accumulation of piles of lapilli and ash with steep slopes, but this will occur only if there is enough fine material deposited along with the coarse to make it impermeable, or sufficiently so to facilitate run-off at least during heavy rains. Being quite unconsolidated, such debris slopes are immediately gullied when streams are formed on them. Some tuff rings contrast strongly in this respect with scoria mounds, which may long remain undissected (Chapter X).

"Craters of Elevation"

There is an old theory which was advocated in the early part of the nineteenth century by Leopold von Buch and strongly supported by Humboldt—though rejected by Scrope, Lyell, and others, and now entirely discarded—that the quaquaversal dips observed in the beds composing central volcanoes result from vertical uplift that has taken place at each centre of eruption. Von Buch found it difficult to believe that lava flows had solidified in attitudes other than nearly horizontal, and he assumed tilting away from a central axis to account for their steep inclination on the slopes of cones.

Though innumerable observations of the relations of inclined lava flows to ash beds have led to the conclusion that their dips are for the most part original, and the "elevation" theory of volcanic up-growth has long ago given place

[22] R. A. Daly remarks of these: "The strong trade wind is the obvious direct cause of the more rapid initial accumulation of ash and cinder on the leeward side of each vent. The initial asymmetry has been increased by the erosion of the windward slopes and drifting of the finer material to leeward. This thinning and weakening of the structure on the windward side has caused the lava, later rising in the vent, to break through or over the rim on that side." (Geology of Ascension Island, *Proc. Am. Ass. Arts & Sci.*, 60 (1), p. 10 1925.)

to the "ejection" theory, it is on record nevertheless that some volcanic outbreaks have been preceded by a certain amount of up-doming of the ground, perhaps as a result of laccolithic injection of magma into the underlying structure (Chapter II). Thus beds and constructional surfaces around craters, though laid down originally with a considerable inclination, must undoubtedly be tilted a little more steeply in some cases by slight up-warping. They are subject, of course, to a reverse effect also, due to collapse.

Another cause of up-tilting of edges around a crater may be explosion, though it seems probable that during most explosive eruptions elevation occurs only to a very limited extent, and it is likely to be more than cancelled by subsequent collapse. On the other hand strong and permanent up-tilting has taken place around some craters due to single explosions which have made the craters but which have quite probably not been of volcanic origin.

Meteor Craters

Apart from mine and bomb craters the only explosion-made craters of non-volcanic origin on the earth which have attracted attention are those which are formed by the impact and explosion of meteorites.

The enormous and very numerous craters on the moon have been given more attention than similar though much smaller features on the earth. In spite of many assertions that lunar craters are volcanoes very serious consideration must be given to the hypothesis that each of them has resulted from a single explosion such as must take place as the result of impact of a high-velocity meteorite.[23] The analogy of some terrestrial forms with the lunar craters seems very close, and it is reasonable to explain these also as the results of meteoritic explosions.

An investigation by Gifford[24] of the ideal form of a crater ring built of material ejected by a single explosion has resulted

[23] A. C. Gifford, The Mountains of the Moon, *New Zealand Jour. Sci. & Tech.*, 7, pp. 129-142, 1924; The Origin of the Surface Features of the Moon, *ibid.*, 11, pp. 319-327, 1930; The Origin of the Surface Features of the Moon, *Scientia*, pp. 69-80, Aug. 1930.

[24] A. C. Gifford, *loc. cit.* (23) (1924).

in development of a deduced form very closely similar to the symmetrical ring-enclosed depression, or saucer-shaped hollow, with central boss which is seen on the moon and has been

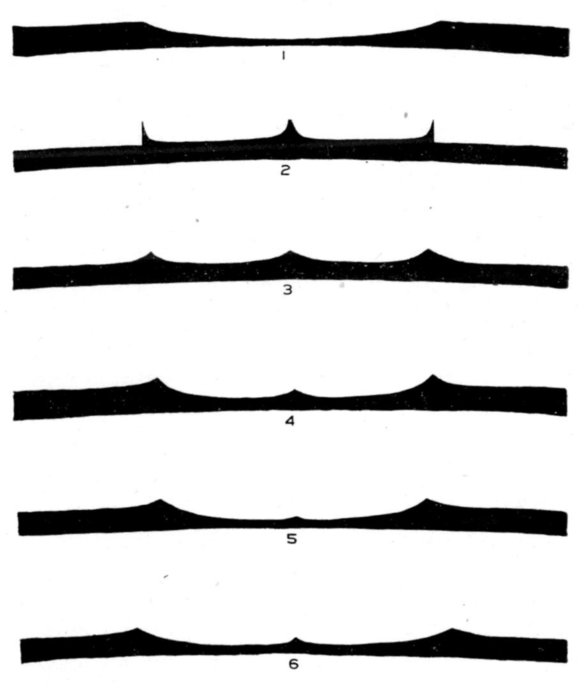

(After Gifford.)

Fig. 146. (1) "Crater formed by the ejection of material."
(2) "Graph showing theoretical density of material returned on to a plane surface."
(3) "The form as modified by material settling at the natural angle of slope."
(4) "No. 3 superimposed on No. 1 gives the theoretical form of a typical lunar crater."
(5) "A section of the crater Copernicus."
(6) "A section of the crater Theophilus."

familiarised by photography as typical of the moon's surface (Fig. 146). The theory of this is summarised as follows:

To take an ideal though of course an impossible case, let us imagine material ejected from a point on the surface . . . and all thrown out with the same velocity but in directions equally distributed through the space above the surface. Let this material return under gravitation and rest where it falls. . . . The resultant structure . . . . will be a saucer-shaped depression with a very steep mountain wall around it and a steep peak in the centre. The actual conditions approximate to this ideal case. The explosion takes place not indeed at one point but in a very limited space. The material is ejected simultaneously with approximately equal velocities in all directions; but a greater quantity will be pushed radially along the surface than thrown vertically upwards.[25]

The velocities of meteors are from 10 to 45 miles per second. A small meteorite

which meets the rare terrestrial atmosphere a hundred miles above the surface with a speed of 45 miles a second loses nearly all of its energy in a few seconds, and, being completely vaporised, disappears at a height of 10, 20, or 40 miles above the earth's surface. If it is so massive as to escape being completely turned into vapour by the friction of the air, its speed is still further reduced while traversing the dense lower layers, and it strikes the earth with so slight a momentum that it can penetrate only one or two yards and may even fail to bury itself.[26] Very different is the fate of a meteorite meeting the moon, which has no protective envelope of air. . . . There is nothing to moderate the meteoric speed before it strikes the solid surface.[27]

It has been pointed out by Wylie,[28] however, that this contrast between lunar and terrestrial conditions is not maintained when very large meteorites fall (especially those over 100 feet in diameter). The terrestrial atmosphere will offer little resistance to the passage of these, and they will strike the earth with almost their original velocities. Thus the fate of such large bodies if they collide with the earth must be the same as that of meteorites striking the moon, that is to say they

[25] A. C. Gifford, *Scientia,* p. 77, Aug. 1930.

[26] "The 820-pound Paragould (Ark.) stone, a witnessed fall, penetrated the clay soil to 8 feet" (E. P. Henderson and C. W. Cooke, The Sardis, Georgia, Meteorite, *Proc. U.S. Nat. Mus.,* 92, pp. 141-150, 1942).

[27] A. C. Gifford, *loc. cit.* (23) (1924), p. 135.

[28] C. C. Wylie, On the Formation of Meteorite Craters, *Pop. Astron.,* 41, pp. 211-214, 1933.

must explode with sudden release of an immense store of energy.[29]

It would be a mistake to assume that a meteorite must fall vertically in order to make a crater of circular outline; for Gifford recalls that—as pointed out by A. W. Bickerton in 1915—"the normal meteoric speed is sufficient to produce an explosive action, and that consequently oblique impact will produce roughly circular rings." He continues:

Any mass moving at the rate of 40 miles a second has kinetic energy equivalent to 494,700 calories per gramme. Now, the energy of dynamite is only about 1100 calories per gramme. So, many of the meteorites which strike the lunar surface must be changed instantaneously into explosives with 400 or 500 times the intensity of dynamite.[30]

Gravity is six times weaker on the moon than on the earth, and for this reason the size of lunar craters will be larger than that of terrestrial craters of similar origin. Lunar craters, moreover, it cannot be doubted, are very ancient; not being subject to erosion of any kind, they are everlasting. They date possibly from a time when both earth and moon were more subject to bombardment by large projectiles than they now are (or, alternatively, they have been formed one by one at very long intervals) and similar scars formed on the earth have been for the most part obliterated by erosion.

It is reasonable to suppose that—due allowance being made also for differences of velocity—the sizes of lunar craters will be related to the sizes of the meteorites which have exploded to produce them. Gifford's calculations have led him to the conclusion that the mass of the exploding body must be

[29] After estimating the energy released by the explosion of a large low-velocity meteorite (estimated to have been a body several hundred tons in weight) in the air over Pennyslvania in 1938, Randolph remarks: "If the big meteorite had come straight down instead of at this long slant, it would have reached the earth with a lot of its structure still intact and a lot of its energy still in it; and if it had landed on Pittsburgh there would have been few survivors. Its kinetic energy of 31,400,000 foot-pounds per pound is more than 20 times as great as the explosive energy of T.N.T. . . . . . But fortunately the energy of this meteorite had all been absorbed by the air before it reached the ground." (Preston, Henderson, and Randolph, The Chicora Meteorite, *Proc. U.S. Nat. Mus.*, 90, pp. 387-416, 1941.)

[30] A. C. Gifford, *loc. cit.* [25], p. 76.

"certainly not less than one four-thousandth part of that of the material ejected."[31]

Terrestrial Meteor Craters

In the case of the pseudo-volcanic Meteor Crater, in Arizona, the edges of limestone strata that were previously

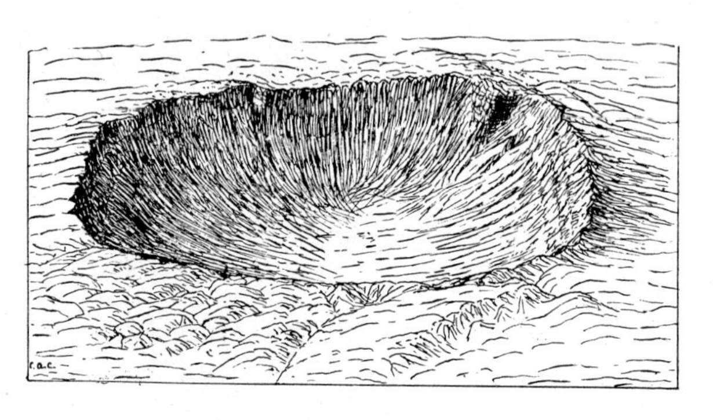

Fig. 147. Meteor Crater, Arizona. (From a photograph.)

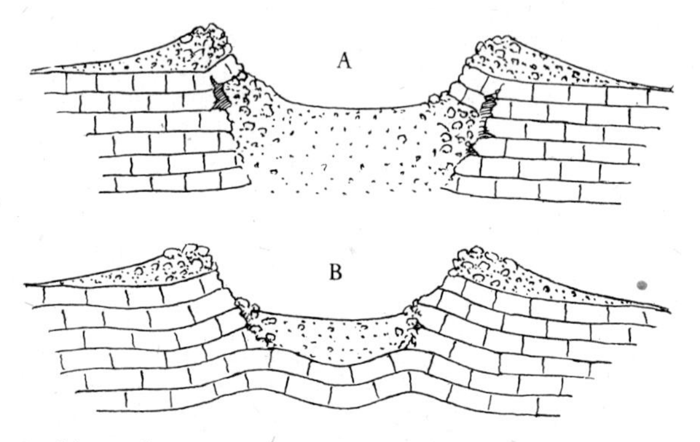

Fig. 148. Up-tilting of strata and formation of a crater ring by explosion. A: Fracturing and tilting by outward explosion. B: Ring anticline formed by percussion. (After Spencer.)

horizontal, as well as the flat surface of a plateau, have been strongly bent upward by an explosion (Figs. 147, 148) so that around the rim of the crater they dip at from 10° to 80°. The crater is 600 feet deep and three quarters of a mile in diameter,

[31] *Loc. cit.*, [25], p. 78.

and the rim around it rises to a height of 150 feet above the surrounding plateau. It has apparently been formed by an explosion resulting from impact of an exceptionally large meteorite with the earth.

The rim is covered with ejected blocks of limestone up to 60 feet in diameter, and large blocks have been thrown to a distance of half a mile from the edge of the crater. It has long been realised that Meteor Crater must be the result of an explosion. So clearly was this apparent to Gilbert[32] that he rejected the theory of meteoritic *impact* in spite of strong corroborative evidence afforded by the presence of very numerous fragments of meteoric iron in the vicinity of the crater.[33] The probability of a meteoritic *explosion* had not been suggested when Gilbert studied the problem of Meteor Crater, and the meteoritic hypothesis in the only form in which he considered it (the impact hypothesis) postulated that an extremely large meteorite had imbedded itself in the floor of the crater. A magnetic survey had failed to indicate the presence there of a large mass of iron, and a topographic survey had shown the volume of ejected material in the rim to be approximately the same as that of the cavity (82 million cubic yards). Hence Gilbert decided in favour of an explosion, which, however, must be related in some way, he thought, to volcanic activity.

A few other examples of craters of like origin are known, though those that have undoubtedly been made by the explosion of meteorites are generally quite small. There are, however, some large hollows which remain in the doubtful category because, though they are very probably explosion pits, the possibility of volcanic explosion as their cause cannot be ruled out altogether. Where such craters are unique features of the regions in which they occur and are not associated with

[32] G. K. Gilbert, The Origin of Hypotheses, *Science*, 3, p. 1, 1896.

[33] By a remarkable coincidence another fall of fragments of a meteorite occurred in 1912 within 60 miles of Meteor Crater. In this case a stony meteorite exploded in the atmosphere. Thousands of fragments of it have been collected, from which it has been estimated that the body weighed about 500 pounds. (W. M. Foote, Preliminary Note on the Shower of Meteoric Stones at Aztec, near Holbrook, Navajo County, Arizona, *Am. Jour. Sci.*, 34, pp. 437-456, 1912.)

any positive indications of recent volcanic activity, however, the hypothesis that they have been made by meteoritic explosions must be entertained.[34]

One such is that which contains the circular lake Bosumtwi, in Ashanti, 6½ miles in diameter and 1000 feet deep. The depression, quite large enough to be termed a caldera, in which the lake lies has a raised rim, but the rim is not built of ejected fragments.[35] Though the basin is much larger than any terrestrial forms definitely known to be of purely explosive origin, this does not necessarily discredit the meteoritic explosion hypothesis as applied to it. Rohleder[36] discards the meteoritic hypothesis but favours that of a single shallow (phreatic) volcanic explosion. There are few signs of volcanic activity in recent times in the surrounding region, but an outcrop of pumiceous agglomerate found in the vicinity of the lake has been made the basis of the theory of volcanic origin.

Other craters for which a meteoritic origin has been suggested by Gilbert,[37] Spencer, and others as possible are the Pretoria "salt pan," in South Africa, which is two-thirds of a mile in diameter and 400 feet deep, and the basin of the Lonar Lake, in India. The Pretoria Salt Pan was initiated, according to Wagner,[38] by a "tremendous phreatic explosion"—in a region showing no other signs of volcanic activity—followed by subsidence.

The Lonar Lake is midway between Bombay and Nagpur in a terrain of ancient dissected basalt (Deccan traps) in which volcanic activity has been extinct for several geological periods. The vertical-walled circular pit, rather more than a mile in diameter and 300 to 400 feet deep (Fig. 149), is a unique feature in the region, for in the whole Indian peninsula there is not a single trace of recent volcanic activity (with this possible exception). On all sides except north and north-east

[34] L. J. Spencer, Meteorite Craters as Topographical Features on the Earth's Surface, *Geog. Jour.*, 81, pp. 227-248, 1933; reprinted *Rep. Smithson. Inst. for 1933*, pp. 307-325, 1934.

[35] M. Maclaren, Lake Bosumtwi, Ashanti, *Geog. Jour.* 78, pp. 270-276, 1931.

[36] H. P. T. Rohleder, Lake Bosumtwi, Ashanti, *Geog. Jour.*, 87, pp. 51-65, 1936.

[37] G. K. Gilbert, *loc. cit.* (32).

[38] P. A. Wagner, *Mem. Geol. Surv. S. Afr.*, 20, p. 27, 1922. See illustration of this feature in L. C. King, *South African Scenery*, Fig. 174, p. 147, 1942.

the "crater" has a raised rim with a maximum external height of 100 feet, which is

composed of blocks of basalt irregularly piled and precisely similar to the rock exposed in the sides of the hollow. The dip of the surrounding traps is away from the hollow but very low. . . . The raised rim is very small and cannot contain a thousandth part of the rock ejected from the crater, but it is impossible to say how much was reduced to fine powder and scattered to a distance or removed by denudation.[39]

Fig. 149. Lonar Lake, India, the basin of which is possibly a crater formed by explosion of a meteorite. (Outlined after Newbold.)

A suggestion of "cauldron subsidence" made by La Touche and quoted by Wadia[40] does not agree particularly well with the description (as quoted above) of the upturned edges of basalt sheets and of the rim with its veneer of ejected blocks; and the explosion hypothesis favoured by Oldham may be retained if supplemented by the suggestion that the explosion was meteoritic. There seems some reason to support this suggestion also in the case of the Pretoria Salt Pan, which is a unique feature in South Africa, just as the Lonar Lake is in India.

An attempt has been made to explain very numerous shallow depressions in the south-eastern United States (the

[39] R. D. Oldham, Medlicott and Blandford's *Manual of the Geology of India*, 2nd ed., p. 19, 1893.

[40] D. N. Wadia, *Geology of India*, p. 23, 1919.

"Carolina bays") as meteorite scars, but the theory has fallen into disfavour as a result largely of exhaustive investigations by Johnson,[41] who has been able to explain the features quite satisfactorily without invoking the aid of processes other than purely terrestrial. This makes it unnecessary to include any description or discussion of the "bays" here. No suggestion has been made that they are of *volcanic* origin.

High Temperatures developed by Meteoritic Explosions

According to calculations made and quoted by Gifford and Wylie[42] even the very low meteoritic velocity of ten miles per second would, on impact, result not only in an explosion which would scatter widely such fragments of the meteorite as remained unmelted and also much terrestrial rock debris, but this would be accompanied by the melting of much, and even vaporisation of some, meteoritic and terrestrial material.

It is clear that the meteoritic explosions which have excavated some small craters have developed temperatures far in excess of any recorded in association with volcanic phenomena or developed by any other natural process on the surface of the earth with the possible exception of lightning discharges, which frequently fuse small quantities of sand, forming "fulgurites." Not only have scattered fragments of iron from exploded meteorites been gathered, but in certain cases—notably in the vicinity of the Henbury craters in central Australia and at Wabar in Arabia—fused quartzitic rock and quartz sand have been found. At Wabar, indeed, some of the fused quartz has been boiled by the heat generated.

> For a distance of about 40 metres from the rim [of a small crater] the outer slopes are thickly strewn with cindery masses of silica glass and smaller complete bombs of the same material ranging in size down to small "black pearls," which were picked up in large numbers. The rims of the craters appear to be built up mainly of this silica glass . . . . Inside they [the bombs] consist of a very cellular white silica glass . . . . These structures suggest that there was a pool of molten and boiling silica (the silica vapour causing

[41] Douglas Johnson, *The Origin of the Carolina Bays,* New York, 1942.
[42] A. C. Gifford, *loc. cit.* (23); Wylie, *loc. cit.* (28).

the highly cellular structure) and that a rain of molten silica was shot out from the craters through an atmosphere of silica, iron, and nickel produced by the vaporisation of the desert sand and part (perhaps a large portion) of the meteorite.[43]

"Cryptovolcanic" Structures

Within a number of roughly circular areas in accurately mapped parts of North America in which the older rocks outcrop Bucher[44] has described geological structures that are extremely complex as compared with those of their surroundings. These he has classed with "cryptovolcanic" structures comparable to the Steinheim basin, in Germany, and has ascribed their origin to explosions ("sudden liberation of pent-up volcanic gases which had accumulated near the surface"). Such explosions, it may be remarked, are quite unlike any known volcanic phenomena and have occurred "at points where there had previously been no volcanic activity." He regards the structures as exemplifying the underground effects of single explosions on a scale capable of opening large craters if they occurred superficially. These have been muffled, however, by the overburden.

In the areas thus affected complexity of structure, which is conspicuous both in plan and in section, has been introduced by the development of numerous faults and some folds in the strata affected. The pattern of the folds "resembles that of damped waves, a central uplift surrounded [in a particular example] by two pairs of down- and up-folds with diminishing amplitude. It is the sort of pattern that results from a sudden impulse such as that of an explosion."[45]

The small central dome usually present may be an indicator of percussion from below, but Boon and Albritton, who have proposed and favour the alternative hypothesis that the structures in question have been produced by the explosion of large meteorites striking the ground above them, do not take this view. On the contrary, while "a series of concentric

[43] L. J. Spencer, *loc. cit.* (34), pp. 314-315.
[44] W. H. Bucher, Cryptovolcanic Structures in the United States, *Rep. XVI Internat. Geol. Cong.*, 2, pp. 1055-1084, 1936.
[45] W. H. Bucher, *loc. cit.* (44), p. 1068.

waves would go out in all directions, forming ring anticlines and synclines" as a result of percussion from above, *"the central zone, completely damped by tension fractures produced by rebound, would become fixed as a structural dome."*[46]

The structures have been exposed by erosion long after the erosive processes have destroyed any craters or other surface forms produced in association with them by the same explosions.

Williams[47] does not mention the meteoritic hypothesis, but accepts the volcanic-explosion theory in partial explanation of these "cryptovolcanic" structures as well as of various large isolated craters and calderas for the origin of which suggestions of meteoritic origin have been made. He is satisfied that most of these and also the surface depressions in Germany known as the Steinheim Basin and the Rieskessel, which latter has the dimensions of a large caldera, originated as the result of, or in association with, explosions—followed, however, by subsidence or collapse of the surface.[48] These German basins are of complex structure and came into existence, in the opinion of most observers, as the result of processes set in train by the injection of laccoliths or sheets of igneous magma at a shallow depth. If this is correct, the meteoritic hypothesis may be ruled out at once as far as the origin of such basins is concerned. Whether the American "cryptovolcanic" structures are of the same nature is another question, however, and the hypothesis of meteoritic explosion seems worthy of consideration as applied to them.

[46] J. D. Boon and C. C. Albritton, Meteorite Craters and their Possible Relationship to "Cryptovolcanic Structure," *Field and Laboratory,* 5, pp. 1-9, 1936.

[47] Howel Williams, Calderas and their Origin, *Univ. Cal. Publ., Bull, Dep. Geol. Sci.,* 25 (6), pp. 303-304, 1941.

[48] Compare E. Suess, *The Face of the Earth,* IV, English ed., p. 568, 1909.

CHAPTER XV

Submarine Eruptions; Pillow Lavas

In the course of geological ages immense quantities of lava have been poured out and vast formations of lava and tuffs have accumulated on the ocean floors. Tyrrell[1] has computed that the crustal strata contain associated with sedimentary rocks of geosynclinal origin a volume of basaltic material "of the same order of size as that of the great flood-basalt [or plateau-basalt] accumulations." These are basaltic rocks that accumulated along with the sedimentary series enclosing them in periods anterior to the folding which they have undergone.

Island Volcanoes

When attention is focussed on more recent volcanism it is clear that submarine emission of lava and of the fragmentary products of magma has taken place on a stupendous scale in order to build up from ocean depths the foundations of the numerous large volcanic islands in the Pacific, Indian, and Atlantic Oceans. Even at the present day there is much evidence of activity on the ocean floors, and Jaggar has remarked that,

> if we could know of all the submarine lava flows which come out through cracks in the bottom of the sea between New Zealand and Samoa, we might discover that [submarine] lava outflow is not so rare as has been imagined.[2]

Little is known with certainty of the effect produced by contact of effusive lava with ocean water, but submarine eruption is obviously explosive in many cases even where the magma concerned is basaltic. During the submarine eruptions of 1811 in the Azores and that at Krakatau (East Indies) in

[1] G. W. Tyrrell, *Volcanoes,* p. 226, 1931; Flood Basalts and Fissure Eruptions, *Bull. Volcanol.,* 1, pp. 89-111, 1937.

[2] T. A. Jaggar, *The Volcano Letter,* 265, p. 3, 1930. See also M. Neumann van Padang, Ueber die Unterseevulkane der Erde, *De Mijningenieur,* 5-6, 1938.

1928, as pictured in sketches and photographs, vertical belches took place. These closely resembled eruptions of the giant geyser (now quiescent) at Waimangu, New Zealand, which habitually threw up stones (Fig. 150), and geyser-like eruptions

Muir and Moodie, photo

Fig. 150. An eruption of Waimangu geyser, New Zealand, which was active on hundreds of occasions between 1900 and 1904, belching to a height of over 1000 feet. Explosive emission of scoria and bombs from submarine vents at a moderate depth closely resembles this activity.

of the same kind in a solfataric crater on the volcano Poas, in Costa Rica. An explosive eruption which took place through the crater lake of the Soufrière of St Vincent and was photographed by Lacroix was very similar to these.

The eruption of 1928 at Krakatau was photographed by Stehn, and later explosions of the same kind which ejected scoria and bombs, together with columns of water, to a great height were described and photographed by Neumann van Padang.[3] Some such explosions may very well be of magmatic

[3] M. Neumann van Padang, Der Krater des Anak Krakatau, *De Ingen. in Ned. Ind.*, 4, pp. 101-107, 1936.

origin, but some of them, notably those taking place where basaltic magma of a kind generally non-explosive is making contact with sea-water, are more probably phreatomagmatic, as defined by Stearns (Chapter XIV). Pumiceous or scoriaceous fragments sufficiently inflated to float rise to the surface. Whether the process ever involves solution of the sea-water in magma as a preliminary to explosion is unknown.

There are abundant lava rocks among the geological strata which without doubt were poured out on the ocean floor without exploding, and lava has been observed to flow quietly into the sea at Matavanu, Samoa,[4] and without violent explosions at Ambrym in 1894 and on some occasions in Hawaii. The Matavanu lava—to the total amount of several cubic kilometres—poured into the sea for five years, and must have flowed away down a submarine slope.

In the case of submarine lava flows observed by Perret at Stromboli in 1915 and by Washington at Santorini in 1925, neither of which exploded, it has been suggested that the lava advanced under the protection of "a water-cooled and flexible sheath."[5] Similar outflow of andesitic lava, which spread without exploding to some distance under Kagoshima Bay, took place during the Sakurajima eruption of 1914.[5a]

On the other hand, rather violent explosions have taken place where a lava from Mauna Loa, Hawaii, has entered the sea, the lava has been shattered, and a quite large cone of the resulting lapilli has been built up.[6] A new island which emerged in 1913, a mile offshore, west of a basaltic volcano on Ambrym Island, New Hebrides,[7] may have originated from lava outflow, as it appeared in front of a lava flow which had

[4] Tempest Anderson, The Volcano of Matavanu in Savaii, *Quart. Jour. Geol. Soc.*, 66, pp. 621-639, 1910; I. Friedländer, Beiträge zur Geologie der Samoa-Inseln, *Abh. K. Bay. Ak. d. Wiss.*, II, 24, pp. 507-541, 1910.

[5] H. S. Washington, Santorini Eruption of 1925, *Bull. Geol. Soc. Am.*, 37, pp. 349-384 (p. 376), 1926.

[5a] T. A. Jaggar, Sahurajima, Japan's Greatest Volcanic Eruption, *Nat. Geog. Mag.*, 45, pp. 441-470, 1924.

[6] H. T. Stearns and W. O. Clark, Geology and Water Resources of the Kau District, Hawaii, *U.S. Geol. Surv. W-S. Paper*, 616, pp. 125-127, 1930; G. A. Macdonald, *The Volcano Letter*, 474, p. 2, 1941.

[7] Dr Bowie, quoted by P. Marshall, The Recent Volcanic Eruptions on Ambrym Island, *Trans. N.Z. Inst.*, 47, pp. 387-391, 1915.

entered the sea a few hours earlier. It may, however, have been built up of scoria over an independent submarine vent.

As has been suggested by Bonney[8] and assumed by N. D. Stearns,[9] the up-building of volcanic piles from the floor of the deep ocean to become islands that are later enlarged as lava domes may take place as a result of the heaping of the fragments into which basalt magma is sometimes shattered when it makes contact with water.[10] H. T. Stearns, on the other hand, pictures the Hawaiian Islands as built up all the way from the ocean floor as lava domes.[11] Accounts of a submarine eruption on the outer flank of Mauna Loa which occurred on February 24, 1877, record that lumps of the porous basaltic lava a foot and more in diameter came to the surface of the sea but rapidly became waterlogged and sank again.[12] Thus it seems to be indicated that in this case fragmentation of the submarine lava was in progress during the eruption.

New Islands built by Submarine Eruptions

Newly-appearing volcanic islands, which have been observed from time to time, are of two kinds. Those consisting of basaltic material seem to be built in all cases of pyroclastic fragments (scoria and lapilli); while those of the more acid lavas have in most cases a solid foundation of the nature of a cumulo-dome.

Falcon Island (Fig. 151) in the Tonga group, South Pacific, which has emerged, has been cut down by marine erosion, and has been rebuilt from time to time, is an example of the

[8] T. G. Bonney, *Volcanoes,* 2nd ed., p. 47, 1902.

[9] "The base is steep-sided and consists of . . . . material chiefly of explosive birth; the capping is plasters of heavier air-born lavas" (*An Island is Born,* p. 104, Honolulu, 1935).

[10] According to P. H. Kuenen's investigations by sonic sounding submarine slopes assumed to consist of volcanic debris are straight and steep and this (foreset) profile, as might be expected, shows no relation to the subaerial slopes of the adjacent volcanic islands (Submarine Slopes of Volcanoes, *Snellius Expedition,* 5 (1), pp. 62-69, 1935).

[11] Geological Map and Guide to Oahu, *T.H. Div. Hydrog. Bull.,* 2, p. 8, 1939. See also Jaggar, *infra* (16).

[12] Coleman Phillips, On the Volcanoes of the Pacific, *Trans. N.Z. Inst.,* 32, p. 202, 1900; J. D. Dana, *Am. Jour. Sci.,* 36, p. 29, 1888.

former kind.[13] As described by Phillips,[14] the lapilli of which Falcon Island had been built, as it existed (still hot) in 1894, seem closely to resemble the material of magmatic origin (but produced by the phreatomagmatic process) in the tuff rings of eastern Oahu (Chapter XIV). Phillips has noted their resemblance to lapilli from the Tarawera eruption of 1886.

Falcon Island was rebuilt of basaltic fragments in 1927 in the form of a crescentic segment of a cone, 365 feet high, to leeward (north-westward) of a crater,[15] but Jaggar[16] has

Fig. 151. Falcon Island, an intermittent volcanic island of changing form in the South Pacific Ocean, as it appeared after being rebuilt by the eruption of 1927. Marine erosion has since reduced it again to a shoal. (Drawn from a photograph by A. Thomson, Apia.)

expressed his belief that a submarine lava dome exists under the island:

While such volcanoes exhibit chiefly fragmental material, it must be remembered that a pile of solid lava in the form of a dome probably exists beneath. How an outflow of liquid lava like one of the Mauna Loa flows would behave under deep, cold water is a matter of theory, not of observation.

Another new island (Fig. 152) built entirely of pumiceous "scoria" and ash with the composition of hypersthene basalt bordering on andesite has made its appearance more or less centrally within the great submarine caldera made by the

[13] For an account of the early history of Falcon Island and also for descriptions of other new islands built by submarine eruptions in the Atlantic and Pacific Oceans and the Mediterranean Sea see A. Geikie, *Textbook of Geology,* 4th ed., pp. 332-335, 1903. See also H. Luke, Tonga: the Last Kingdom of the South Seas, *Scot. Geog. Mag.,* 59, pp. 49-54, 1943.

[14] Coleman Phillips, The Volcanoes of the Pacific, *Trans. N.Z. Inst.,* 31, p. 513, 1899.

[15] J. E. Hoffmeister and H. S. Ladd, Falcon, the Pacific's Newest Island, *Nat. Geog. Mag.,* 54, pp. 757-766, 1928; also Ladd and Alling, Falcon Island, *Am. Jour. Sci.,* 1929.

[16] T. A. Jaggar, *loc. cit.* (2).

Plinian eruption of Krakatau, in the Strait of Sunda, in 1883. The new island, Anak Krakatau, was first seen in 1928 after the occurrence of submarine eruptions. It was soon afterwards destroyed by marine erosion, but was later rebuilt several times, growing eventually in a series of eruptions in 1931-1935

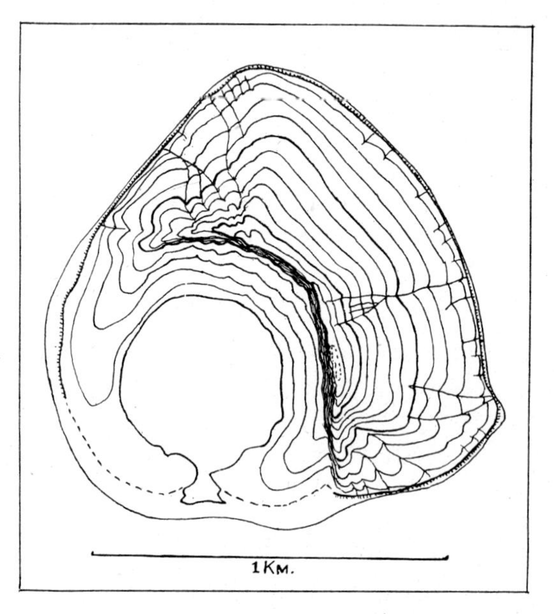

Fig. 152. Anak Krakatau, the new island built over the eruptive centre of Krakatau, East Indies, as it was shortly after the eruption of July, 1935, which had increased the area of the island from 0·792 to 1·155 square kilometres and had spread over it 5 to 12 metres of new ash; but had reduced the height from 88 to 63 metres by enlarging the crater. Contour interval: 5 metres.

to a height of 90 metres in crescentic form to north and east of a lagoon-filled crater hollow. This crescentic form has been assumed as the result of progressive marine erosion, due to exposure to rollers from the Indian Ocean, which prevents accumulation south-west of the crater. Since 1935 (Fig. 152) there have been several vigorous eruptions, which have been closely observed as a precautionary measure because of the possible danger of another outbreak comparable to that of 1883. The island changes in form as affected alternately by rapid constructive and destructive processes, but the net result has been increase in size and after the eruption of 1939 the area had grown to 1·5 square kilometres.

Only shallow ravining of the surface by a few consequent streams had taken place prior to 1937, but after the eruptions of August and September of that year a deep fine-textured dissection of the gentle slope of the cone was observed, photographed, and mapped by the Netherlands Indies Volcanological Survey. Sharp-edged ridges and spurs separated an immense number of branching gullies. The sudden development of this dissection pattern during or immediately after an eruption can be explained only as the result of a shower of ash finer and much less permeable than that of former ejections. Later eruptions buried and obliterated the spur pattern.

New Islands formed by Lava Protrusion

In the case of some islands which have emerged as a consequence of submarine eruptions a doubt remains as to whether the debris of which they appear to be composed has a core of solid rock. Such a doubt has been expressed by Washington[17]

Fig. 153. New volcanic island west of Mahangetang, Sangi Islands, from south-south-west, November, 1919. (From a photograph.)

in regard to Graham Island, which appeared in the Mediterranean Sea, near Pantelleria, in 1831; but the material of which this island was built up rapidly to a height of 200 feet above the sea was basaltic scoria and the pile was reduced by marine erosion to the condition of a shoal in a few months. It is generally regarded, therefore, as of the same kind as Falcon Island.

The new islands, on the other hand, at the Santorini volcano (eastern Mediterranean) and at Bogoslof (Aleutian Islands)

[17] H. S. Washington, The Submarine Eruptions of 1831 and 1891 near Pantelleria, *Am. Jour. Sci.*, 27, pp. 131-150, 1909.

and another (Fig. 153) which arose west of Mahangetang, Sangi Islands, East Indies, are emergent cumulo-domes of lava of non-basaltic composition.[18] Within the great caldera of Santorini cumulo-domes have emerged from the sea on several occasions, and sluggish outflow of viscid lava (Fig. 49) has occurred also in shallow water in association with these.

Even more spectacular is the succession of events in the Bogoslof Islands, for not only have high craggy islands rapidly emerged, adding themselves to the group, but these have been subject to partial or complete destruction by explosion within a few months of their appearance. Cumulo-domes, moreover, owing to their fissured condition, generally offer only a poor resistance to wave attack, and thus the one member of the Bogoslof group, Castle Rock or Old Bogoslof, which has largely escaped destruction by explosions has been cliffed and considerably reduced in size by marine erosion in a short time. It was formed in 1796, and thirty years later it still covered an area of nearly two square miles and was 350 feet high; but it dwindled appreciably in size during the nineteenth century. Its form in 1907 is shown in Fig. 154. By that time Grewingk,

Fig. 154. Bogoslof Island (or group) in July, 1907, viewed from north-east. From south-east (at left): Castle Rock; McCulloch Peak and remnant of Metcalf Cone (steaming); Grewingk. (From a photograph.)

or New Bogoslof, which had emerged in 1883, had assumed a flat-topped form, wave-built spits of broken rock had been formed so as to bridge the space of about a mile originally separating Old and New Bogoslof, and within a warm lagoon enclosed by bars of wave-drifted boulders and explosively projected blocks two more new domes had appeared (1906-1907).

One of these, Metcalf Cone, 400 feet high, was a steeply conical mound of talus surrounding a spine, while the other,

[18] Howel Williams, The History and Character of Volcanic Domes, *Univ. Cal. Publ., Bull. Dep. Geol. Sci.,* 21 (5), pp. 51-146, 1932.

McCulloch Peak, 500 feet high, was of domed or tholoid form. Half of Metcalf had been blown away before McCulloch arose (Fig. 154).

Bogoslof was now a continuous island two miles long. . . . . McCulloch Peak was three-quarters surrounded by steaming salt water [of the lagoon] at 90° F. . . . . Immediately adjacent to it was the Metcalf half-dome, with the central spine wonderfully revealed. . . . . Bogoslof in 1907 was at its maximum above the waves. . . . Everything indicated that the huge pressure inside the submerged volcano, which was pushing up McCulloch Peak, was also lifting the volcano on its back and carrying the chain of islets with it.

On September 1, 1907,

a dense black cloud rose from Bogoslof, ash and sand fell at Unalaska. . . . McCulloch Peak had blown itself up. A steaming lagoon was left in its place; the rest of the island was piled high with fallen debris. . . . In July, 1908, the remains of Metcalf Peak had subsided, and there had probably been another explosion the previous winter.[19]

Explosions, erosion, bar-building, and occasional appearance of small new islands of lava in the lagoon continued; the composite island was divided in two by the opening of a navigable channel; and then, in 1926, a new cumulo-dome appeared, which was surrounded in 1927 by an extensive warm lagoon at 70° F., this being enclosed by a complete ring of boulder banks and sand bars 10 feet above the sea. The surviving remnants of Castle Rock and Grewingk were again connected by this means. The new central island was then 200 feet high, 1000 feet in diameter, and similar in form to other tholoids of recent growth (Chapter XI).

Submarine Lavas: Pillow Structure

The appearance of many lavas exposed in association with marine sediments in sections of crustal strata has been likened to piles of "pillows, bolsters, cushions, or filled sacks"[20] (Figs.

[19] T. A. Jaggar, Recent Activity of Bogoslof Volcano, *The Volcano Letter,* 275, pp. 1-3, 1930; Evolution of Bogoslof Volcano, *ibid.,* 322, pp. 1-3, 1931.

[20] G. W. Tyrrell, *Volcanoes,* p. 50, 1931.

155, 156). This is, indeed, the most striking peculiarity of lavas which have been poured out under water, and "pillow structure," as it is termed, has resulted apparently from very rapid chilling. Basaltic lavas especially (which, as noted earlier in this chapter, are extremely abundant intercalated with sedimentary strata of geosynclinal and commonly of marine origin) exhibit pillow structure.

Professor J. A. Bartrum, photo

Fig. 155. A pillow lava of Tertiary age exposed in section in sea cliffs at Muriwai, west coast of Auckland, New Zealand.

Internally the rock in some "pillows" exhibits a radial arrangement of columnar shrinkage jointing[21] (Fig. 155). The most characteristic feature of a typical pillow, however, is a skin which contains it and has obviously been the first part of

[21] J. V. Lewis, The Origin of Pillow Lavas, *Bull. Geol. Soc. Am.*, 25, pp. 591-654 (p. 650), 1914; J. A. Bartrum and F. J. Turner, Pillow Lavas, Peridotites, and Associated Rocks of Northernmost New Zealand, *Trans. N.Z. Inst.*, 59, pp. 98-138, 1928; J. A. Bartrum, Pillow Lavas and Columnar Structure at Muriwai, Auckland, *Jour. Geol.*, 38, pp. 447-455, 1930.

it to become solid.[22] This skin is glassy and vesicular, and thus differs distinctly in texture from the stony (more crystalline) rock within it. The solidification of the skin, which is very much like that on pahoehoe flows, and which when first formed has obviously been very tough and elastic, is generally ascribed to rapid chilling in contact with water or mud. "The pillows mutually indent one another, as if they had been soft at the time of their formation. They are frequently elongated like bolsters, and show a marked parallelism of their longer axes" (TYRRELL).

An example of actual change from pahoehoe solidification to pillow formation as a result of contact with water has been observed at the Matavanu volcano, Samoa.[23] A small flow of lava entering the sea

> extended itself into buds or lobes. The process was as follows: An ovoid mass of lava still in communication with its source of supply, and having its surfaces, though still red hot, reduced to a pasty condition by cooling, would be seen to swell or crack into a sort of bud with a narrow neck . . . . and thus would rapidly increase in heat, mobility, and size until it became a lobe as large as a sack or pillow. . . . Sometimes the neck supplying a new lobe would be several feet long and as thick as a man's arm before it expanded into a full-sized lobe; more commonly it would be shorter, so that the fresh-formed lobes would be heaped together. They looked white-hot, even in daylight, and as waves washed over them the water seemed to fall off unaltered, without boiling, owing probably to its being in the spheroidal condition. (TEMPEST ANDERSON.)

Obviously the slender connecting tubes may easily be severed, allowing the inflated lobes to become discrete ellipsoids. It appears probable that in some cases these ellipsoids, after becoming partially solidified, have cascaded down subaqueous slopes to accumulate as talus deposits;[24] but such an occurrence must be rare, and most pillows remain in the positions in which

[22] An example of an old lava pillow has been figured from which a portion of the contents had run out while still liquid after a thick skin had become rigid. R. E. Fuller, The Aqueous Chilling of Basaltic Lava on the Columbia River Plateau, *Am. Jour. Sci.*, 21, pp. 281-300 (Fig. 3), 1931.

[23] Tempest Anderson, The Volcano of Matavanu in Savaii, *Quart. Jour. Geol. Soc.*, 66, pp. 621-639, 1910; Volcanic Craters and Explosions, *Geog. Jour.*, 39, pp. 123-132, 1912.

[24] As assumed by R. E. Fuller in explanation of the origin of some pillow structures. *Loc. cit.* (22).

they were formed, as is shown by absence of evidence of rolling and mutual abrasion.[25]

Lava pillows are commonly three to four feet in diameter. Occasionally, however, flattened pillows are as much as 12 feet long.[26]

Some pahoehoe surfaces of lava flows that have solidified on land exhibit bulbous swellings (Fig. 39, B), which are seen in rather rare instances to follow one another as bead-like expansions of narrow pahoehoe streams, or "toes"; and it is possible that in published descriptions of ancient lavas there has been in a few cases confusion of the bulbous surfaces and projecting toes of successive thin pahoehoe flows with true pillow lavas. Too great stress laid on such resemblances has led to some announcements that the formation of pillows is not generally a result of water-cooling;[27] but according to other competent opinion pillow lava, in the properly restricted use of the term, is very different from any bulbous subaerial pahoehoe, and there seems little doubt that sufficiently rapid chilling to cause lava to solidify in this distinctive way occurs only under water, or at least in close contact with water.

In the Snake River basalt field (Chapter VIII) pillow structure has been formed abundantly in the lavas which have poured into and filled river canyons, but in no other flows. Though the basalt in the canyons has solidified probably with-

[25] J. V. Lewis, *loc. cit.* (21).

[26] J. A. Bartrum, *loc. cit.* (21); Bartrum and Turner *loc. cit.* (21).

[27] J. V. Lewis (*loc. cit.* (21)) regards the formation of pillow lava as a variant of pahoehoe solidification (bulbous budding), and considers that subaqueous cooling, though conducive to the assumption of pillow structure, may not be essential. See also M. G. Hoffman, Structural Features in the Columbia River Lavas of Central Washington, *Jour. Geol.,* 41, pp. 184-195, 1933; J. T. Stark, Vesicular Dike and Subaerial Pillow Lavas in Borabora, Society Islands, *Jour. Geol.,* 46, pp. 225-238, 1938; *ibid.,* 47, p. 205, 1939; H. E. McKinstry, "Pillow Lavas" of Borabora, Society Islands, *Jour. Geol.,* 47, p. 202, 1939.

T. A. Jaggar (*The Volcano Letter,* 7, 1925) has observed that pahoehoe surfaces in Hawaii have taken on a resemblance to pillow lava if formed "during torrential rainstorms."

H. T. Stearns (Pillow Lavas in Hawaii, *Proc. Geol. Soc. Am. for 1937,* pp. 252-253, 1938) notes that among the basalts of the Hawaiian Islands "genuine pillow lavas" are extremely rare. They have been formed only in contact with water, though "many spheroidal forms have been mistaken for pillow lava. . . . . . The various bulbous forms of pahoehoe in Hawaii . . . . . are definitely not pillow lavas, but are elongated structures made during the normal process of subaerial pahoehoe spreading."

out actual immersion in water, ponded river water must have percolated under the lava, which has thus been permeated during its solidification by steam from such water as well as from saturated gravel in the bottom of the canyon.[28]

In most pillow lavas "the interspaces are filled with fragments spalled off from the pillows or with sediments of the sea floor."[29] A submarine origin is most obvious in those cases where the spaces have remained open until filled by infiltration of sea-bottom siliceous or calcareous ooze[30] or other recognisable sediment of marine origin (Fig. 156).

Professor James Park, photo

Fig. 156. Cliff section showing calcareous ooze filling in the spaces among pillows in a pillow lava of Tertiary age at Cape Wanbrow, Oamaru, New Zealand. (Courtesy of Professor James Park, from *The Geology of New Zealand.*)

[28] H. T. Stearns, Origin of the Large Springs and their Alcoves along the Snake River in Southern Idaho, *Jour. Geol.,* 44, pp. 429-450, 1936.

[29] G. W. Tyrrell, *Principles of Petrology,* p. 38, 1926.

[30] H. Dewey and J. S. Flett, British Pillow Lavas and the Rocks Associated with them, *Geol. Mag.,* 48, pp. 202, 241, 1911; J. Park, On Marine Tertiaries, *Trans. N.Z. Inst.,* 37, pp. 489-551 (p. 531), 1905; *The Geology of New Zealand,* p. 146, 1910; *N.Z. Geol. Surv. Bull.,* 20, p. 39, 1918.

In the spilitic pillow basalts of the Franciscan series of California a few of the spaces are hollow, while others "are occupied by red chert or red limestone. The red chert . . . is generally indistinguishable from the ordinary chert of the Franciscan." (N. L. Taliaferro, Franciscan-Knoxville Problem, *Bull. Am. Assoc. Petrol. Geol.,* 27, pp. 109-219, 1943.)

Though it has been thought to be peculiar to basaltic lavas,[31] much perfectly developed pillow structure has been found also in the ancient (Keewatin) andesites of western Quebec, and some is developed even in rhyolitic lavas associated with these.[32] Andesitic pillow-lavas of submarine origin are reported also in Alaska in formations ranging in age from Ordovician to Triassic.[32a]

Structural Significance of Pillow Forms

As Capps[33] and Wilson[34] have observed, the forms assumed by individual pillows in lava flows with pillow structure may be indicators of the attitudes and upper surfaces of the ancient lavas and therefore also of other strata enclosing them. Such evidence is particularly valuable in some cases where overturning of folded strata may have taken place. Some of the "bolsters" or "mattress-like" pillows are found to have considerable extension in the plane of bedding, which almost certainly has been nearly horizontal, and these are commonly flat on both upper and lower surfaces. A layer of perfectly flat-based "mattresses" lies, for example, at the bottom of the pile of pillow lava in the cliff section at Muriwai (N.Z.) described and figured by Bartrum.[35] Being slightly convex above, these flat-based pillows conform to the "bun" type described below.

Most smaller pillows are strongly convex above, making "a succession of domes resembling the surface of a magnified cobblestone pavement" (Capps), while their under surfaces are either flat—the "bun" type of Wilson—or are extended down in V form between pillows of an underlying layer—the "balloon" type[36] (Fig. 157A). The balloon form has been assumed only by very plastic pillows, and perhaps is developed in greatest perfection in non-basaltic lavas, which do not

[31] G. W. Tyrrell, *The Principles of Petrology,* p. 38, 1926.

[32] M. E. Wilson, Structural Features of the Keewatin Volcanic Rocks of Western Quebec, *Bull. Geol. Soc. Am.,* 53, pp. 53-70, 1942.

[32a] A. F. Buddington, *Jour. Geol.,* 34, pp. 824-828, 1926.

[33] *Fide* J. V. Lewis, *loc. cit.* (21).

[34] M. E. Wilson, *loc. cit.* (32).

[35] J. A. Bartrum, *loc. cit.* (21).

[36] M. E. Wilson, *loc. cit.* (32), p. 63.

crystallise at such high temperatures as basalt and which, if originally very hot, will therefore remain longer plastic. According to Lewis[37] "the degree of adaptation" varies "inversely with the promptness with which a restraining integument is developed," but it seems more likely to depend on the degree of plasticity retained by the contents after the pillow is in place. In many cases—and it may be in the case of most basalt pillows—the globes have been less yielding and, though indented at some contacts, have remained rather rotund. It is in this way that open interspaces are left, such as those (Fig. 156) that have been afterwards filled with marine sediment of non-volcanic origin.[38] Even basaltic pillows thrust

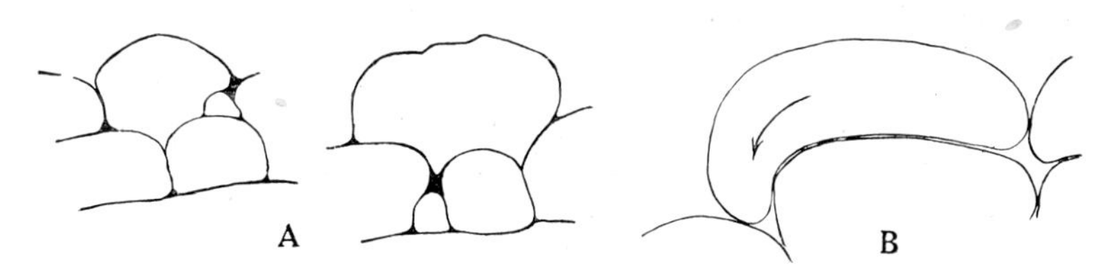

Fig. 157. A: Bun- and balloon-type pillows in Amulet Hills andesite, Western Quebec (from photographs by M. E. Wilson). B: Ideal diagram of a lava pillow overfolded, or "overturned," in the direction of flow (from a sketch by Professor J. A. Bartrum).

down some V-shaped protrusions, however, into the spaces between pillows of an underlying layer in such manner as to make the under surface of a layer recognisable. These are seen in the Muriwai (N.Z.) pillow lava already referred to, which has been figured by Bartrum.[39]

In more ancient pillow lavas at the northern extremity of New Zealand, perhaps of Cretaceous age, which are inclined at 40° to 50°, the plane of bedding has been determined[40] from the forms of the pillows. These are described as "decidedly overturned and elongated in the direction of flow." Fig. 157B

37 J. V. Lewis, *loc. cit.* (21), p. 648.
38 "The adjacent pillows have roughly adjusted their shape to fit together, but their surfaces in detail show only a slight tendency to coincide. A maze of open cavities therefore exist between their curving margins." (R. E. Fuller, *loc cit.* (24), p. 288.)
39 J. A. Bartrum, *loc. cit.* (21), Figs. 3, 4.
40 By Bartrum and Turner, *loc. cit.* (21).

shows this overturned type of flattened pillow in its original attitude—i.e. before it has been tilted by earth movement. To avoid ambiguity this modification of the pillow form might be otherwise described as overfolding of an overlapping front.

CHAPTER XVI

Craters and Calderas; Volcanic Depressions and Lakes

TRUE CRATERS ARE MERE DIMPLES IN VOLCANIC SUMMITS, OR AT most are somewhat enlarged funnels; "calderas" are contrastingly larger geomorphic depressions which also are generally concentric with foci of long-continued central activity. Such calderas are common among major features of the relief of regions of andesitic volcanoes. Most of them contain within their boundaries at least one and generally more younger volcanoes, each with its own crater; they are ringed around, however, by more or less well-preserved scarps—in many cases high and steep—which have an obsequent (anti-dip) relation to the inclination of the ash beds and lava tongues of large ancient volcanic mountains the summits of which have been broken down or destroyed in some way in the process of formation of the calderas. Large depressions of volcanic origin which are of irregular outline and are not obviously centred on ancient large volcanoes are excluded from the category of calderas. As comprehensively defined by Williams,[1] calderas are "large volcanic depressions more or less circular or cirque-like in form, the diameters of which are many times greater than those of the included vent or vents, no matter what the steepness of the walls or form of the floor." A crater, on the other hand, besides being relatively small, is distinguished as "typically a constructional apparatus from which ejecta escape to build up positive landforms, whereas . . . . calderas are passive forms of destruction."

"Crater" is a common English word defined by *The Pocket Oxford Dictionary* as "bowl-shaped cavity," and in this broad

[1] Howel Williams, Calderas and their Origin, *Univ. Cal. Publ., Bull. Dep. Geol. Sci.*, 25 (6), pp. 239-346 (p. 242), 1941.

sense the word must be sometimes employed (Chapter XIV). As defined in this chapter in a restricted sense, "crater" is to be regarded, on the other hand, as a technical term.

Craters

The smaller volcanic hollows which are by definition craters are either of bowl-like form as a result of the simple process of up-building of a ring of volcanic ejecta around a small orifice (Fig. 158A) or have a more funnel-like shape, if

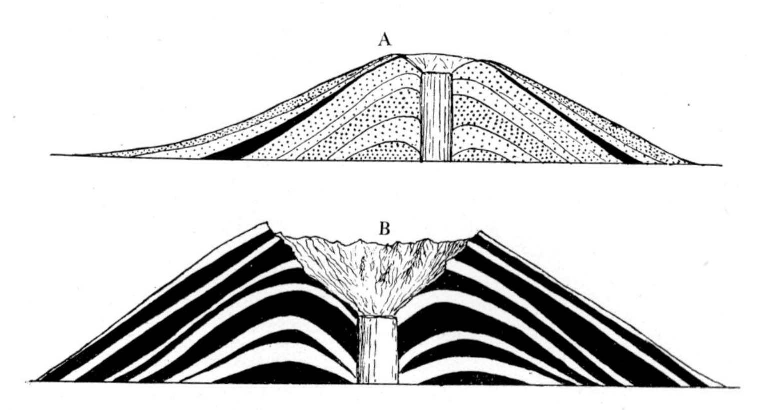

Fig. 158. A: A cone built by mild activity imitated by blowing sand out a vertical tube. (After Kuenen.) B: Enlarged crater opened by a stronger current of sand-laden air from a tube opening at a lower level. (After Kuenen.)

situated over vents that have been somewhat cored out by more violent explosions (Fig. 158B). The up-building process (easily imitated in small-scale experimental cone-building—Figs. 141B, 158) results in interior slopes of moderate steepness (at the angle of rest) within a broadly convex rim such as is found on scoria mounds (Fig. 56) and small growing ash cones (Fig. 109). Craters of this kind are related in size to the dimensions of the feeding channels or pipes beneath them, but these are never very large. Something is known of the dimensions of such pipes from study of the ancient pipe-fillings (necks) which traverse some geological formations.[2] Precipitous walls around a crater (Figs. 8, 111, 159) indicate that it has been enlarged by explosion—a clearing of the throat of the volcano. Such clearing is accompanied and followed by internal landsliding of much ash and coarser debris, which

Fig. 159. A precipitous-walled crater, Mt Ubinas (17,400 feet), in Peru. (From a photograph.)

opens the vent to a funnel shape. Material thus sliding down may be explosively rejected by the volcano, or, in intervals between explosive bursts of activity, it may be engulfed or slowly incorporated in lava. A similar result may be produced by landsliding that follows close upon a quiet sinking of the lava level, and this may expose a new scarp around the crater just as it has done in the Halemaumau pit at Kilauea, Hawaii (Chapter IV). Such processes truncate the summit of a cone sufficiently to expose the edges of the outward-dipping layers in the walls of the crater in a manner which has been imitated experimentally by Kuenen[3] (Fig. 158B). By ejecting sand rather forcibly from a tube opening considerably below the apex of a sand-built cone this experimenter has succeeded in reproducing on a small scale the phenomena accompanying eruptions of considerable intensity.

[2] Very numerous agglomerate necks exposed by erosion in Fifeshire, Scotland, which have been described by A. Geikie (Geology of East Fife, *Mem. Geol. Surv.*, 1902) are of various sizes from a few yards to three quarters of a mile in diameter. There are many "diatreme" fillings comparable in size with these in Arizona (p. 257). Daly states that the average diameter of pipes is well under 300 metres and that those under large volcanoes are not necessarily wider than those under small ones, but in some cases very small pipes or fissure vents expand upward, funnel-wise, to form throats beneath the larger craters (R. A. Daly, *Igneous Rocks and the Depths of the Earth*, pp. 159, 381, 1933).

[3] P. H. Kuenen, Experiments on the Formation of Volcanic Cones, *Leidsche Geol. Meded.*, 6 (2), pp. 99-118, 1934.

Basaltic Pit Craters

On basalt domes there are generally a number of pits in which lava rises occasionally or stands for long periods exposed to the atmosphere as lava lakes (Chapters II-IV), but these present no close analogy with true craters, which are features of the truncated summits of mounds and cones built in the main by accumulation of projected material. The pits, "pit craters" of Dana and most other Hawaiian investigators, occur in some cases (Figs. 160, 169) within much larger, axial, and in the main fault-bounded depressions (described later in this chapter) sometimes included in the category of calderas, but also termed "sinks," which seem to be normal features of dome summits; but other pit craters are scattered rather widely. There are a number, for example, south-eastward from the summit depression on the Kilauea dome, and many on the broad ridge (situated over a rift zone) that extends south-westward from the summit of the East Maui or Haleakala dome[4] (Fig. 214). In some groups there is a tendency to alignment which seems to indicate an origin in relation to underlying fissures or to the rift zones as a whole.

Though some pit craters are attributed to "collapse only" (Stearns), those which have contained lava lakes seem to have been initiated by gas-fluxing at points on fissures at which activity is concentrated while other parts become sealed by congealed lava.[5] "The development of an outlying vent which is at first a crack leads to a concentric structure about a vertical shaft, owing to the tendency of lava to freeze in the narrower portions of the crack, and to enlarge the vent circularly in the wider portions."[6]

Lava may occupy the floor of a pit crater for a time as a lava lake, the level rising and falling; and falls of rock from

[4] H. T. Stearns and G. A. Macdonald, Geology and ground-water Resources of the Island of Maui, Hawaii, *T. H. Div. Hydrog. Bull.*, 7, Plate 6A, 1942. These pits have been enlarged by phreatic explosions (*loc. cit.*, p. 101).

[5] R. A. Daly, The Nature of Volcanic Action, *Proc. Am. Ac. Arts & Sci.*, 47 (3), p. 69, 1911: "As the vent is enlarged by gas-fluxing the magma rises within it, and, kept fluid by the emanating gas, permits of further upward blowpiping." See also J. B. Stone, The Products and Structure of Kilauea, *Bishop Mus. Bull.*, 33, 1926.

[6] T. A. Jaggar, The Crater of Kilauea, *The Volcano Letter*, 364, p. 3, 1931.

the walls, such as occur at times when the lava has sunk to a low level, open the pit more widely with nearly circular ground plan in the manner which has been observed over a period of many years at the active Halemaumau pit at Kilauea. The majority of such pit craters on the Kilauea dome have apparently been formed rather rapidly and have afterwards become inactive; some have very recently contained lava lakes,

Fig. 160. Pit craters in a sink on the summit of the basalt dome, Nira Gongo, central Africa. (From a photograph.)

however. Only rarely has lava overflowed, and explosions in the pits have been of rare occurrence and have apparently contributed little to their development. In the nearly vertical walls the edges of horizontal lava flows are exposed. These are the lavas of which the domes have been built and through which the pit craters have been drilled. Their relation to new lava which rises in the pits is that of a perilith.

The Halemaumau pit (Fig. 169), which is a very large example, has grown rapidly to its present dimensions as a result of rock-sliding and crumbling of the walls of perilith (though there have been episodes of reduction to a smaller area when outflow of lava has occurred). In 1913 the pit was 1300 feet in diameter, having grown to that size since its initiation as a small pit less than a century earlier. In 1923 it was oval, with the longer diameter 2000 feet; and a period of rapid enlargement during low ebb of lava began in 1924, the longer diameter increasing to 3500 feet in 1930.

Calderas

Paroxysmal eruptions of the Vulcanian kind in which gas escapes with high velocity (as at Vesuvius in 1906, especially in the "gas" phase of the eruption) enlarge the pipes beneath some craters by a sort of corrasion which has been termed "coring-out," and Escher[7] has supposed it possible for pipes to be cored out to a great depth by Plinian eruptions, thus attaining diameters of one to two kilometres. This would greatly facilitate the opening out of larger funnel-shaped orifices by internal landsliding.

Coring-out took place on a relatively small scale at Vesuvius in 1906, and according to Escher's theory of the origin of calderas there is no distinction except that of size between explosion craters formed in this way and calderas. In surface form at least, apart from theory of origin, it is not possible to draw a sharp line of distinction between craters and calderas, and some volcanologists have maintained that there is therefore no justification for the use of the term "caldera." Though a crater may be several times as broad as the root portion of the pipe beneath it, however, a caldera is generally many times larger. It is commonly upwards of a mile, and may be many miles, in diameter.

The eruption of Vesuvius in 1906 truncated the cone (Fig. 161) and replaced the former smaller summit crater by a gaping funnel (see also Fig. 8) due to explosion and contemporaneous and subsequent internal landsliding. This new crater was a third of a mile in diameter. It was scarcely of the dimensions of a caldera; and it is doubtful whether such an eruption, in which a simple explosive blowing-off of ash-laden gas plays the principal part, ever truncates a volcanic

[7] B. G. Escher, On the Formation of Calderas, *Proc. IV. Pac. Sci. Cong.,* 2, pp. 571-589, 1930.

Fig. 161. Vesuvius, showing the large composite cone built since A.D. 79 "nested" within a caldera which dates from that year (wall of Monte Somma at right). The crater was enlarged and the cone was re-surfaced in 1906.

summit or opens a cavity to the extent of making a full-sized caldera.

The cone of the modern Vesuvius stands partly within a larger ring (of which Monte Somma is a part—Figs. 1, 161, 162) and this is a remnant—called by some of the older volcanologists a "basal wreck" and by Jaggar and Perret a "Somma ring"—of the great volcanic mountain of ancient times in

Fig. 162. Part of the wall of the caldera of Vesuvius preserved in the surviving part of the ring of Monte Somma. (From a photograph by Tempest Anderson.)

which the Plinian eruption of A.D. 79 opened a caldera $2\frac{1}{2}$ miles in diameter. After an examination of the material distributed by the explosive paroxysm, especially of the pumice which overlies the ruins of Pompeii, Lacroix[8] has found no reason to doubt the correctness of the time-honoured explanation of the destruction of the "symmetry of the ancient cone of Somma" by explosion, and so he accounts thus for the partial replacement of the ancient mountain by a caldera. It is permissible, however, to suspect that the great explosive belches of pumiceous debris which took place during the Plinian eruption of Vesuvius were accompanied or followed by collapse and engulfment of the summit.

Though many volcanologists and geomorphologists[9] have been content to explain calderas, Somma rings, and "basal wrecks" as the results (on a grand scale) of simple explosion due to escape of pent-up gases, this theory has been very generally discredited by the failure of investigators to discover the scattered fragments of destroyed mountains in sufficient quantities. As a substitute for this theory one of collapse and subsidence along faults has been proposed, but has been found to be of far from universal application. Based on an analogy with sinks on Hawaiian domes, which may be thus explained, the theory seems inapplicable to the case of the pumice volcanoes in which the majority of typical calderas are found, and it has therefore been abandoned as an explanation of these.

It has been shown in some cases (where quantitative estimation has been possible) that the volume of ejecta during Plinian, caldera-making eruptions has been indeed enormous, but the materal ejected has been largely pumiceous and of magmatic origin. This abundance of new ejecta, together with the necessity of explaining the disappearance of former high volcanic mountains by collapse or subsidence rather than their direct destruction by explosion, have brought into favour the "explosion-collapse" hypothesis as it has been termed by Williams after its recent adoption by Van Bemmelen, Van den

[8] A. Lacroix, *La Montagne Pelée après ses éruptions, etc.*, p. 122, Acad. Sci., Paris, 1908.

[9] See, for example, E. de Martonne, *Traité de géographie physique*, 5th ed., p. 735, 1935.

Bosch, and others.[10] This hypothesis, as formulated by Dana[11] in explanation of the phenomena of the Krakatau eruption of 1883, was stated as follows:

> The explosive eruption blew to great heights fragments of the liquid lavas in the shape of scoria and sand or ashes, but did not blow off the solid rocks of the mountain. The disappearance of these and the making of the cavities are explained by the engulfment or down-plunge of material *to fill the space left empty* by the projectile discharges.

It will be clear to the reader that the kind of "explosion" referred to here is very different from the detonation of a mine. Like most of the other truly volcanic phenomena classed as "explosive" it is much more continuous than that, and it may be likened rather to the blowing off of steam from a boiler or, still better perhaps, to the exhaust of an internal-combustion motor. Van Bemmelen terms the process not "explosion" but "emptying-out."

Van den Bosch emphasises the importance of collapse as a consequence of such emptying-out:

> What becomes of the summit of the volcano and of the matter that formerly occupied the calderic space? "Thrown out," it is said. I do not think so. As the huge gas stream shoots up ashes and pumice originating from the lava by the gas-producing process in the magma "body," the walls of the vent collapse and the pieces are thrown away, ... but the greater portion of the volcano and the underground matter sinks more or less vertically downward.[12]

Imagined stages in the caldera-forming process, as worded by Williams in explanation of a series of diagrams based on those of Van Bemmelen (Fig. 163) are:

1. First mild explosions of pumice. Magma stands high in the conduits.
2. Explosions increase in violence. Magma level falls into main chamber.

[10] R. W. van Bemmelen, Het Caldera Probleem, *De Mijningenieur,* 1929; C. A. van den Bosch, The Problem of Calderas, *Proc. IV Pac. Sci. Cong.,* 2, pp. 233-235, 1930; Howel Williams, *loc. cit.* (1).

[11] J. D. Dana, *Am. Jour. Sci.,* 36, p. 107, 1888.

[12] *Loc. cit.* (10).

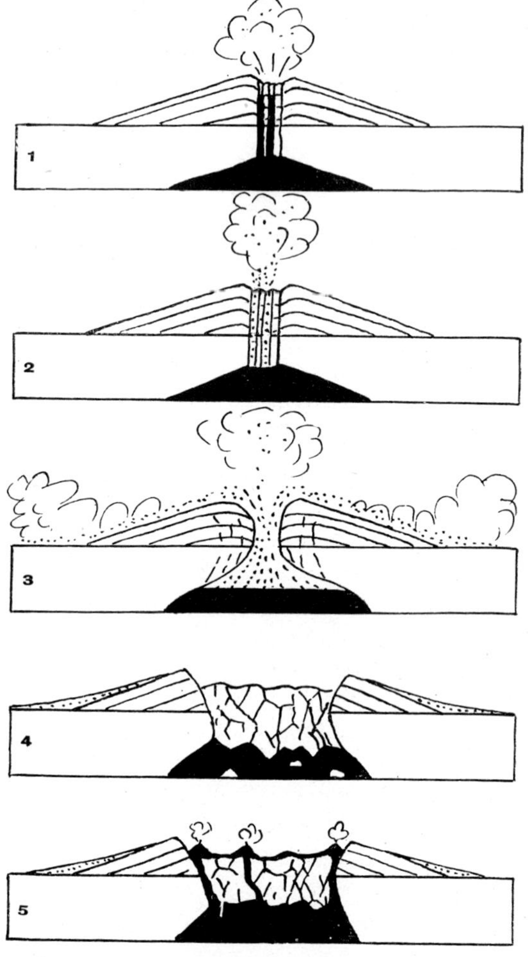

Fig. 163. Successive stages in the development of a caldera, according to R. W. van Bemmelen. (After Williams.)

3. Culminating explosions. Part of the ejecta is hurled high above the cone, but most of it rushes down the flanks as *nuées ardentes.* Magma level deep in chamber. Roof begins to crack.
4. Lacking support, the top of the cone collapses into the magma chamber.
5. After an interval of quiescence and erosion new cones appear on the caldera floor, especially near the rim.

Dana's explosion-collapse theory, though devised originally in explanation of the eruption of Krakatau, has been found applicable in general to caldera-making eruptions. At Santorini in the eastern Mediterranean, where, as at Krakatau, a submarine caldera has been formed, an enormous quantity of pumice has been emitted. This forms the material, termed "pozzolana," which is spread as a thick sheet over the island of Thera (a long arcuate remnant of a broad volcanic cone which was destroyed by Plinian eruption, and from which steep obsequent cliffs descend into a submarine caldera). Fouqué[13] found the vast pozzolana deposit to be 99 per cent pumice; Washington[14] has attributed its origin to the "foaming head of a lava column" and has described the method of its formation as a "more or less sudden boiling up of the lava in the throat of the volcano."

Crater Lake Caldera

Of the various calderas which remain in a youthful condition as conspicuous landscape features that occupied by Crater Lake, in Oregon, is perhaps the most perfect and is certainly the best known example; and it may be described, therefore, as a type form. Fortunately it has been exhaustively studied by experienced investigators.

The diameter of the Crater Lake caldera, which is nearly circular, is 5½ miles; the depth of the lake in it is 2000 feet; and the surrounding scarp rises 500 to 2000 feet above the surface of the lake.

A test of the applicability of the theory of simple explosion in explanation of calderas is, as Diller[15] has clearly stated in his discussion of the problem presented by Crater Lake, that fragmentation and distribution by an explosion of the rocks composing a mountain must spread coarse fragmentary material as a thick layer immediately around the rim of the

[13] F. Fouqué, *Santorin et ses éruptions,* Paris, 1879.

[14] H. S. Washington, Santorini Eruption of 1925, *Bull. Geol. Soc. Am.,* 37, pp. 349-384, 1926.

[15] J. S. Diller and H. B. Patton, The Geology and Petrography of Crater Lake National Park, *U.S. Geol. Surv. Prof. Paper,* 3, 1902; J. S. Diller, Did Crater Lake, Oregon, originate by a Volcanic Subsidence or an Explosive Eruption? *Jour. Geol.,* 31, pp. 226-227, 1923.

cavity opened by the explosion. No such accumulation of the debris of the former mountain surrounds Crater Lake; but the explosion theory has recently returned to favour as a competitor of Diller's subsidence theory,[16] largely because the walls of the caldera do not appear to be fault scarps.

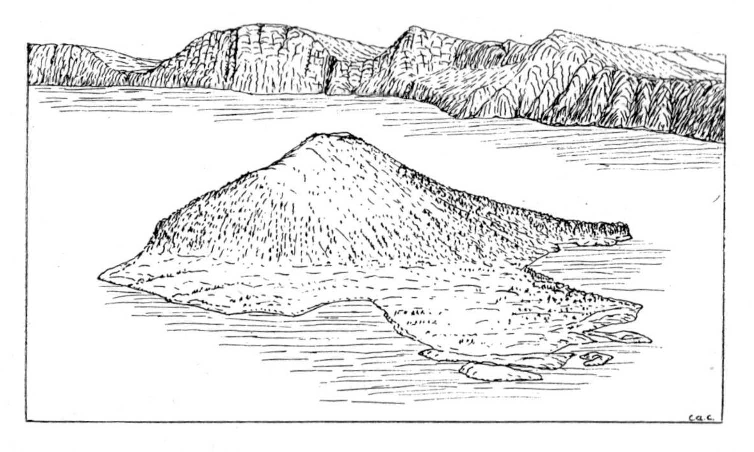

Fig. 164. Crater Lake caldera, Oregon, and Wizard Island. (From a photograph.)

The Crater Lake caldera (Figs. 164, 165) has, as all observers agree, taken the place of centre and summit of a formerly high volcanic cone; and to this vanished volcano the name Mount Mazama has been applied. Not more than a few thousand years can have elapsed since the destruction or disappearance of the mountain summit. Notches in the rim of the caldera (Fig. 164) are quite clearly not due to erosion which has taken place since the opening of the caldera, for such erosion is so slight as to be negligible and the condition of the walls is infantile. The initial rim of a caldera opened in a smooth, undissected cone would have a simple crest-line; so notches which are present in the crest-line show that the flanks of the vanished Mount Mazama were seamed by large consequent valleys.

[16] W. D. Smith and C. R. Swartzlow, Mount Mazama: Explosion versus Collapse, *Bull. Geol. Soc. Am.,* 47, pp. 1809-1830, 1936.

Glaciers partly covered the former Mount Mazama. Not only are there U-shaped glacial troughs on the ring formed of the flanks of Mazama (which are seen in cross-section where they are beheaded by the walls of the caldera—Fig. 164), but there is also much glacial morainic debris interbedded at

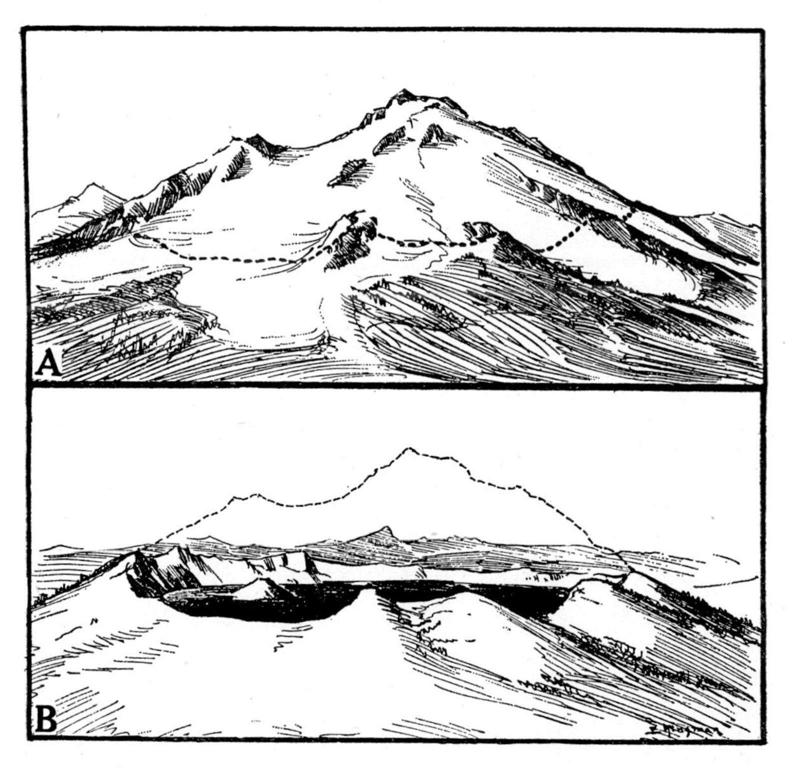

Fig. 165. A: Restoration of Mount Mazama, with its glaciers. The dotted line is at the rim of the Crater Lake caldera. B: Crater Lake caldera. The profile of Mount Mazama is restored by the dotted line. (Reproduced by permission of Professor W. W. Atwood Jr.)

several levels with the lava and ash beds of the ancient volcano where these are revealed in the inward-facing scarp. Discovery of these and other signs of vigorous glaciation confirms earlier estimates of the height of Mount Mazama (about 12,000 feet); they show also that the volcano grew slowly and that active growth was in progress during a glacial epoch, probably the latest.[17]

Within the caldera ring, though close to one side of it, a recent revival of volcanic activity has taken place which has resulted in the building of a small new cone and in outflow of lava; but this activity has ceased again after the formation of Wizard Island (Fig. 164).

The evidence for and against the explosion-collapse explanation of the caldera containing Crater Lake has been summed up by Williams.[18] About 17 cubic miles of Mount Mazama have disappeared. Though there is no such piled-up ring of coarse ejected blocks and fragments on the rim of the caldera as is required by the hypothesis that it has been opened up as the result of a simple explosion, and certainly not more than two cubic miles of such debris have been ejected, yet an abundance of pumiceous ash of very recent origin is spread around the base of the former mountain, and this has been ejected without doubt more or less simultaneously with the opening of the caldera. The volume of inflated material in this fringe or ring is estimated at from 10 to 12 cubic miles. Racing as *nuées ardentes* down the slopes and valleys on the dissected flanks of Mount Mazama during the culminating phase of paroxysmal activity, it has been distributed very widely. The pumiceous material, which has been derived obviously from new magma, has carried with it a relatively small quantity of blocks and fragments which have come from the demolished mountain. The volume of the magma from which the pumice has been formed would be about five cubic miles, the removal of which would develop a potential vacuity beneath the mountain, inducing collapse.

Some space has been made available in this way for a portion of the collapsed mountain, but about ten cubic miles remain to be accounted for, and this Williams does by postulating that "collapse of the peak into the magma chamber was

[17] W. W. Atwood Jr, The Glacial History of an Extinct Volcano, Crater Lake National Park, *Jour. Geol.*, 43, pp. 142-168, 1935; reprinted *Smithsonian Rep. for* 1935, pp. 307-320, 1936.

[18] Howel Williams, *loc. cit.* (1); *The Geology of Crater Lake National Park, Oregon, etc.*, Carnegie Inst., Washington, 1942.

brought about not only by the drainage resulting from eruption but also by the drainage consequent on some form of deep-seated intrusion."[19] To the generalisation of Van Bemmelen,[20] "the volume of breakdown calderas must be about the same as the volume of the blown-out magma," Williams has, therefore, added the qualification, "unless eruption has been accompanied by intrusions at depth."

The discrepancy between the space made available by pumice-ejection and the volume of rock which has disappeared owing to collapse may be exceptionally great in the case of the Crater Lake caldera, and it may not be necessary in every case to which the explosion-collapse explanation is applied to postulate the draining away of a vast quantity of magma into some hypothetical subterranean receptacle. The question, however, commonly remains unanswered as to whether there is some such discrepancy, for, as Williams has pointed out, the difficulties in the way of making reliable estimates of the volumes of rock involved may be insurmountable.

Modern Caldera-making Eruptions

Several Plinian explosions in recent centuries have opened calderas, destroying at the same time cones which not only had been long ages in building but commonly also had ceased to evince signs of vigorous activity. Hence the theory has been proposed that "the paroxysmal explosions leading to caldera-making are an old-age symptom."[21] A long period of dormancy may be a necessary preliminary to Plinian explosion; but some volcanoes have renewed their youth after such an event. Vesuvius, for example, shows no signs of approaching extinction, and Krakatau, which has been reduced to a caldera by

[19] Howel Williams, *loc. cit.* (1), p. 275.

[20] R. W. van Bemmelen, *loc. cit.* (10).

[21] Howel Williams, *loc. cit.* (1), p. 335. During a long suspension or slowing down of activity "slow crystallisation of magna in the hearth permits the building of high gas pressure and an accumulation of lighter, acid melt in the upper parts. . . . . The nature of volcanic eruptions depends chiefly on the degree to which the magma has crystallised, for this affects not only the chemical composition and viscosity of the residual melt, but also in large measure controls the vapour tension consequent on retrograde boiling" (*loc. cit.*).

Plinian eruptions on more than one occasion, is now busily at work building a new cone—Anak Krakatau (Chapter XV).

Among the most notable caldera-making eruptions of modern times is that of Coseguina, in Nicaragua, which took place in 1835. "During the eruption the height of the cone was considerably reduced, but to what extent is not certainly known, probably by at least one-half, for it is now a crater four miles in diameter and only 3600 feet above the sea."[22] The great eruptions of Krakatau, 1883, already referred to, and of Katmai, Alaska, in 1912, both formed calderas, though not of the largest size.

The sequence of events postulated, as above, to account for the features and history of Crater Lake is closely parallel to the observed phenomena of the Krakatau eruption as they have been interpreted by Verbeek and Stehn.[23] The eruption left a great submarine hollow in the Strait of Sunda on the site of a former island. Williams describes the hollow as "a caldera produced by collapse following the evisceration of a magma chamber by explosions of pumice," and includes all calderas formed in this way under the head of the "Krakatau type."

For the eruption of Katmai, Alaska, in 1912, an explanation which includes a theory that the summit of the mountain collapsed is preferred to the theory of simple explosion, because, just as is the case at Crater Lake, an insufficient quantity of debris consisting of projectiles of solid-rock origin has fallen round about to justify adoption of the latter theory. This seems to have been an instance of piecemeal subsidence, however, not sudden collapse, and a crater of normal dimensions probably expanded thus to become a caldera as its sides caved in. The view that an enormous magma-filled cavern had been formed by solution of the old volcanic materials in new magma in the interior of the volcano and was already in existence prior to the actual collapse of the summit has not been fully accepted by Williams, though such a theory of caldera-formation has

[22] T. G. Bonney, *Volcanoes*, 2nd ed., p. 254, 1902.

[23] R. D. M. Verbeek, *Krakatau*, 1886; C. E. Stehn, The Geology and Volcanism of the Krakatau Group, *Guidebook IV Pacific Sci. Cong.*, Batavia, 1929; H. Williams, *loc. cit.* (1), pp. 253-265. J. W. Judd has ascribed the opening of the caldera to explosion only (*The Eruption of Krakatoa*, pp. 1-56, Roy. Soc. London, 1888.)

been entertained and has been regarded by some geologists as of general application. The process seems rather to have been in the main a "caving-in of the crater walls into a pool of unusually hot and gas-rich rhyolite" (WILLIAMS). Melted rock of mixed origin was then blown out as pumice.[24]

The caldera which was opened by this means at Katmai was three miles in diameter and 3700 feet deep, and during its formation two cubic miles of the mountain were disposed of. The total volume ejected (mainly as pumice) from the Katmai vent itself supplemented by that which issued from other orifices in the vicinity and accumulated in the Valley of Ten Thousand Smokes (Chapter XII) has been estimated at not less than $6\frac{1}{4}$ cubic miles, however. The two cubic miles of old rock material if melted and inflated as pumice would expand to make more than double that volume, and so there is a reasonable measure of agreement between the size of the caldera formed and the potential vacuity which may be supposed to have been the cause of engulfment. A distinction which Williams makes between a "Katmai type" of caldera (in which engulfed rock is dissolved in magma and ejected with the pumice) and the Krakatau type, to which the majority of calderas—including Crater Lake as described above—are referred, though of great interest to petrologists is of little geomorphic significance unless the forms of caldera walls formed as a result of progressive engulfment (Katmai type) are found to differ recognisably from those which result from cataclysmic collapse in the Krakatau type.

OTHER EXPLOSION-COLLAPSE CALDERAS

Most of the other large hollows and "basal wrecks" which have resulted from the partial destruction of large pumice-built and composite cones may be reckoned in the class of explosion-collapse calderas.

A particularly large example is that of Aniakchak, in

[24] C. N. Fenner, The Katmai Region, Alaska, . . . Eruption of 1912, *Jour. Geol.*, 28, pp. 569-606, 1920; R. F. Griggs, *The Valley of Ten Thousand Smokes*, Nat. Geog. Soc., Washington, 1922; Howel Williams, *loc. cit.* (1), pp. 296-300.

Alaska[25] (Fig. 166). It is enclosed by a nearly continuous and approximately circular wall, 19 miles in circumference, which is precipitous and in most parts unscalable and ranges in height from 1200 to 3000 feet (average 1600 feet). This encloses a lowland with an area of 30 square miles containing two lakes; and the larger lake, which is 2½ miles long and 1100 feet above sea-level, overflows through the rim of the

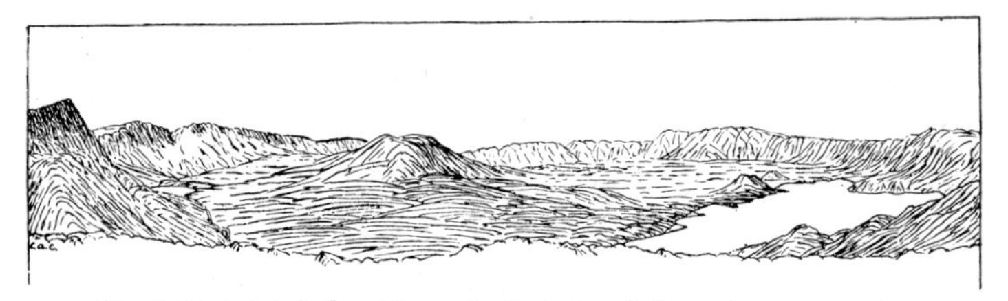

Fig. 166. Aniakchak caldera, Alaska, viewed from the eastern rim. (From a photograph.)

caldera by way of a narrow gorge, for the most part river-cut. Several ash cones stand within the caldera, but only one of these is of considerable dimensions, containing about a cubic mile of ejecta. The whole floor is ash-covered to an unknown depth, and a new paroxysmal eruption broke out in 1931.

Though sedimentary strata and other prevolcanic rocks are exposed in the scarp of the Aniakchak caldera, some beheaded lava flows on the rim indicate the former presence of a large volcanic mountain on the same site, and as a conservative estimate of the altitude of its summit 6000 feet has been proposed. Destruction of a cone of such dimensions and the opening out of the caldera must have involved removal of about 16 cubic miles of material.

Galtes caldera, discovered by Hubbard at the south-western end of Alaska Peninsula, is reported to be larger than Aniakchak.[25a]

Among other large and well-known calderas are four in

[25] Described and figured by W. R. Smith, who does not discuss the explosion-collapse hypothesis, but regards the feature as probably explosion-made (Aniakchak Crater, Alaska Peninsula, *U.S. Geol. Surv. Prof. Paper,* 132, pp. 139-145, 1925).

See map of the caldera in O. D. von Engeln, *Geomorphology,* opp. p. 602, 1942.

[25a] B. R. Hubbard, *Cradle of the Storms,* pp. 152-153, 1936.

southern Japan. Of these Kikai is submarine and the others are on the island of Kyusyu, two of them Aira and Ibosuki drowned to form parts of Kagoshima Bay, while the fourth, Aso, is inland, forming a basin about 150 square miles in extent. Italian examples include the basin Bolsena, about half the area of Aso.

Both Aso and Bolsena calderas have been the sites of continued or renewed volcanic activity on a relatively small scale subsequently to the paroxysms which opened the great hollows. Bolsena caldera contains the lake of that name, which overflows as a surface stream; Aso has developed into a valley system drained through a gorge, and parts of the surrounding scarp are maturely dissected. Those calderas (Bolsena is described as a typical example) in which the average descent of the interior walls is gradual—owing to maturity of dissection, step faulting, or other cause—have been distinguished in Italy as "concas." The eastern wall in particular of the Bolsena caldera descends as a succession of steps separated by gravity faults. Consequent drainage follows these splinters lengthwise. There are concas in Japan also, where that of Inawasiro in particular has been compared with Lago di Bolsena.[26] Its area is about 400 square kilometres.

Similar in a general way to the large calderas of southern Japan are those of New Britain. The wall of one of these encloses extensive lowlands at Rabaul, and within it are also several younger volcanoes, including the newly built cone Vulcan (Chapter XIII).

A small caldera (Fig. 167) on the large obsidian dome Mayor Island, New Zealand, may be explained by the explosion-collapse hypothesis because of the abundance of pumice which surrounds it[27] (Fig. 95) without admixture of lava blocks or fragments such as would indicate a partial wrecking of the volcano by simple explosion—unless, indeed,

[26] H. Tanakadate, The Problem of Caldera in the Pacific Region, *Proc. IV Pac. Sci. Cong.,* 2B, pp. 730-744, 1930.

[27] J. A. Thomson, Geological Notes on Mayor Island, *N.Z. Jour. Sci. & Tech.,* 8, pp. 210-214, 1926.

such blocks have been ejected by lateral explosion so as to come to rest beyond the shores of the island.[28]

C. A. Cotton, photo

Fig. 167. The caldera on Mayor Island, Bay of Plenty, New Zealand.

Volcanic Rents and Sector Grabens

True calderas, such as have been described and cited on the foregoing pages as examples, are more or less closely concentric with large central volcanoes the ancient cones of which have been partly demolished. The symmetry of some other cones, however, which have escaped central or total demolition is marred by the development of depressions which seem to be true grabens, being collapsed or subsided strips and polygons bounded by nearly straight faults. Other graben-like hollows ("volcanic rents," bounded by "horse-shoe" faults) are arcuate in plan and have been interpreted as gaping rents left open as the result of lateral down-sliding movements of considerable

[28] C. A. Cotton, Some Volcanic Landforms in New Zealand, *Jour. Geomorph.*, 4, pp. 297-306, 1941.

magnitude. Such a tearing apart of the land surface is said to have taken place within the last few thousand years in Java as a result of lateral movements (still in progress) due to sliding of heavy masses of volcanic accumulation down underlying regional slopes, other evidence of such movement being found in the crumpling of young strata into folds by "secondary tectogenesis," according to the "undation" theory; and this has been claimed as the cause of the formation of the great Soenda, Tengger, and other "calderas," which are regarded as the broadest parts of such volcanic rents.[29]

A conspicuous graben which completely transects the extinct Ringgit volcano (also in eastern Java) has been explained as another feature of purely tectonic origin, which has developed as a burst-open anticlinal crest during late Pleistocene folding and upheaval.[30] This rift has opened during folding because of the presence of a resistant body of solidified magma in a shallow volcanic hearth under an ash-built cone.

Eccentric and lateral "calderas" which appear to be bounded in part by faults on the flanks of volcanic mountains are classed by Williams[31] as "sector grabens." One such amphitheatre, on the flank of the dome of Etna, is the great hollow of rather irregular outline known as the Val del Bove. Though, in common with most other hollows on volcanoes, this has been regarded by some volcanologists as a cavity opened by an explosion, it has been accounted for by others as the result of subsidence. There is a somewhat similar amphitheatre,

[29] Howel Williams, *loc. cit.* (1), pp. 311-315. See also R. W. van Bemmelen, Ein Beispiel für Sekundärtektogenese auf Java, *Geol. Runds.*, 25, pp. 175-194, 1934; Examples of Gravitational Tectogenesis from Central Java, *De Ingenieur in Ned. Ind.* (iv), pp. 55-65, 1937; The Volcano-tectonic Structure of the Residency of Malang (E. Java), *ibid.*, pp. 159-172, 1937.

The theory that the Tengger "caldera" is a tectonic feature is in competition with the two explanations put forward by B. G. Escher (*loc. cit.* (7)) and M. E. Akkersdijk (Caldera of the Tengger Mountain: Excursion E2, *IV Pacific Sci. Cong.*, Java, 1929), which have postulated explosive emptying-out, and one by H. Kuenen (Contributions to the Geology of the East Indies from the Snellius Expedition, I: Volcanoes, *Leidsche Geol. Meded.*, 7 (2), pp. 273-331, 1935) which appeals to the ring-dyke mechanism of cauldron subsidence.

[30] R. W. van Bemmelen, De Ringgit-Beser, *Nat. Tijds. Ned. Ind.*, 98, pp. 171-194, 1938.

[31] Howel Williams, *loc. cit.* (1), p. 246.

Mohokea, which seems to be of the nature of a sector graben, on the southern flank of Mauna Loa, in Hawaii.

Many supposed examples of sector grabens have been considerably enlarged and modified by erosion, some so much so that it is now difficult or even impossible to decide whether erosion has or has not been guided initially by a sector graben. This is the case, for example, in the Caldera of La Palma, Canary Islands, which has given to geomorphology the term "caldera" (Chapter XVIII); and the great hollow in the crown of the basalt dome Haleakala, in the island of Maui, Hawaiian Islands, generally regarded until very recently as either a volcanic rent or a sector graben, is now confidently explained as the product of erosion unaided by guidance other than that afforded by the surface of the dome (Chapter XVIII).

Basaltic Sinks

The large summit depressions, often termed craters and sometimes calderas, which occupy axial positions on the active Hawaiian domes Mauna Loa and Kilauea, are centred over the main lava columns, or rift zones containing magma. Their immediate cause has been gravity subsidence along superficial faults which may perhaps be related to the rift zones within and beneath the volcanoes,[32] though independently of these some fissuring of the summit of a dome must take place when magma rises and swells within it, and the shrinkage which follows a subsequent withdrawal has been generally considered to be the cause of the collapse of portions of the nearly level summit. In general, faults formed during such gravity slumping require little or no guidance.[32a] To explain withdrawal of magma Williams[33] postulates "intrusion of dykes at depth" as well as the mechanism of outflow at low levels mainly relied upon by Jaggar to account for fluctuations of the lava level in the course of the cycle of activity (Chapters II-IV). Whatever may be the main reason for withdrawals, the fact

[32] H. T. Stearns, Geology and Water Resources of the Kau District, Hawaii, *U.S. Geol. Surv. W-S. Paper,* 616, 1930.

[32a] C. R. Longwell, Discussion: Classification of Faults, *Bull. Am. Ass. Petrol. Geol.,* 27, pp. 1633-1640, 1943.

[33] Howel Williams, *loc. cit.* (1), p. 289.

that they occur and the possibility that they cause summit collapse have long been realised.[34]

Stearns, however, while recognising that lateral outflow of lava has taken place in all stages of dome-building, has been led by study of the structure in dissected domes to adopt the view that such outpourings have not caused extensive central collapse in the earliest phase of rapid up-building, but that collapse and graben subsidence occur later and "proceed at an accelerated rate near the end of the very hot primitive-basalt phase." Hence he seeks an explanation of such collapse rather in a postulated undermining or "stoping" of the summit by a body of magma rising as an intrusive stock in the core of the dome. In fact, he states positively that "the Hawaiian caldera is the Glen Coe type,"[35] i.e. of the nature of cauldron subsidence.

The Hawaiian summit hollows, though they are commonly described along with true calderas as related forms, are best referred to as "sinks."[36] Possibly this term is applicable to all "calderas of the Kilauean type" (as defined by Williams), for which the "outflow-collapse" theory has been suggested in explanation, and of which examples from various volcanic regions have been cited;[37] but the larger Hawaiian sinks are distinctive geomorphic forms with recognisable fault scarps around them. Most pit craters are of the nature of sinks, at least in part, and are enlarged as their walls cave in by landslide faulting and simple rock-sliding; but the largest sinks, Mokuaweoweo on the summit of Mauna Loa (Fig. 168) and the Kilauea sink (Figs. 169, 170), undoubtedly owe their large dimensions in great part to integration of smaller sinks and pit craters (Fig. 160) as the strips of the summit between these

[34] C. E. Dutton, Hawaiian Volcanoes, *U.S. Geol. Surv. Ann. Rep.*, 4, pp. 214-217, 1884.

[35] H. T. Stearns, Four-phase Volcanism in Hawaii, *Bull. Geol. Soc. Am.*, 51, pp. 1947-1948, 1940; Geology and Ground-water Resources of Maui, Hawaii, *T. H. Div. Hydrog. Bull.*, 7, p. 184 and Pl. 12, 1942.

[36] T. A. Jaggar (Seismometric Investigation of the Hawaiian Lava Column, *Bull. Seism. Soc. Am.*, 10 (4), 1920) has adopted the term "sink," using it in a more restricted sense than R. A. Daly's "volcanic sink," which was introduced with the intention of including all calderas formed by collapse (*Igneous Rocks and their Origin*, pp. 150-153, 1914).

[37] Howel Williams, *loc. cit.* (1), pp. 292-294

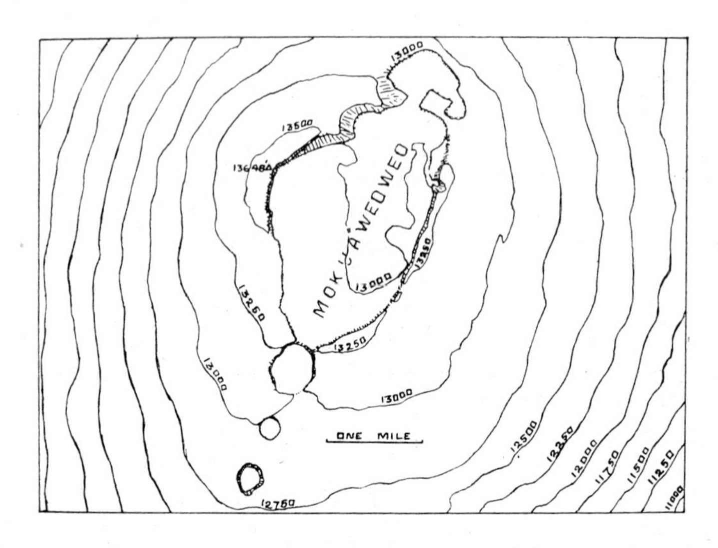

Fig. 168. The great sink Mokuaweoweo, on the summit of the Mauna Loa volcanic dome, Hawaii. The sink has been developed, as its outlines indicate, by enlargement and integration of a row of smaller sinks and pit craters.

Fig. 169. Kilauea sink, Hawaii, showing the splintered fault scarp forming the western wall. The inner pit crater, Halemaumau has recently grown to large dimensions by internal landsliding. Around it the floor of the main sink is formed of recent lava flows. (From a photograph, 1929.)

Fig. 170. Northern wall and floor of the Kilauea sink, Hawaii. This part of the floor was flooded in 1919 by lava which has solidified as pahoehoe.

have broken down,[38] though they have been further enlarged by progressive collapse of marginal fault splinters around their walls. These sinks (of caldera dimensions) are perhaps exceptionally large examples of their kind, for other Hawaiian domes are without similar large summit hollows. Sinks have, no doubt, been opened by subsidence and filled in again with lava on their summits, however, for this appears to be part of the normal mechanism of dome growth (Chapter VII).

There are sinks on the summits of the Central African examples of domes of the Hawaiian type—Nyamlagira and Nira Gongo (Fig. 160). The Nyamlagira sink is two miles in diameter.[38a]

The existing large sinks are of recent origin, for they are collapsed areas breaking through the upper members of the piles of lava sheets which have built up the domes. It is possible that no such caldera-like sinks existed in earlier stages of more vigorous dome-building, when there may have been much more voluminous output of superfluent lava. More time is available, that is to say, for gradual collapse and the coalescence of minor pits to form major hollows now that summit outflows have become small and infrequent.[39] It is not unreasonable to suppose, however, that smaller ancestral sinks of ever-changing aspect have existed, or that sinks have been formed and refilled intermittently as the domes have passed through those stages of growth described by Jaggar[40] as "pericentric about a central sink."

Though large perhaps as compared with those of former times, the Hawaiian summit hollows are small when the grand dimensions of the domes under them are considered, and they do not interrupt the visible summit profiles of the domes except from certain points of view. The sink on the Kilauea dome is about three square miles in extent and that (Mokuaweoweo—Fig. 168) on the great dome Mauna Loa is rather larger. The boundaries of both are for the greater part fresh fault scarps, and down-trailing splinters from these (Fig. 169)

[38] T. A. Jaggar, *loc. cit.* (36), p. 194; H. T. Stearns, *loc. cit.* (32), p. 49.

[38a] J. Verhoogen, *Nat. Geog. Mag.*, 76, p. 519, 1939; see also L. C. King, *South African Scenery,* Fig. 157, 1942.

[39] H. T. Stearns, *loc. cit.* (32), p. 49.

[40] T. A. Jaggar, *loc. cit.* (36), p. 187.

indicate enlargement by progressive collapse of former walls in successive strips. The original floors and subsided strips are hidden beneath very recent accumulations of lava (Fig. 170).

The Ring-dyke Mechanism

As mentioned in Chapter IV some of the observed modern phenomena of summit collapse on basalt piles have been thought to resemble the process known to geologists as "cauldron subsidence" of the Glencoe type, which has become familiarised by study, chiefly in Scotland, of structures associated with ring dykes. A large cylindrical block of the terrain—consisting perhaps, but not necessarily, of a pile of basalt sheets—sinks by gravity faulting and, as either cause or effect, magma ascends around it, resulting in the injection of intrusive rock as ring dykes and related forms and giving rise also to superficial volcanic activity.

A modern example of cauldron subsidence axially situated in a basaltic dome structure is found perhaps in the lake-filled caldera on the Pacific island Niuafo'ou (Fig. 7). Evidence of the presence of ring dykes around it is found in the high level at which the water of the lake stands, and peripheral volcanic activity, consisting mainly of outflow of lava, has taken place very recently from fissures which are concentric with the rim of the circular caldera instead of being radial as is the case on the growing basalt domes Mauna Loa and Etna.

A somewhat similar subsidence of the arched roof over a laccolith—giving rise in this case also to volcanic phenomena—has been claimed as the cause of formation of the circular Morro depression, three kilometres in diameter, in Argentina.[41]

Another large "caldera" which is attributed by Tanakadate, who has described it, to subsidence over a laccolith is the conca of Bolsena, in Italy. It has been mentioned on an earlier page under the head of explosion-collapse calderas, being so classed by Williams; but Tanakadate[42] doubts the existence of a single large volcano at any time on its site. Rather "a volcanic plateau

[41] H. Gerth, Der Morro von San Luis (Argentinien), ein "Erhebungskrater," *Leidsche Geol. Meded.*, 2, pp. 242-250, 1928; Howel Williams, *loc. cit.* (1), p. 307.
[42] H. Tanakadate, *loc. cit.* (26), p. 738.

[over the hypothetical laccolith] was crowned with many small volcanoes." Definite fault scarps and step faulting around Bolsena seem to indicate gradual development of the basin and to support this view.

Volcano-tectonic Depressions

Besides more or less symmetrical calderas and large Hawaiian sinks, each of which is centred on a single volcanic mountain, or occupies the site of its ruins, there are other large but less symmetrical areas of volcanic subsidence which, though less strikingly conspicuous as unit landforms, are nevertheless of great geomorphic interest in that they control and modify drainage and the development of initial and sequential forms of relief.

Withdrawal, outflow, or blowing off of magma which has underlain considerable areas of surface seems to have been rather commonly the cause of collapse in which normal faults have been developed with the production of downwarped basins and complex grabens of irregular outline.

Examples of such "volcano-tectonic" depressions are conspicuous in the island of Sumatra. Notably there is a very large depression of recent (late Pleistocene) formation in northern Sumatra which contains Lake Toba. The cause of the collapse in this case has apparently been the blowing out of some 2000 cubic kilometres of rhyolitic ash, much of which has accumulated to form a plateau around the lake, while some has travelled as far away as the Malay Peninsula.

Van Bemmelen[43] accounts for the collapse accompanying the great tuff eruption by supposing that "as a result of the viscosity of the acid magma the supply of magma from below could not keep pace with the emptying of the chamber. Large breakdowns, possibly a whole series, belonging to the different points of eruption of the Toba region, took place, partially following the older fault systems." Subsequent volcanism and upheaval of the lake floor have somewhat reduced the capacity

[43] R. W. van Bemmelen, The Origin of Lake Toba (North Sumatra), *Proc. IV Pac. Sci. Cong.*, 2, pp. 115-124, 1930; The Volcano-tectonic Origin of Lake Toba (North Sumatra), *De Mijningenieur,* 6, pp. 126-140, Batavia, 1939.

of the basin. It is described as a "giant caldera" 100 kilometres long and 31 kilometres wide.

Along the Semangka Valley also, in southern Sumatra, there are several fault-bounded sunken areas separate from one another.[44]

Other depressed areas that have been explained as the result of volcanic collapse include a number in Japan, notably the Towada and Toya basins, 117 and 133 square kilometres in area respectively.[45] In general these are "deeply sunk in a mountain region or in a plateau of volcanic origin" and are "steep-sided," but with "no distinct rim." They may have "rather irregular form in the beginning, because the structure of their foundations is more complicated than that of a regular cone. The actual form is the result of the subsequent water erosion, which depends very much upon the structure of the terrain." [46]

It is worthy of note that in many cases distinct fault scarps have been identified in the boundary walls of volcano-tectonic depressions and of concas and "calderas" of like origin, which seems to indicate that these are of much slower growth than are calderas which are formed as the consequence of Plinian eruptions, being in this respect more nearly related to Hawaiian sinks. Such scarps include those bounding the faulted steps or splinters in concas, the eastern wall of the Inawasiro conca (dissected into facets), a scarp on which the presence of slickensides is recorded at the Akaigawa depression,[47] and fault scarps of great length and height facing Lake Toba in Sumatra.[48]

THE ROTORUA-TAUPO BASIN (N.Z.)

A large complex of warped surfaces and relatively low-lying faulted blocks in the North Island of New Zealand occupies a strip with north-north-easterly trend which includes

[44] R. W. van Bemmelen, Vulkano-tektonische depressies op Sumatra, *Handel 25 Nederl. Nat. en Gen. Cong.*, 1935; Howel Williams, *loc. cit.* (1), pp. 319-324.

[45] H. Tanakadate, *loc. cit.* (26).

[46] H. Tanakadate, *loc. cit.* (26), p. 729.

[47] H. Tanakadate, *loc. cit.* (26), pp. 733-734.

[48] R. W. van Bemmelen, *loc. cit.* (43).

at the northern end the basin that contains Lake Rotorua and at the southern end the large Lake Taupo. Following earlier authors[49] Williams[50] attributes this relatively low-lying block complex to volcano-tectonic collapse.

By far the largest subterranean void must have been made when the incandescent pumiceous sand was emitted (nearly 2000 cubic miles of it) to make the ignimbrite sheet the surface of which forms the surrounding plateaux as well as underlying the whole or at least the greater part of this volcanic field (Chapter XII; Fig. 100). Many of the tectonic features are of much later date than the ignimbrite, however, and some of them must have resulted from subsidence following the Vulcanian ejection in much later times of an enormous quantity of pumice and ash from vents within and closely adjacent to the district (Chapter XII). Subsidence along faults which intersect the land surface north of Lake Taupo continues at the present day.[51]

As regards the origin of the basin as a whole, viewed as a major tectonic feature, some consideration must be given to the hypothesis that it is of compressional tectonic formation like most of the other major features of New Zealand relief. It may have been developed in a late phase of the late-Tertiary orogeny. Some evidence in favour of this view is afforded by the fact that the western boundary of the suspected volcano-tectonic depression is the limb of a broad anticlinal warp of the land surface and of the ignimbrite sheet (which forms the Mamaku or Patetere Plateau) while westward the descent from this plateau is a gentle monoclinal flexure which significantly passes at its northern end into one of the main tectonic scarps of New Zealand, that separating the upheaved

[49] See J. Marwick and H. Fyfe in L. I. Grange, The Geology of the Rotorua-Taupo Subdivision, *N.Z. Geol. Surv. Bull.,* 37, p. 47, 1937.

[50] Howel Williams, *loc. cit.* (1), pp. 315-319.

[51] L. I. Grange, Taupo Earthquakes, *N.Z. Jour. Sci. & Tech.,* 14, pp. 139-141, 1932.

horst of Hauraki Peninsula from the Firth of Thames graben.[52]

It cannot be claimed, however, that the deposition of the ignimbrite sheet occurred prior to the main late-Tertiary deformation. Information as to the form of the surface on which the ignimbrite accumulated is now available along the frayed western margin of the sheet, where it overlies a terrain which had been broken into fault blocks and subsequently subjected to considerable erosion. In this district the late-Tertiary emergence and deformation were complete apparently about the end of the Miocene period, and erosion of a new land surface was in progress in the early Pliocene prior to the eruption of ignimbrite in the late Pliocene.[53] Ignimbrite here occupies grabens (Fig. 179) and enters eroded embayments in horsts. Unlike the central Rotorua-Taupo area this district does not afford evidence of post-ignimbritic faulting.

Marwick suggests that most (perhaps all) of the broad, smooth, gently inclined, and apparently warped surfaces of the plateau are in reality constructional surfaces of ignimbritic accumulation with fan-like forms. These must, however, be supposed to conform to regional slopes of a buried land surface already deformed by warping, faulting, and monoclinal flexure (and subjected to some erosion) in the Miocene-Pliocene series of events referred to above.

That a tectonic basin was already in existence prior to the accumulation of the ignimbrite is shown by the fact (noted by Marwick and Fyfe[54]) that the fault scarp bounding the Kaimanawa Range (horst) formed an eastern wall against which the ignimbrite lapped and now abuts.

[52] J. Henderson (The Structure of the Taupo-Rotorua Region, *N.Z. Jour. Sci. & Tech.*, 6, pp. 270-274, 1924) remarks: "The Hauraki uplands [horst] grade southward into the Patetere Plateau. The great fracture zone along the western edge of the Hauraki upland . . . . has been definitely traced south. The nearly straight western edge of the Patetere Plateau . . . . evidently marks its southerly continuation." He notes also: "Most writers regard the volcanic plateau of the North Island . . . as a broad dome" (The Faults and Geological Structure of New Zealand, *ibid.*, 11, pp. 93-97, 1929). See also Petroleum in New Zealand, *ibid.*, 19, pp. 401-426 (esp. pp. 407, 416), 1937.

[53] J. Marwick, The Geology of the Te Kuiti Subdivision, *N.Z. Geol. Surv. Bull.*, 41, in press.

[54] *Loc. cit.* (49).

It is, of course, quite possible that pre-ignimbrite faulting and such later deformation as involves the ignimbrite sheet around and between the Taupo and Rotorua lakes have been successive spasms of the late-Tertiary orogeny, for in New Zealand these earth movements were spread over a considerable period. Whatever volcano-tectonic subsidence occurred in association with the ignimbrite eruption might be expected to follow, in part at least, the older lines of dislocation, and these may have guided also later tectonic deformation perhaps coeval with differential movement that separated the Hauraki horst from the Thames graben.

Though there may be considerable uncertainty as to the extent to which the Rotorua-Taupo basin as a whole may be attributed to volcano-tectonic subsidence, there can be little doubt that large areas within the basin have subsided rather recently—in response not to exhaustion of magma by the relatively ancient ignimbrite eruptions but to the formation of actual or potential vacuities as a result of the Vulcanian emission of the pumice and ash now present as the soil-making "shower" deposits of subaerial tuff in the extensive region around Lake Taupo. Features of the lakes in the Rotorua district and the faulted outline of the large Lake Taupo basin (which will be described more fully later in this chapter) indicate subsidence, and a vent somewhere within Lake Taupo has been confidently assumed as the source of an immense quantity of ejected pumice.

Volcano-tectonic Uplifts

The reverse of a volcano-tectonic depression is the arched or up-faulted roof over an intrusive body of magma at a shallow depth. As explained by Van Bemmelen in accordance with his "undation" theory of mountain building, such upheaval over batholiths has been responsible for producing much of the initial relief of Sumatra and Java. Thus, for example, "the Barisan mountain range . . . forms the backbone of the island of Sumatra. . . . The definite arching up of this undatory uplift or geanticline occurred in Plio-Pleistocene times." It is along the crestline of this geanticline that the

Lake Toba volcano-tectonic area of collapse and those of the Semangka Valley are aligned. "In the case of Lake Toba probably a migmatitic batholith rose till quite near the surface, arching up the upper crust," prior to the collapse.[55]

On a smaller scale are the more or less symmetrical domes over laccoliths of conventional form, which have been demonstrated not so much by study of landforms as by interpretation of the up-arching and tilting of stratified and volcanic formations lifted by the intrusive magmas.[56]

Little has been written of the details of the patterns of relief produced by volcano-tectonic upheavals. As interpreted by Rittmann,[57] however, the island of Ischia, near Naples, has been unheaved as a complex "volcano-tectonic horst" over the magma in a volcanic hearth, which has assumed the form of a broad laccolith at a moderate depth. The upper part is now solid, but the magma is still differentiating below and has supplied lava of various kinds for recent outbreaks by way of fissures between displaced fault blocks above the laccolith. This is itself injected beneath an extensive pile of ancient lavas. Component blocks of the horst tend to sink back, owing to shrinkage of the laccolith, generating earthquakes.

Crater Lakes

Some examples of maars (Chapter XIV) and of lakes contained in calderas have already been cited. There are also numerous lakes that occupy craters of moderate dimensions, and many of these are situated high on mountain summits, especially in cases where craters have been enlarged by explosive volcanic throat-clearing (Fig. 171). In some cases the existence of a lake at a high level seems to imply the presence of dykes which intersect and seal the aquifers provided by the outwardly-dipping layers in a volcanic cone.

[55] Quotations from R. W. van Bemmelen, *loc. cit.* (43) (1939).

[56] G. K. Gilbert, *Geology of the Henry Mountains,* Washington, 1877. For descriptions of other examples in western North America and in Iceland see H. H. Robinson, The San Franciscan Volcanic Field, Arizona, *U.S. Geol. Surv. Prof. Paper,* 76, pp. 70-86, 1913; L. and H. K. Hawkes, The Sandfell Laccolith and "Dome of Elevation," *Quart. Jour. Geol. Soc.,* 89, pp. 379-400, 1933; A. K. Lobeck, *Geomorphology,* pp. 507, 530-531, 541, 1939.

[57] A. Rittmann, *Vulkane und ihre Tätigkeit,* p. 83, 1936.

The existence of crater lakes on active volcanoes is only compatible with almost impervious floors and walls. The water is generally allowed to rise until the lowest point of the brim is reached, or until more permeable strata outcropping along a part of the wall drain off all surplus of water (rainfall minus evaporation). (N. WING EASTON.)

On the summit of the volcano Keli Mutu, on the East Indian island of Flores, there are three lakes in separate deep craters, one red (containing precipitated iron oxide), one blue (clear), and one green (with precipitated sulphur).[58]

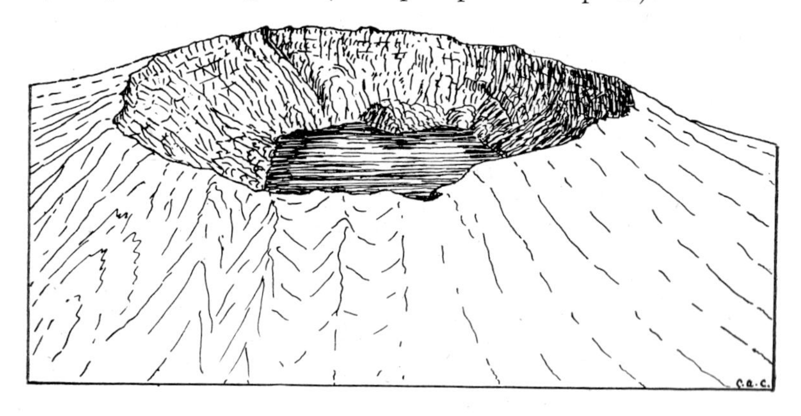

Fig. 171. Crater lake on Coseguina Volcano, Nicaragua. (From a photograph.)

Subterranean damming-up of water by dykes is indicated especially where lakes are present on volcanoes built of basalt lava flows, which generally offer so little resistance to flow and escape of water that the ground-water level, unless thus held up, is at an insignificant height above the base of the cone or dome, or above the level of the sea in the case of island volcanoes. Thus ring dykes probably hold up the level of the water of the lake in the central sink on Niuafo'ou Island, as mentioned on an earlier page.

In the case of volcanoes built of material containing a large proportion of fine ash—and it is in the craters of such volcanoes that lakes are commonly present—the cones are much less permeable. The inner walls of craters in such

[58] G. L. L. Kemmerling, Vulkanen op Flores, *Vulk. Meded.*, 10, 1931.

volcanoes (if they have been enlarged by explosion) reveal also numerous dykes and sills (Chapter XIII) capable of holding up ground water at a high level. Clearing out of the crater of the Soufrière of St Vincent (Fig. 172) by the eruption of 1902 (Chapter XII) was followed within a few months by the reappearance of a lake in the crater standing at the same level

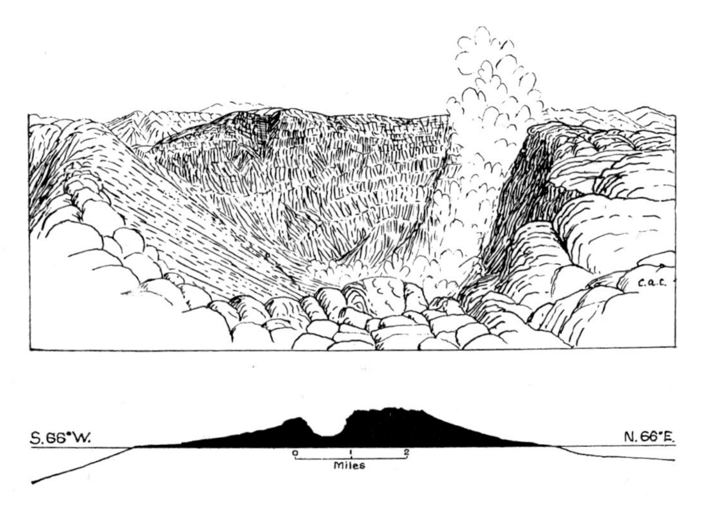

Fig. 172. The Soufrière of St Vincent, in the Lesser Antilles. Above: The crater after the eruption of 1902, when the crater lake had temporarily disappeared. (From a photograph by Dr T. A. Jaggar.) Below: A true-scale profile of the island, showing the size of the crater. (After Jaggar.)

as that which had been in existence before the eruption—that is to say, at the normal level of ground water in the mountain.

Commonly high-level crater lakes do not overflow, and this is the case also with many other lakes in permeable volcanic terrains. Subsidence or the excavation of a completely enclosed (undrained) hollow to below ground-water level by any process other than normal erosion results in the formation of a lake, and such a lake is a natural well. Generally if small it is both fed by and drains into the ground-water system without overflow. Such lakes naturally, if at all deep, are more stable landscape forms—unless, of course, destroyed or modified by a repetition of or continuation of volcanic processes which have formed them—than are those lakes which are subject to filling with sediment brought in by streams and to draining or lowering of level by erosion at an overflow outlet.

In addition to lakes contained within hollows of purely volcanic origin there are many impounded by dams of volcanic construction. These will be referred to again in Chapter XVII.

Lakes of the Rotorua-Taupo Volcanic Field

The lakes of the Rotorua-Taupo volcanic district (Fig. 173) in New Zealand[59] are a group presenting considerable variety

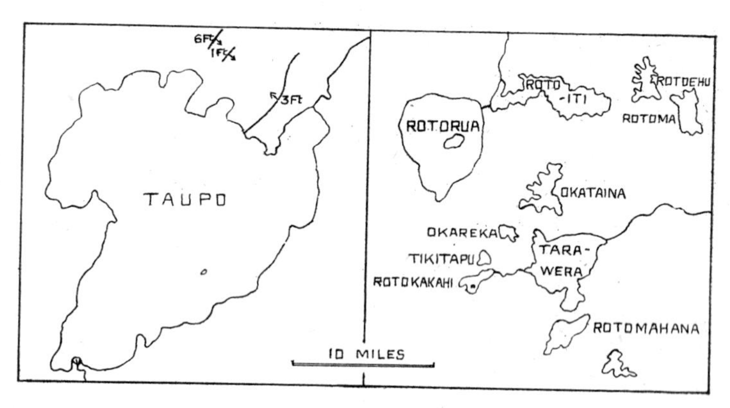

Fig. 173. Lakes of the Rotorua-Taupo basin. Fault scarplets were formed at lines indicated north of Lake Taupo in 1922.

of form and origin and exemplify most of the types characteristically associated with volcanic activity with the exception of mountain-summit crater lakes (of which there are examples, however, on the adjacent volcanoes Ruapehu and Tongariro—Chapter XIII).

In addition to the large Lake Taupo, 238 square miles in extent without including the area of an extensive shallow arm which existed when the lake level was 115 feet higher than it now is, this volcanic field includes many smaller lakes ranging in size from Lake Rotorua (27 square miles) and Lake Tara-

[59] For further details see the descriptions and interpretations of these lakes by J. Marwick and H. Fyfe in *loc. cit.* (49), pp. 26-39, 49.

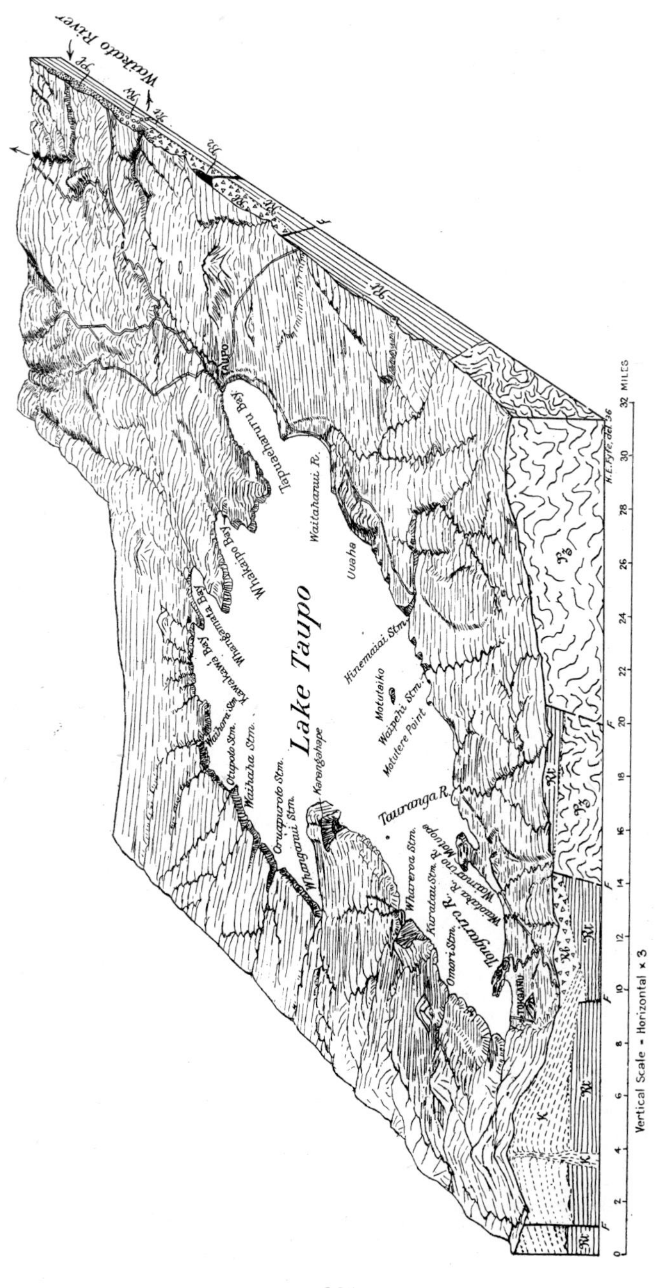

Fig. 174. Lake Taupo, showing structure. (After Grange, Marwick, and Fyfe.)

wera (17 square miles) to a number of lakelets of less than a square mile in extent.

Lake Taupo (Figs. 173-175) is an area of collapse; but within it vents may be situated from which great quantities of pumice have been projected. The lake is bounded nearly all around by fault scarps, though those of the eastern side are much lower than the western scarps, where the "Waihi" fault scarp (running inland at the southern end of the lake—Fig. 175) is unmistakable and farther north the high vertical cliffs

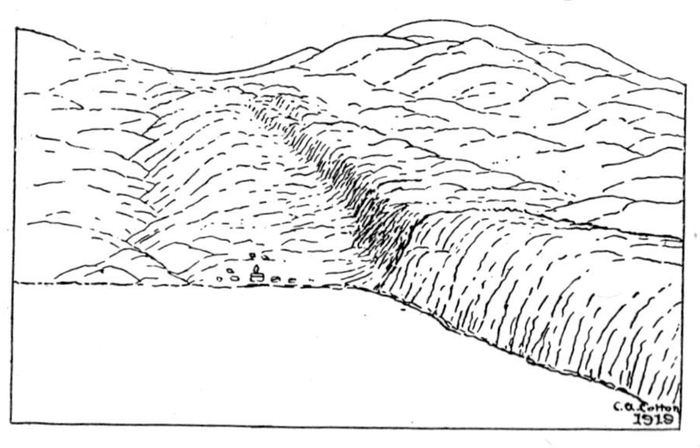

Fig. 175. The western scarp of Lake Taupo, showing the fault scarp running inland in a southerly direction—the Waihi fault scarp of the Geological Survey.

of Karangahape Bluff are a well-known feature of the scenery. The western scarps are from 700 to over 1000 feet high (some 300 feet below the level of the lake being included). There has been some recent activity with emission of basaltic lava, apparently in association with the fault at Karangahape Bluff. The apparent absence of any continuation across the lake floor in the case of the scarps bounding the salient at Karangahape Bluff and those which separate the northern bays of Lake Taupo (Fig. 174) has been the subject of some speculation,[60] but continuity of normal faults due to local subsidence is not to be expected. In volcanic terrains such faults are notoriously "irregular and discontinuous."[60a]

In the northern bays of Lake Taupo differential subsidence, amounting at one place to 12 feet, took place as late as 1922,

[60] *Loc. cit.* (49), p. 28. [60a] C. K. Leith, *Structural Geology,* p. 72, 1923.

when a strip of the surface sank between faults parallel to scarps bounding the bays[61] (Fig. 173). The history of the development of the lake basin may be a very long one, however, and those of its boundaries that suggest very recent marginal inbreak may have been made only by the latest of a long series of block subsidences.

In the contours of the lake floor, so far as they are known, there is no very definite indication of the presence of a crater or craters. Several depressions have been suggested as possibly the vents from which eruption of the pumice of the Taupo shower (Chapter XII) took place,[62] but these hollows are only about 100 feet deep and are several miles in diameter. In spite of this it seems an almost unescapable inference that the vast quantities of pumice, both fine and coarse, which mantle the surrounding landscape were blown out through some part or parts of the floor of the lake. If the latest showers had this origin it is quite probable that earlier ejections of pumice and ash from the same source accompanied or initiated the first collapse which made a caldera-like hollow later enlarged to form the present volcano-tectonic depression.

The smooth floor and rather shallow depth of the lake (maximum, 543 feet) suggest that there is a thick accumulation of lake-floor sediment in the basin. The lake water is very clear and silt accumulation must now be slow, but in the recent past, during paroxysmal eruptions of the volcanoes of the Tongariro group to the south-west, an immense quantity of ash must have been carried into the lake by rivers in addition to air-borne ash. Much pumice from the Taupo shower also must have formed a widespread floating raft perhaps of great thickness, which as it became waterlogged and sank would be spread over the whole floor.

Whether or not there is anything of truth in the suggestion that some of the material of the basal ignimbrite sheet in this region was emitted from vents within the limits of the present Taupo Lake, there is no indication that any crater-like vent

[61] L. I. Grange, *loc. cit.* (51).

[62] L. I. Grange, Physiography and Geology of the Taupo-Rotorua District, *A. & N.Z. Ass. Adv. Sci Handbook for New Zealand*, pp. 22-26, 1936.

has since remained open continuously as a nucleus of the existing lake basin. The lake lies, however, within a broad and probably downwarped hollow in the surface of the ignimbrite sheet—a part of the larger Rotorua-Taupo basin of problematical and complex origin which has already been referred to.

Marwick and Fyfe have regarded the development of a large basin by warping around Lake Taupo as part and parcel of the origin of the inner lake basin. They have described the apparently warped surface of the ignimbrite sheet as follows:

> On the long eastern and western sides the surface of the land for ten or twelve miles back slopes gently down towards the lake. These slopes, though modified by later pumice deposits and by erosion, are formed in general by the ignimbrite, and . . . . seem definitely due to warping and do not represent slopes of deposition. Since these slopes if extended would bring the land surface below the present lake level, warping must be recognised as a factor in forming the lake.[63]

Lake Tarawera is ascribed to collapse along fault boundaries, though the extreme flatness of the lake floor in the midst of a district of considerable relief has been found difficult to explain. Perhaps the partial filling of the basin may be ascribed, as in the case of Taupo, to the sinking of rafts of pumice.

Lake Rotorua is somewhat ancient and has been reduced in area after two rather lengthy episodes of terrace-making and cliff-cutting at 300 feet and 200 feet respectively above the present lake. The shrunken lake occupies only the centre of a basin-shaped depression in the plateau of ignimbrite, which has been attributed to subsidence following volcanic activity. The little-dissected ignimbrite plateau and the rock sheet of ignimbrite under it dip gently into the basin from north-west, west, and south-west, but the south-east boundary appears to be a fault scarp. An island (Mokoia) in the lake is a rhyolite mound apparently of recent growth.

[63] Marwick and Fyfe, *loc. cit.* (49), p. 27. It may be noted that slopes similar to those that lead towards Lake Taupo, but leading down in this case westward towards the fringe of the ignimbrite sheet, have been assumed more recently by Marwick to be fan-like surfaces of *nuée ardente* deposition (see Chapter XVII). The theory that the centripetal slopes around Lake Taupo are the result of warping seems unassailable, but the warping may not be all post-ignimbritic.

While the largest lake basins in this volcanic field are thus regarded as due to volcano-tectonic subsidence, some of the smaller lakes are, in part at least, drowned valley systems, and only some of the smallest seem to be explosion pits (e.g. Tikitapu, Fig. 91). The largest lake of which the present

C. A. Cotton, photo

Fig. 175A. Lake Rotomahana, New Zealand. The hollow it occupies was filled gradually after the expulsion of the earlier lake on the same site by the explosive eruption of 1886. View looking south-west.

outlines are attributable—in part, at least—to explosion is Rotomahana, now three square miles in extent. In its present form Lake Rotomahana is the youngest lake in this region, as it dates from the explosive fissure eruption of Tarawera (1886), a phreatic phase of which blew out the floor of the former Lake Rotomahana and enlarged its basin (Fig. 175A) so as to make of it what has been sometimes described as an "explosion caldera."[64] It is only by stretching the definition, however, that this hollow can be called a caldera.

[64] R. A. Daly, *Igneous Rocks and the Depths of the Earth,* Fig. 80, p. 165, 1933; H. Williams, *loc. cit.* (1), pp. 248-249.

The drowning of valley systems may obviously result either from obstruction by lava dams or from irregular warping (with which may be associated deformation by faulting) of the land surface. Lake Okarita, with an area rather more than a square mile, is a stream valley blocked by rhyolite extrusion, and Okataina (four square miles) with "four radiating arms that appear from their tapering and irregular shape to be drowned stream-valleys" is ponded in the same way.

A group, the "Three Lakes" (Rotoiti, Rotoehu, and Rotoma) well known for their beauty, exhibit considerable variety and none of them resembles their neighbour Rotorua (Fig. 173). The overflow river from Rotorua traverses a corner of Rotoiti (13 square miles) the largest of the three. These lakes, "linked genetically as well as geographically," are of somewhat complex origin, but "the alignment of the lakes, . . . all of which have been centres of volcanic activity, indicates that a powerful east-west fracture underlies [Three Lakes fault]. This suggests that the scarp to the south is a fault scarp and that the lakes occupy a fault angle." The individual features of the Three Lakes are summarised as follows:

Rotoiti: The west end [Fig. 176] is a drowned stream valley in a fault angle offset to the north from the Three Lakes fault; the east end is an explosion crater in the Three Lakes fault angle.

Fig. 176. View looking north across the west end of Lake Rotoiti (compare Fig. 173); inlet at left, outlet left of centre; fault scarp at right descends to drowned valley system. (Sketch by Dr J. Marwick.)

Rotoehu: The south end is an explosion crater in the fault angle [Three Lakes fault]; the north end is a system of confluent drowned stream valleys.

Rotoma: Explosion craters in the fault angle.[65]

[65] Quotations from Marwick and Fyfe, *loc. cit.* [49].

The lake Rotoaira, south of Lake Taupo, which lies at the foot of the volcano Tongariro, has some appearance of being ponded by lava flows from that source, which reach its shores. The discovery of a low fault scarp,[66] however, which faces the lake and is so situated as to reverse the flow of a former stream, makes it appear that the lake basin is due to local volcano-tectonic subsidence.

---

[66] J. Henderson, Earthquake Risk in New Zealand, *N.Z. Jour. Sci. & Tech.*, 24B, pp. 195-219 (p. 198), 1943.

CHAPTER XVII

Dissected Basalt and Ignimbrite Plateaux; Basalt Plains

MANY REGIONS WHICH EXHIBIT THE STRUCTURAL PLATEAUX, benches, and terraces characteristic in a general way of terrains containing horizontally bedded competent formations are ancient lava fields. The very extensive Deccan "trap" region (Chapter VIII) and the plateaux of Abyssinia are often cited as examples.[1] The last remnants of a volcanic plateau to survive may still stand out from the surrounding landscape as prominent mesas and buttes (Figs. 177, 178). The Færoe Islands and the eastern and north-western districts of Iceland (Fig. 30) are examples of lava plateaux that have been submaturely or maturely dissected and subjected also to erosion by valley glaciers.

ANCIENT VOLCANIC PLATEAUX

The more ancient and most extensive basalt plateaux are not only maturely dissected but have been affected by warping and tilting, and in some cases also by faulting, to such an extent that in general they no longer exhibit only landscapes developed directly by erosion from initial constructional volcanic forms; and strictly landscapes developed in a late postvolcanic cycle from tectonic forms on such terrains must be excluded from the category of volcanic landscapes. The basalts of eastern Iceland dip westward at an inclination of 7°.[2] Those of the western part of the Columbia River region

[1] "An enormous area of this country [Abyssinia] seems to be composed of volcanic rocks chiefly in the form of sheets of basalt lava, which rise into high plateaux and break off in steep—sometimes precipitous—mural escarpments along the sides of the valleys" (E. Hull, *Volcanoes Past and Present,* p. 191, 1892).

[2] L. and H. K. Hawkes, The Sandfell Laccolith and "Dome of Elevation," *Quart. Jour. Geol. Soc.,* 89, pp. 379-400 (p. 382), 1933.

of North America are quite strongly folded and other parts of it are converted into ranges and basins by block faulting. The Columbia River itself, where it takes its course through the plateau-basalt region, "is generally believed to have been integrated during the later stages of outpouring of the Miocene

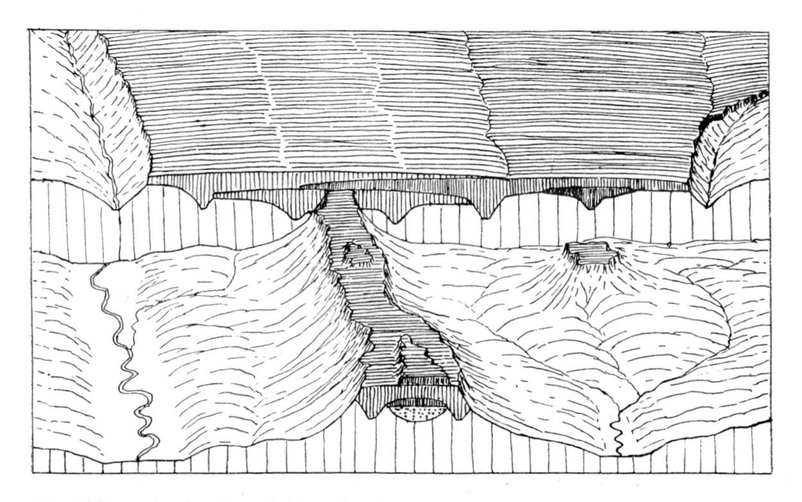

Fig. 177. Early (rear) and late (front) stages in the erosional development and destruction of a lava plateau. (Mainly after W. M. Davis.)

Fig. 178. A small residual mesa of a basalt plateau at Polignac, near Le Puy, France. (Drawn from a photograph.)

Columbia River basalts, to have assumed its westward course at that time, and to have maintained its position, as an antecedent, during post-Miocene uplift which is supposed to have raised the Cascades to their present height."[2a]

[2a] J. H. Mackin, review of Geology of the Lower Columbia River (by E. T. Hodge), *Jour. Geomorph.*, 3, pp. 70-75, 1940.

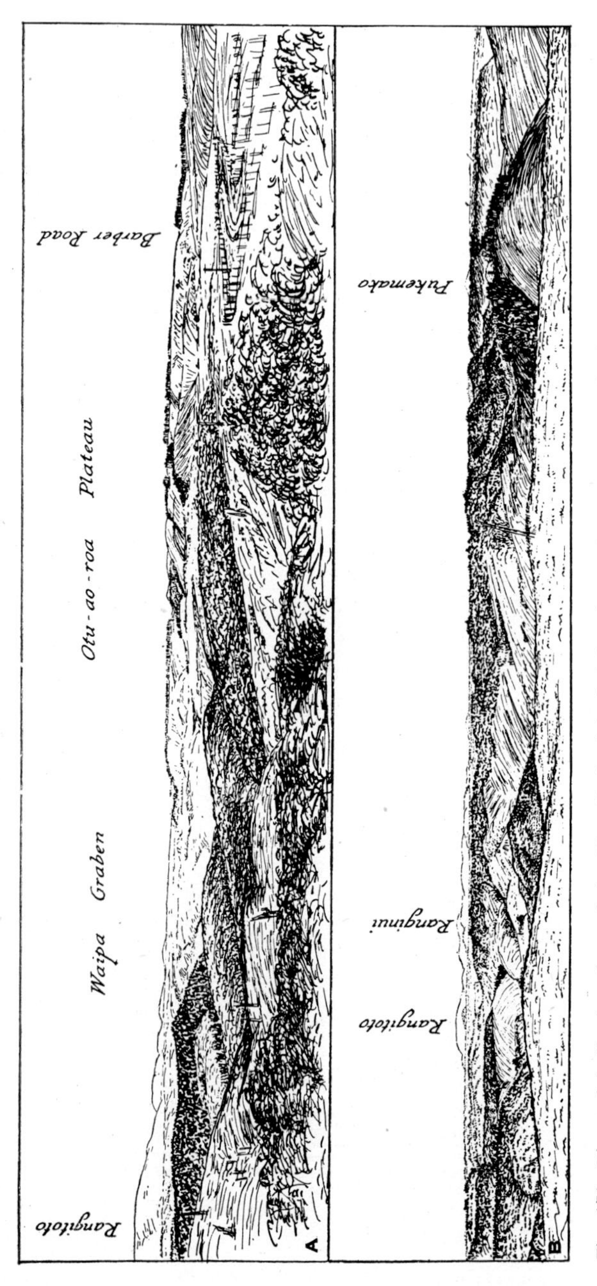

Fig. 179. Dissected inclined surfaces of the New Zealand ignimbrite plateau, in the Te Kuiti district. The regional slopes are perhaps constructional, but are possibly due to warping. Rangitoto, Ranginui, Pukemako, and the Otu-ao-roa Plateau, which are dissected horsts, are portions of a prevolcanic landscape surrounded, like islands, by the constructional plateau surface of ignimbrite.
A: View looking south, showing westward slope of the ignimbrite surface.
B: View looking east up the slope of the ignimbrite plateau, which descends also north and south. This convex surface may perhaps be a constructional fan of *nuée ardente* deposition.

(Drawings by Dr J. Marwick from photographs.)

In the ignimbrite plateau district of New Zealand the surface of the ignimbrite sheet, though in most parts not yet deeply dissected (Fig. 179), has been perhaps upheaved and has been broadly warped as well as locally broken by faults (Chapter XVI) prior to the development of the current landscape cycle.

The Ignimbrite Plateau of New Zealand

For the reason noted above the drainage pattern on the ignimbrite plateau of New Zealand is not as a rule consequent on original slopes of deposition. The constructional surface was in all probability more nearly level than are the arched forms which now exist. It is true that Marwick's investigation of the western margin of the ignimbrite has led to the suggestion that a broadly convex slope at about 3° towards the west which is there present on the plateau is not due to warping but is a fan of *nuée ardente* deposition[3] (Fig. 179). This interpretation may have to be rejected, however, for it implies that all the ignimbrite material has come from the direction of Lake Taupo, whereas it seems probable that much of it has issued locally from fissure vents like that which has been discovered a few miles to the south at Ongarue (Chapter XII).

Dissection of Ignimbrite

Detailed study of the extensive ignimbrite plateau of New Zealand will undoubtedly yield some information regarding the development of drainage and also of modifications due to selective or subsequent erosion.

It is unlikely that agglutination ever affects all parts of a body of ignimbritic tuff quite equally. If it does not, it will make a terrain which is heterogeneous as regards the resistance it will offer to erosion; and differential or selective erosion will isolate some of the more resistant parts as mesas, while adjustment to structure will develop a pattern, though perhaps a very irregular pattern, of subsequent river courses. Complications will be introduced where subsequents guided by

[3] J. Marwick, The Geology of the Te Kuiti Subdivision, *N.Z. Geol. Surv. Bull.*, 41, in press.

patches or strips of incoherent sandy tuff at high levels become superposed across hard cores of well-agglutinated ignimbrite below them. Professor J. A. Bartrum in a letter to the author has described an example of superposition near Mokai, New Zealand, which results from diverse consolidation of ignimbritic material. "A stream," he writes, "flows from a plain-like area through a narrow gap in a ridge of typical ignimbrite about 200 feet in relief, and this ridge dies out shortly into the peripheral plain. This is evidently a case of superposition from upper layers of ignimbrite formation."

The general course of the cycle of dissection may thus diverge considerably from that on a lava plateau. So little system is there likely to be in subsequent stream patterns, however, that they will rarely be recognisable as such and thus distinguishable from consequent and insequent systems in a maturely dissected landscape, especially at the stage where only mesas and buttes of the volcanic terrain survive. In the case of the ignimbritic terrain of the North Island of New Zealand there are large areas of plateau still undissected, the presence of which indicates a high degree of continuity of the well-agglutinated and resistant ignimbrite; and large tectonic features made by earth movements of considerable magnitude are well preserved. In the case of some large mesas, however, an origin by circumdenudation in a heterogeneous terrain is suspected. Firm bodies of ignimbrite may have been isolated by the removal of the looser tuff from intervening areas. Such mesas "terminate rather abruptly [in escarpments]; though originally they may probably have passed laterally into softer and more incoherent matter which was an easy prey to the agents of erosion and was soon removed."[4]

Remarkable pinnacles and a few somewhat broader buttes (Fig. 180) which are conspicuous features on the Patetere Plateau not far from Rotorua, New Zealand, perhaps owe their preservation to superior hardness developed locally during the consolidation of the ignimbrite at points where a high temperature has been maintained by fumaroles, which

[4] P. Marshall, Acid Rocks of the Taupo-Rotorua Volcanic District, *Trans. Roy. Soc. N.Z.*, 64, pp. 323-366 (p. 332), 1935.

may have been of the "rootless" kind (deriving gas from the ignimbritic tuff itself) such as were observed on the "pumice flows" of Komagatake, in Japan (Chapter XII).

C. A. Cotton, photo

Fig. 180. A butte of ignimbrite on the Patetere Plateau, Mamaku, New Zealand.

Lava Plains

Many lava fields of comparatively recent origin have as yet suffered little or no deformation or disturbance by earth movements and, though they may have undergone sufficient superficial alteration to destroy the initial features of their minor relief (Chapter IX), still remain as young landscapes on which one may observe processes of erosion and dissection characteristic of the volcanic variant of the geomorphic cycle. Some of these are of small area, but the plains of the Snake River basalt field, in southern Idaho, are of regional extent (Chapter VIII). These plains are almost uniformly level for thousands of square miles except for some minor undulations of the surface developed during consolidation. The basalt field is undissected and unaffected by erosion except in the vicinity of the Snake River, which has cut a young consequent canyon across it. The surface is a complex of level-surfaced lava flows, some old enough to have developed thin weathered soils, while others, some of them up to 100 square miles in extent, are of such recent origin that they are still fresh and unweathered. The undissected condition of even the less recently built parts of the Snake River plains and of similar plains elsewhere, and the delayed dissection of lava plains in general, even of those that have a considerable measure of available relief, is only in part explicable as a consequence of the resistance offered to erosion by massive bodies of lava rock. It is due in great part to the extremely permeable nature of piles of basalt flows which incorporate little ash or fragmentary material of any kind. Pahoehoe basalt flows are traversed by tunnels, and aa flows have scoriaceous (clinkery) surface layers which are extremely permeable. Thus a pile of basalt lava flows swallows water and there is generally no run-off from its surface. The scoriaceous layers and lava tunnels provide easy passages for water in a lateral direction through the flow complex, or water may pass down through broken or permeable sheets to flow beneath the lava along former water-courses now buried.

Thus very wide spacing of major river valleys and an almost complete absence of minor tributary valleys or ravines are characteristic features of young lava plains and plateaux. For a long distance the Snake River, for example, receives from the lava plains bordering and overlooking its canyon no tributaries in the ordinary sense; but its volume is augmented very considerably by the outflow from voluminous springs which issue from the lava beds forming the steep side of the canyon. These are fed by ground-water streams collected in the more permeable layers of basalt, for the lava-flow complex swallows not only the water that falls as rain on its surface but also the rivers that enter the Snake River basin from the north.

Snake River Alcoves

Some of the largest of the springs that drain from the lava field north of the Snake River now discharge into the heads of short, rather deep, amphitheatre-headed and steep-sided canyons,[5] known as the Snake River "alcoves," as a result of a remarkable process of headward erosion, still in progress, by means of which the springs have sapped back into the basalt terrain. The rock is progressively undermined at the head of each alcove (Fig. 181) so that it collapses and falls as talus. Progressive removal of this canyon-head talus, which is necessary if the headward sapping process is to continue in the alcove, is effected by chemical disintegration of the basalt

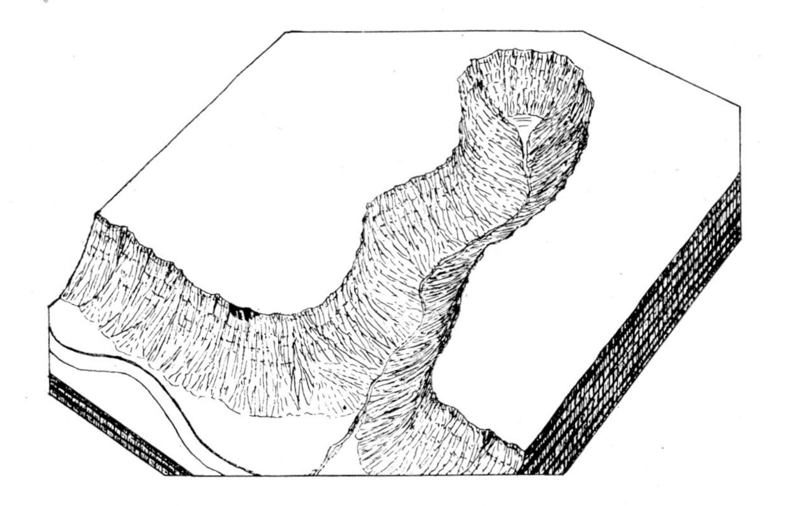

Fig. 181. Headward development of a Snake River alcove.

in the spring water, after which a voluminous stream fed by the spring and draining out through the alcove removes the products of this disintegration, partly in solution.[6] The Blue Lakes alcove, which is a typical example (Fig. 182), is a canyon two miles long and 300 feet deep. Rock talus is abundant along the side walls, but scanty at the head. Neither the steep heads of the alcoves nor their precipitous walls can be attributed to ordinary escarpment-making processes, for it requires a weak stratum under the cap rock to develop a

[5] I. C. Russell, Geology and Water Resources of the Snake River Plains of Idaho, *U.S. Geol. Surv. Bull.*, 199, pp. 26-28, 127-130, 161-168, 1902.

[6] H. T. Stearns, Origin of the Large Springs and their Alcoves along the Snake River in Southern Idaho, *Jour. Geol.*, 44, pp. 429-450, 1936.

structural escarpment, and this is not present in the Snake River basalts. Though spring sapping has rarely been recognised elsewhere except in some very weak terrains,[7] this "alcove" type of headward erosion, albeit slow in action, may have operated very generally as the predominant process developing insequent valleys in the protracted youthful stage of the dissection of lava plateaux.

Fig. 182. The Blue Lakes alcove, excavated below the Snake River basalt plains and occupied by a spring-fed insequent tributary of the Snake River, Idaho. (From photographs.)

An analogy has been suggested between the Snake River alcoves and the amphitheatre heads of maturely opened radial valleys (the "Oahu" valley type—Chapter XVIII) in the basalts of Hawaiian domes[8]; but the cases are different, for the extension of these Hawaiian valleys headward takes place above ground-water level[9] and their heads are not in all cases occupied by springs.

Consequent River Courses over Lava Plains

The majority of lava flows display some convexity of surface and even the flattest flows may have convex margins, which

[7] Examples in Florida, described originally by Sellards, are cited by Douglas Johnson, *Submarine Canyons,* p. 91 and Fig. 4, 1939.

[8] H. T. Stearns, *loc. cit.* (6).

[9] H. T. Stearns, Geology and Water Resources of the Island of Ohau, *T.H. Div. Hydrog. Bull.,* 1, p. 24, 1935.

tend to be wall-like. Even the lake-like lava sheets of the Snake River plains which Russell has described (Chapter VIII) generally terminate abruptly, as though the ponded lavas had been walled in by solidified barriers, instead of lapping with thin edges like lakes of water against the land slopes around which they contour. Smaller flows, more especially those of non-basaltic lavas, solidify in most cases with visibly convex surfaces, and some such convexity is generally taken for granted in making deductions regarding probable initial courses of consequent streams over lava fields in cases where conditions permit of a certain amount of run-off. Obviously even flat-surfaced lava flows with steep margins will enclose potential channels between them where they meet, and the inclined surfaces of interfingering flows from different sources make re-entering angles with one another, which may guide surface streams of water. Originating in this way there are many lakes and swampy tracts of consequent origin on the low-lying lava plains of south-eastern Australia.[10] So permeable are most basalt flows, however, that few consequent rivers originate in lava fields, and basalt plains which are high enough to have some available relief are eventually dissected in the main by insequent streams. (This does not apply, however, to domes—see Chapter XVIII.)

On the Australian lava plains, as described by Hills[11] and others, the development of consequent drainage governed by initial consequent slopes is interfered with not only by the generally absorbent and permeable nature of the basalt lavas but also by local subsidences, which are explained as due to the collapse of the rather thin basalt cover over areas of cavernous limestone as these are progressively dissolved by ground water. There is a possibility also of shallow lake basins being formed owing to the collapse of lava tunnels and caverns, but trenches so formed are generally discontinuous and rarely become water-courses. Where there are open underground channels the ground water is generally at a consider-

[10] E. S. Hills, *The Physiography of Victoria,* p. 197, 1940.
[11] E. S. Hills, *loc. cit.* (10).

able depth, and so hollows formed by collapse are commonly dry and end blindly.

Basaltic Scablands as an Example of Glacial Sculpture in Lava Fields

Investigations made by Hobbs[12] indicate a probability that a considerable area of the previously water-worn and maturely developed surface of the Columbia River plateau of basalt of Miocene age has suffered glacial erosion of a peculiar kind under an extensive marginal lobe of the Pleistocene northern ice sheet. This is the region known as the "scablands," or channeled scablands, in which a unique assemblage of landforms has long been recognised as failing to conform to any of the well-known types of landscape attributable to fully understood processes of erosion.

> This surface . . . bears little resemblance to the accepted features from operation of any of the known geological agents. In general it may be said that the scablands surface is made up of anastomosing channels within the basalt, channels which are arranged on the plan of braided streams. . . . Separating the channels are residual low mesas—black and barren—which are the "scabs" [Fig. 182A]. These are bounded by precipitous walls, often terraced, sometimes a hundred feet or more in height.
>
> Scattered on the floors of the channels are hundreds of rock basins, large and small, generally filled with water but sometimes with gravel. Along the sides of the channels are local deposits of gravel, which Bretz has called "bars," and gravel is found also clinging in niches of the walls. This gravel is clearly the residue of much larger deposits which have been in great part swept away.
>
> On the larger residual islands between the channels are heavy deposits of dark-brown loess in character identical with that which surrounds the scablands on east, south, and west and constitutes the rich wheat lands. These islands are scabs masked by their loess cover. (Hobbs)

12 W. H. Hobbs, Discovery in Eastern Washington of a New Lobe of the Pleistocene Continental Glacier, *Science,* 98, pp. 227-230, 1943.

According to Flint:[13]

It is generally agreed that the valleys constitute the normal preglacial drainage pattern of the plateau, but that many of them were widened, deepened, and modified by the making of subsidiary channels and closed depressions during their occupation by proglacial melt-water, so that to-day they are scabland tracts rather than normal valleys.

Various special explanations of the origin of the peculiar features of the scabland tracts have been proposed, all of which

Fig. 182A. Basalt buttes, or "scabs," on the floor of one of the scabland valleys of Washington. (Drawn from a photograph by Junius Henderson.)

have assumed excavation by voluminous streams of water derived from the melting of the Pleistocene ice.[14] The gist of these hypotheses, as summarised by Worcester,[15] is as follows: Of three different explanations for the origin of the features one "requires the catastrophic liberation of glacial water" and the other two

invoke normal stream erosion that was brought about by an unusual event, that is, the blocking of a great stream. This was by no means impossible through lava flows, landslides, ice jams, earth move-

[13] R. F. Flint, Origin of the Cheney-Palouse Scabland Tract, Washington, *Bull. Geol. Soc. Am.,* 49, pp. 461-524 (p. 465), 1938.

[14] J. Harlen Bretz, The Channeled Scablands of the Columbia River Plateau, *Jour. Geol.,* 31, pp. 617-649, 1923; Alternative Hypotheses for Channeled Scablands, *ibid.* 36, pp. 193-223, 312-341, 1928; *Grand Coulee,* Am. Geog. Soc., New York, 1932; I. S. Allison, New Version of the Spokane Flood, *Bull. Geol. Soc. Am.,* 44, pp. 675-722, 1933; R. F. Flint, *loc. cit.* (13).

[15] P. G. Worcester, *Geomorphology,* p. 216, 1939.

ments, or combinations of these. While most physiographers seem to prefer to accept the more normal erosion by glacial melt-water, the scouring of deep, discontinuous basins and the lifting out of great blocks of basalt certainly strongly imply torrents of water.

The new hypothesis of glacial differential erosion favoured by Hobbs is supported by the discovery of a line of terminal moraines bordering the scablands area which he assumes has been covered by a glacial lobe, while beyond this again is the peripheral widespread aeolian loess-like deposit known as the "Palouse soil" of the wheat lands. Though few examples of the special features produced by ice-sheet erosion on lava fields are known, Hobbs points out that "the area overridden by the Icelandic glaciers [as described by Kjartonsson[16]] supplies a close parallel to the topographic development of the scablands."

Against the view of Bretz that the channels and their rock basins have been excavated by torrents of water [Hobbs cites] the recognised fact that no such features have been produced by running water in any other part of the world. The rock basins of the scablands are found in the wider channels particularly, and rock basins are an almost universal feature of glaciated regions. It is the channels with their included low mesas which are the unique feature of the scablands. If such features seem to be unique, it is almost unique for glaciers to have advanced over lava plains. Nowhere in the world except in Iceland has an ice sheet, either an ice cap or a continental glacier, been known to have overridden flat lava beds. The lava flows of the Columbia plains reveal within the same bed the widest differences of character. In one place the structure may look like a sponge: it is coarsely vesicular or scoriaceous. Elsewhere it is hard and firm; and the residual mesas have this character. The vesicular lava is especially vulnerable to the glacier's initial attacks by plucking, and after it had been quarried from the rock bed the rock must have been crushed and so made incapable of effective abrasion of the bedrock. The till of the moraines should therefore be poor in clay and characterised by smaller rock nodules, perhaps of pebble size. The vesicular facies of the lava flows are extended in the direction of flow, and

[16] G. Kjartonnson, *Med. fra Dansk Geol. Foren.*, 9, pp. 426-458, 1939.

if tapped from the preglacial valleys might well yield a channeled rock surface when the glacier had overridden it.

Ponding and Diversion of Rivers

Where volcanic mountains of whatever form are built up, and also where lava flows or lahars spread out to make plains which occupy valleys either wholly or in part but are not sufficiently extensive to bury the whole prevolcanic landscape, some characteristic effects are produced. Such interference with the course of the normal geomorphic cycle must be classed as a "volcanic accident." Diversion of rivers has occurred, in some cases on a very large scale, where volcanoes, perhaps chains of volcanic mountains, have established new consequent divides. The large Lake Van, in Armenia, is said to be held up by a lava dam formed by a flow from Nemrut volcano. So also Lakes Nicaragua and Managua have been ponded on the western side by a line of volcanoes (Fig. 106) some of which are still active, and drainage is now diverted so that it overflows eastward across a former continental divide. The largest of several lakes of this kind in Central Africa is Lake Kivu, ponded, with other smaller lakes, by the recently arisen volcanoes that form the Mufumbiro or Virunga Mountains, among which Nyamlagira is at present active.

> Formerly the drainage from this region flowed northward along the western rift-valley zone to join the Nile *via* Lake Albert. The volcanoes of the Nyamlagira group, however, erected a barrier across the rift valley and dammed back the waters to fashion a drowned valley system unique in Africa. The impounded water expanded as Lake Kivu and eventually escaped at the southern end into the Ruzizi, whence it now flows to Lake Tanganyika and . . . . to the Congo. Thus the Kivu drainage, even in an area of high relief, was switched from the Nile and Mediterranean to the Congo and the Atlantic.[17]

Another example is afforded by the volcanoes of the Roman Campagna, which have obstructed rivers formerly flowing directly from the Apennines to the sea.

[17] L. C. King, *South African Scenery*, pp. 153-154, 1942.

Hence their waters were constrained to gather in the longitudinal depression between the limestone mountains and the row of volcanoes. It happened that the lowest outlet was found between the third and fourth volcanoes, probably because they are farthest apart. There the gathered waters, forming a good-sized consequent river, ... the Tiber, ... eroded a valley . . . . and so reached the sea.[18]

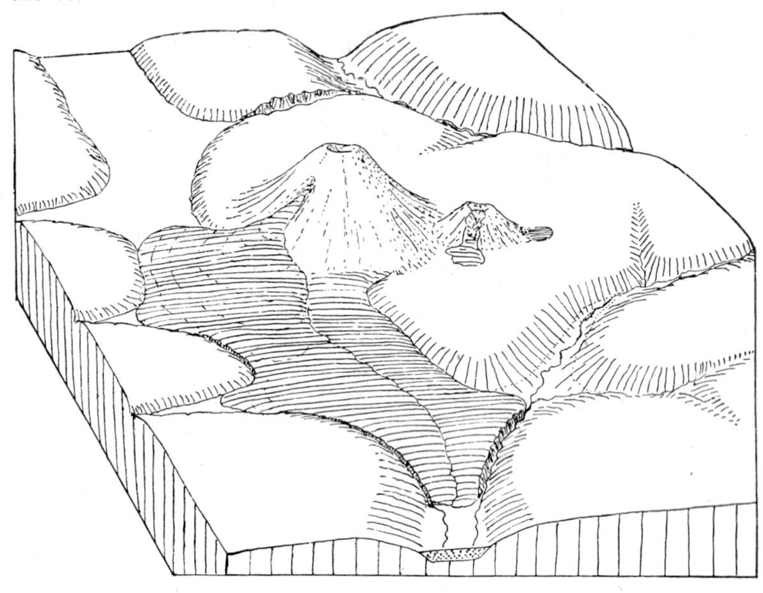

Fig. 183. A lava field in a valley ponds and diverts rivers. (After W. M. Davis.)

Many smaller pondings and diversions by lava-flow dams of the kind illustrated in Fig. 183 have taken place. In northern New Zealand, the upper Waitangi valley, for example, has been converted into Lake Omapere (Fig. 184), which has a westward overflow taking the place of the former eastward drainage by way of the Waitangi River.

In prehistoric (but legendary) times in Java a great lahar from the volcano Tangkoeban Prahoe impounded the water of a river with the formation of a temporary lake which found a new outlet.[19]

[18] W. M. Davis, The Seven Hills of Rome, *Jour. of Geog.*, 9, pp. 197-202, 1911.
[19] R. W. van Bemmelen, Ein Beispiel für Sekundärtektogenese auf Java, *Geol. Runds.*, 25, pp. 175-194 (179-180), 1934.

C. A. Cotton, photo

Fig. 184. Lake Omapere, formed by ponding of the former headwaters of the Waitangi River, New Zealand. The lava dam, a basalt plain surmounted by Te Ahuahu scoria cone, is at the right.

A lake ponded as a part of the process of stream diversion will eventually, like every other lake, disappear again from the landscape when it has become either filled in with sediment or drained by lowering of the outlet, and the fact that a river has here changed its course will become less obvious. The volcanic dam may survive, however, and long remain a conspicuous ridge or mesa, and the landscape may preserve traces—such as a silt-floored plain and possibly also shoreline cliffs and terraces —of the temporary existence of even a short lived lake. The diverted river also will still present in its new course some of the features of youth or, at any rate, will occupy a valley less maturely developed than others in the vicinity which have been in existence since before the diversion.

A lava-invaded valley in south-eastern Australia, that of Morass Creek, which is illustrated in Fig. 185, affords examples

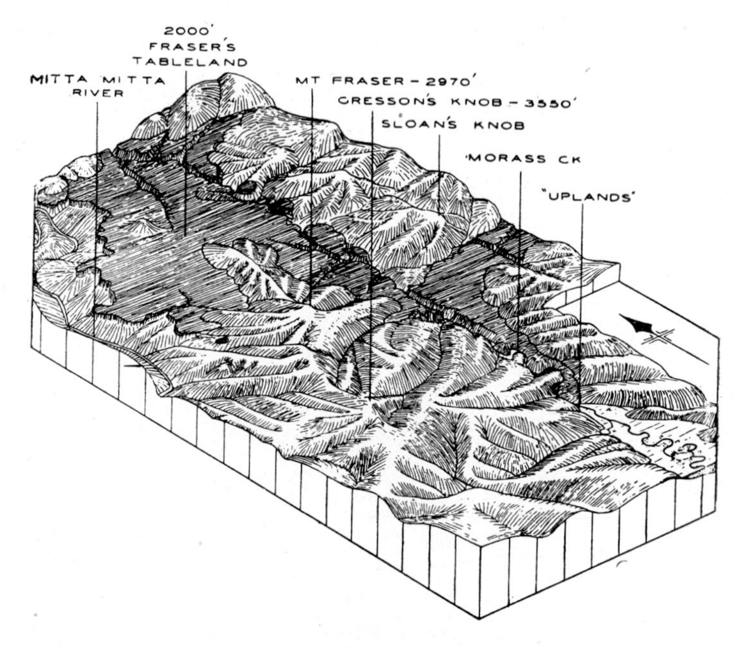

Fig. 185. The relation of the Morass Creek basalts, Victoria, to the prevolcanic landscape. (After E. S. Hills.)

of various features characteristic of ponded rivers and lava plains. As described by Hills:[20]

The walls of the pre-basaltic valley rise high above the lavas, which flowed both up and down the valley of the pre-existing stream. . . . . They also flowed up the valleys of tributary streams. . . . . [On resulting lava plains] the stage of dissection of the flows is youthful, [and] over extensive areas the initial surfaces of the upper flows are but little modified by erosion. . . . [Morass Creek, however, has cut a gorge] mainly along the eastern edge of the basalts.

A lake ponded by the lava in the main valley has been since drained by deepening of the outlet gorge through the lava dam, but the lake-floor deposits still remain as an extensive plain (Fig. 185, right). Large parts of this plain are still marshy, though dissection is beginning as a result of progressive deepening of the outlet gorge.

Volcanic Inversion of Relief

"Inversion of relief" in general is the replacement of valleys by divides and of divides by valleys. Volcanic inversion of relief[21] can take place where somewhat convex-shaped lava flows invade and usurp river valleys and become divides.

This case of invasion of river valleys, or of only partial burial of the landscape as a whole under lava, might be treated at some length as part of a series ranging from complete flooding to sporadic splashing, as it were, of the land surface with lava. It is not proposed here, however, to make a deductive picture of the volcanic landscape mosaics which may result from erosion after various degrees of discontinuous lava flooding. It will suffice to mention the useful Hawaiian term "kipuka" which has been introduced into geomorphology and used, especially by Thomson and by Stearns,[22] for the strips or

[20] E. S. Hills, The Cainozoic Volcanic Rocks of Victoria, *Trans. Roy. Soc. Vic.*, 51, pp. 112-139, 1939.

[21] So called by E. de Martonne (*Traité de géographie*, p. 527, 1909; 5th ed., p. 744, 1935). W. M. Davis has used the expression "inverted topography" (*loc. cit.* (18), p. 200; *Die erklärende Beschreibung der Landformen*, p. 325, 1912).

[22] J. Allan Thomson, The Geology of Western Samoa, *N.Z. Jour. Sci. and Tech.*, 4, pp. 49-66 (p. 52), 1921; H. T. Stearns, Geology and Water Resources of the Kau District, Hawaii, *U.S. Geol. Surv. W-S. Paper*, 616, p. 45, 1930. *Dagala*, the designation employed for kipukas on Etna is used as a synonym by F. A. Perret (*The Vesuvius Eruption of 1906*, p. 75, 1924).

patches of a land surface which escape burial by lava. In Savaii and Hawaii these consist of older lavas, but in general they expose the surface of any terrain, volcanic or prevolcanic, that has escaped burial under the most recent flows. A kipuka may or may not stand at a level above the surfaces of adjacent lava flows which more or less surround it; in some cases it is in a hollow between flows, a patch which they have capriciously avoided flooding. Convex kipukas surrounded by lavas so that they stand out like islands (Fig. 186) have been called "steptoes" by Russell.[23]

Lava Dams

Some small floods of mobile basalt that have solidified as lava plains in river valleys have formed low dams so as to impound shallow lakes (soon converted into marshy flats) but have failed to divert the overflow streams. Where such a flow

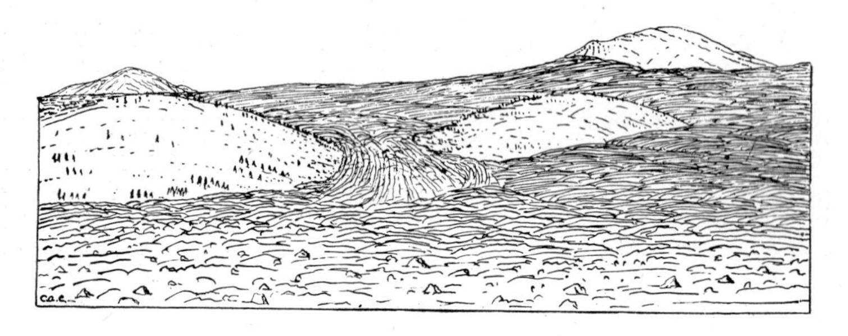

Fig. 186. Kipukas, dagalas, or steptoes of a granite terrain surrounded by basalt lava flows, McKenzie Pass, Washington. (From a photograph.)

has been insufficiently convex after solidification to displace the water stream by thrusting it over towards the valley side the ponded water overflows the dam, as has been the case rather commonly in northern New Zealand. A single flow of this kind, filling a narrow valley to a considerable depth, is generally massive (i.e. without pahoehoe tunnels) and holds up water. The river which takes its course across such a dam may develop falls of the Niagara type, which as they gnaw

[23] I. C. Russell, Preliminary Report on the Geology and Water Resources of Central Oregon, *U.S. Geol. Surv. Bull.,* 252, p. 78, 1905.

back headward develop a gorge through the obstruction (Fig. 187). Commonly the cap rock of such falls is an upper layer of the basalt which is free from joints, but beneath this there is a zone of columnar jointing, the rock of which is readily excavated by plunge-pool erosion.

C. A. Cotton, photo

Fig. 187. The Wairua Falls, developed where a river crosses a dam of basaltic lava, North Auckland, New Zealand.

Lateral and Twin-lateral Streams

It is a different matter when a convex-surfaced flow of non-basaltic lava, or even of basalt of the less fluent kind, invades a valley, for in this case the lava may become not merely a dam but also a divide immediately on its solidification. A stream of water flows beside instead of over the lava flow, or it may now separate two parallel consequent streams ("lateral" and "twin-lateral" streams of Hills[24]) in the re-entering angles between the convex lava surface and

[24] E. S. Hills, *loc. cit.* (10), p. 194.

the adjacent convex kipukas or pre-existing valley walls (Fig. 188). Courses adopted in this way may be deepened in pre-volcanic rocks (if these are less resistant than the lava) and eventually lateral streams in adjacent parallel valleys may coalesce as these are enlarged (one capturing the other by abstraction). Thus eventually valleys come to take the places

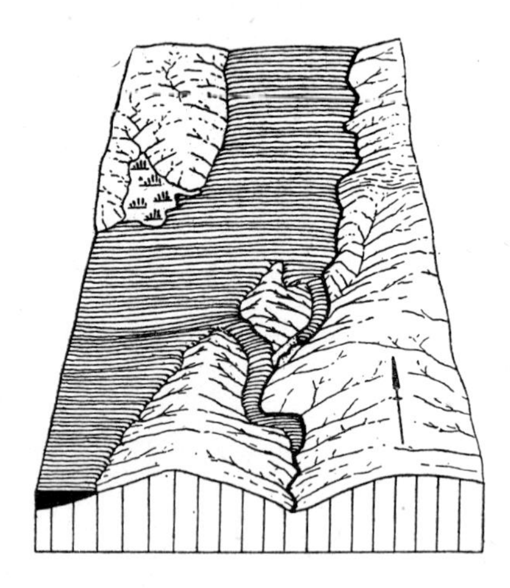

Fig. 188. A lateral stream after being ousted by a lava flow from the axis of a valley is eventually crowded out of it altogether and diverted, Plenty River, Victoria, Australia. (After Jutson and Hills.)

of former dividing ridges, while the lava flows which have taken possession of the former valleys are left standing above the general level as dividing ridges: inversion of relief is complete (Figs. 177, 189).

The common occurrence of alluvial gravels under lava residuals isolated by erosion is readily understood when it is remembered that the thickest parts of lava sheets and the strips likely to survive longest as landscape features are those which have flowed into and along valleys and filled river channels. Auriferous gravels buried in this way under lavas which have flowed into valleys and have later been reduced to ridges and mesas have been mined extensively in California and Australia, being termed "deep leads." Such cases of inversion of relief in

Tuolumne County, California, were long ago figured by Whitney.[25]

Residual lava ridges made prominent by inversion of relief, like those parts of extensive lava sheets that survive in mesas where general lowering of the surrounding land surface is in

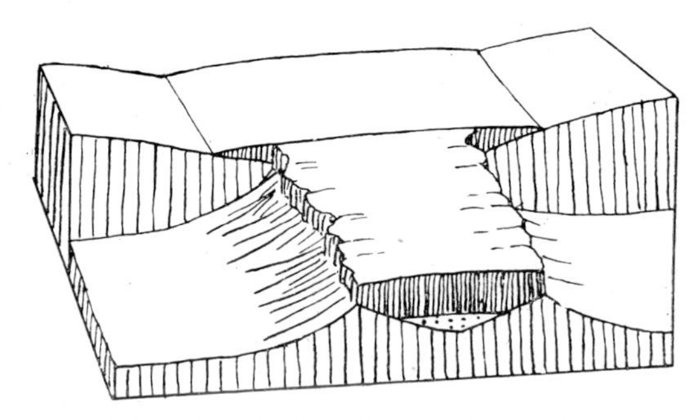

Fig. 189. By volcanic inversion of relief a lava flow in a valley has become a ridge. (After W. M. Davis.)

Fig. 190. Lava-capped ridge, illustrating volcanic inversion of relief, Kazbek, Caucasus. (After Abich and Suess.)

progress, are generally bounded by escarpments (Fig. 190). The cap rock of a basalt escarpment may be a massive (unjointed) layer among jointed or broken flows, a lava sheet overlying scoria, or the whole basalt formation (Fig. 189) where it overlies weaker rocks. Retreat of escarpments narrows each ridge, and eventually the last remnants of all high-perched

[25] J. D. Whitney, The Auriferous Gravels of the Sierra Nevada of California, *Mem. Mus. Comp. Zool. Harv.*, 6, 1879.

lava residuals will be destroyed in this way. Their imminent disappearance is indicated in Fig. 177. Much reduced lava residuals on a divide in south-eastern Australia, which are aligned along the course of an ancient river and thus afford an example of inversion, are shown in Fig. 191.

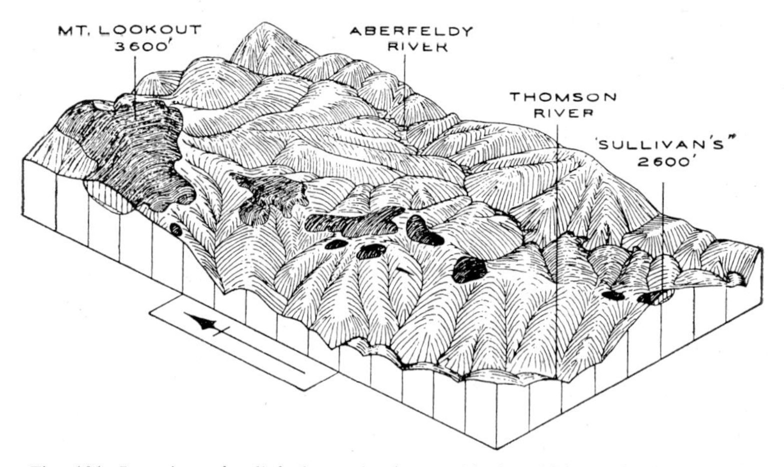

Fig. 191. Inversion of relief shown by lava residuals which mark the line of a former valley, Victoria, Australia. (After E. S. Hills.)

CHAPTER XVIII

Erosion and Destruction of Volcanic Mountains

THE DISSECTION OF VOLCANIC MOUNTAINS BY EROSION—TO BE followed, it may be, by their complete destruction—pursues the same general course as that of other large positive landscape features.

SMOOTH CONES

Some cones conserve the initial volcanic constructional form in an infantile, or as yet almost undissected, condition; and among these there are a few still coated by smooth-surfaced ash layers recently deposited as Vulcanian showers. Vulcan (New Britain) as it appeared when newly built (Fig. 109) exhibited such a smooth surface. Russell has described almost perfect preservation of the original constructional form on the cone of Agua, in Guatemala, an extinct volcano more than two miles high (Chapter XIII). Fujiyama, in Japan, is another high cone the smooth surface of which has often been remarked upon (Fig. 108), especially as the volcano has not been in eruption for more than two centuries. Though this immunity of the surface from gullying has been attributed to induration owing to an assumed cement-like quality of the ash,[1] it is quite probably due in the main, if not entirely, to coarse texture and permeability, as in the case of scoria mounds (Chapter X). In contrast with Fujiyama, Mayon in Luzon (Philippine Islands) assumes a smooth form when newly coated with ash after an eruption but soon becomes deeply furrowed as numerous consequent gullies are cut in the fine and relatively impermeable ash.

[1] G. F. Becker, The Geometrical Form of Volcanic Cones, *Am. Jour. Sci.*, 30, pp. 283-293, 1885.

Such furrowing of ash slopes generally begins with the first rain, and the course of consequent dissection on cones is similar to that which may be observed on shower deposits veneering and conforming to any slopes (Fig. 96). The drainage, distributed at first in innumerable rills, is later gathered into a relatively small number of the consequent streams which gain the mastery (Fig. 192) and cut deep ravines—"barrancos" of

Stewart and White, photo

Fig. 192. The small active volcano White Island, largely ash-built, showing mature dissection of outer slopes of the cone, enlarged and breached crater, and lahar hummocks on the floor of the crater.

German authors, "barrancas" of some American authors (Fig. 193). The pattern of consequent stream-cut ravines is still a very simple one—radial and centrifugal on the cone—following the initial slopes. The barrancas are bottle-necked, and remnants of the constructional slope survive longest between their shallow debouchures. These are features characteristic of a water-eroded slope.

Fig. 193. Deeply-cut ravine (barranca) on the flank of the large active volcano Popocatapetl, Mexico, above the timber line.

Initial Relief on Ash-cone Slopes

The parasol pattern of radial gashes and ribs, as developed on some growing cones by hot avalanching of ash has been described in Chapter XIII. These gashes maintain a more uniform depth than stream-cut ravines (p. 362). Ravines are excavated also on some cones during eruptive episodes by the erosive action of block flows and glowing avalanches which originate as *nuées ardentes.*

Clean-swept and striated rock surfaces, which Perret[2] has observed and described, testify to the abrasive power of the coarser debris swept along by the glowing avalanches of Martinique. The gap through which innumerable *nuées ardentes* have issued from the ring crater of Mont Pelée has been enlarged by this erosion until further enlargement has been checked by extremely resistant lava masses at either side of it which Perret has termed the "gate-posts" (Fig 81, at left). Perhaps most of the gaps in crater rings, largely ash-built, through which glowing avalanches escape, have been initiated by lateral thrust exerted by spreading tholoids in the manner indicated by Wing Easton (Chapter XII), who describes the gaps as the heads of what he terms "pressure valleys."

[2] F. A. Perret, *The Eruption of Mt Pelée 1929-1932,* p. 76 and Fig. 53, 1935.

Long and deep gashes have been eroded by glowing avalanches on the steep upper slopes of the sharp cone of Merapi (Java) during the eruptions of recent years (Chapter XII). Confined as narrow streams in these ravines, the successive glowing avalanches are thus concentrated so as to emerge at certain points commanding the lower slopes, over which they then fan out widely. Some ravines eroded in this way are later choked by hot-rock slides and ladus (block flows).

All landslide and avalanche gashes that remain on a cone after activity ceases will provide consequent courses in which drainage is concentrated, and they will be rapidly deepened thereafter by water erosion.

Inversion of Relief on Cones and Domes

On volcanoes which occasionally emit lava to flow down and solidify as tongues in radial consequent valleys the pattern of consequent streams changes from time to time during slow volcanic growth. Such changes in pattern afford an instance of volcanic inversion of relief (Chapter XVII); for the ravines of one stage become divides in the next if they are then

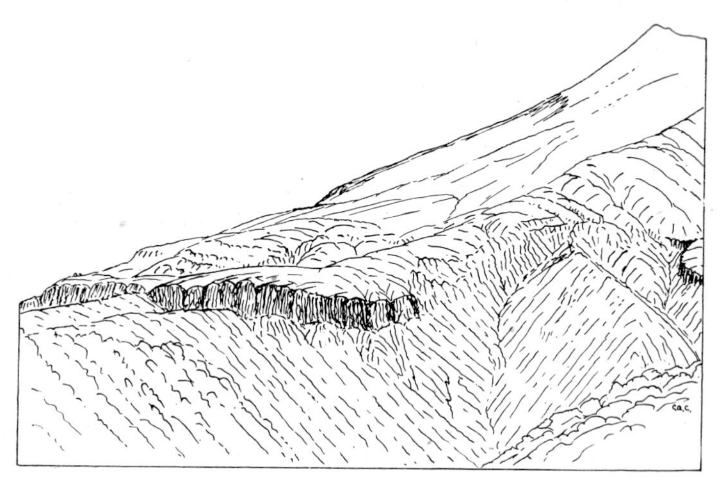

Fig. 194. Convex lava spurs on Egmont volcano, New Zealand, showing the development of inversion of relief.

occupied by convex lava tongues with sufficient relief to form ridges, and these separate new sets of valleys which are deepened by new consequent streams on the sites of former divides (Fig. 194).

A classic example of such reshuffling of radial consequent features is found on the flank of the ancient Alban Hills volcano, near Rome. As described by Davis, numerous individual cases of inversion are seen:

> The railroad crosses the new valleys on embankments and passes through the ridges in cuts. . . . On entering and leaving a cut the thin border of the firm columnar lava is seen lying on the red weathered tuff, while in the middle of the cut the base of the lava is not revealed; that is, the depth of the cut is there less than the thickness of the lava stream; hence the lava stream occupies an ancient valley.[3]

Inversion of relief takes place as a result of flank outflows of lava on Hawaiian basalt domes, but in such cases generally a new flow, which takes the form of a very broadly convex ridge, is guided not by an eroded valley but merely by the shallow depression left uncovered by other recent flows or the re-entering angle made where one laps over the edge of another (Fig. 47).

Planeze Stage of Dissection

At the submature stage of dissection of a cone or of a basalt dome dwindling sectors of the constructional surface survive on the ridges between deeply-eroded major consequent valleys (Figs. 195, 196). Such sectors are termed "planezes" (from the French *planéze*[4]).

[3] W. M. Davis, The Seven Hills of Rome, *Jour. of Geog.*, 9, pp. 197-202, 1911.

[4] E. de Martonne: "Nom commun qui merite d'être retenu" (*Traité de géographie physique*, p. 529, 1909; 5th ed., pp. 746-748, 1935). "Planeze" conveniently replaces such expressions as "short-sector ridge," "conic remnant of the constructional surface," and "sloping triangular facets which are little dissected remnants of the constructional surface of the original lava dome," which are used by Davis, Hinds, and Wentworth respectively—W. M. Davis, *The Coral Reef Problem*, pp. 244, 254-255, 1928; N. E. A. Hinds, The Relative Ages of the Hawaiian Landscapes, *Univ. Cal. Publ., Bull. Dept. Geol. Sci.*, 20 (6), p. 254, 1931; C. K. Wentworth, Soil Avalanches on Oahu, Hawaii, *Bull. Geol. Soc. Am.*, 54, pp. 53-64 (p. 55), 1943.

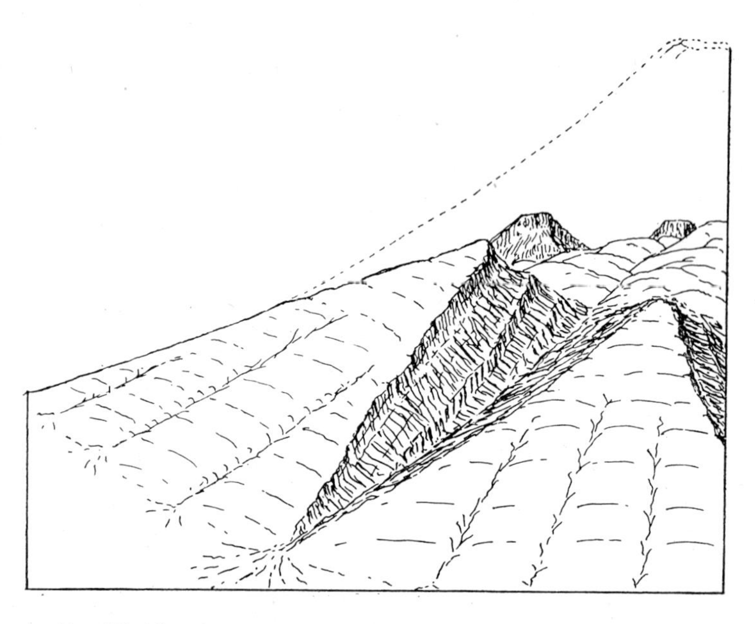

Fig. 195. The planeze stage of dissection of a composite volcanic cone.

Fig. 196. Submature dissection, western slope of Mont Pelée, Martinique, showing a planeze at the left. (After a photograph by Tempest Anderson.)

At an early stage of planeze development, on either a cone (Fig. 195) or a dome (Fig. 197) the planezes may assume a prominent flat-iron form bounded on two sides by escarpments, while the constructional or dip slope of each such sector runs up to a point, beyond which the planezes are either

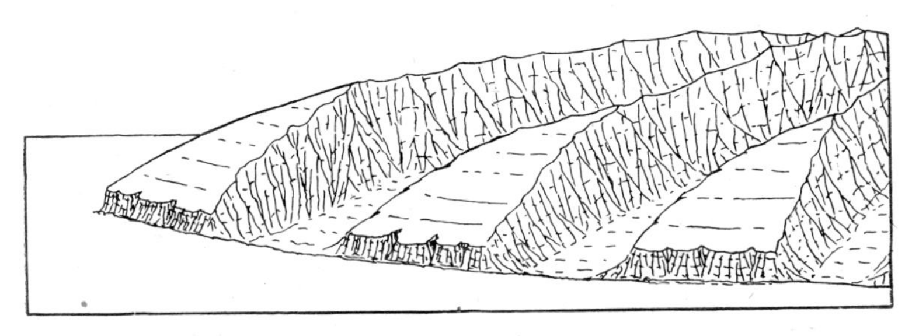

Fig. 197. Submature, or planeze, stage of dissection of an island which has been initially a basalt dome.

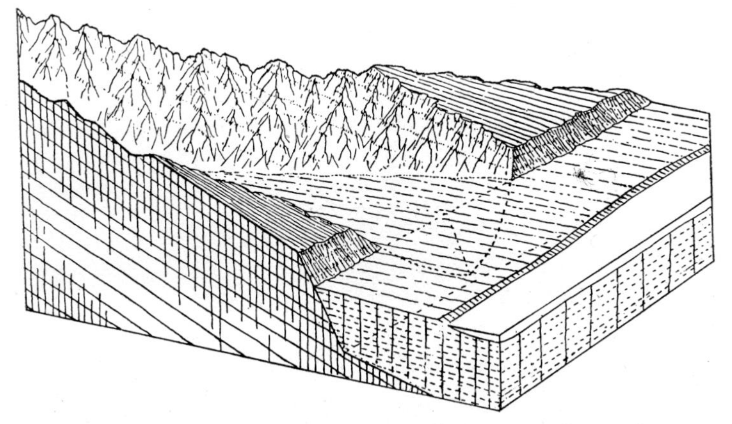

Fig. 198. A planeze (cliffed by marine erosion) bordering Wailupe valley, eastern Oahu, Hawaiian Islands. (After W. M. Davis, *The Coral Reef Problem*, Am. Geog. Soc. N.Y.

continued in sharp-edged ridges towards the volcanic centre (as in the dissected volcano of the Cantal, France, the type locality for planezes[5]) or may overlook an interior hollow developed by more prolonged erosion of the volcanic centre (as deduced in Fig. 195).

[5] E. de Martonne, *loc. cit.* (4) (1935), Fig. 294.

Examples of planezes in all stages of development and destruction are present on dissected basaltic domes of the Hawaiian type (Fig. 198).

Volcanic Skeletons

At full maturity of dissection the initially conical or domed shape of a volcanic mountain has been destroyed by erosion and it has been reduced to a group of hills. In this hill group,

Photo by Geo. Grant; by courtesy of W. W. Atwood Jr.

Fig. 198A. Dykes that intersect the ancient volcano Mazama are beginning to stand out as hogbacks from the crumbling wall of Crater Lake caldera, Oregon.

however, an arrangement of radial ridges and a centrifugal pattern of consequent valleys may still survive, as is the case in the Euganean Hills, in north Italy[5a]; and it is possible for such a pattern eventually to be superposed, at least in part, on a basement or undermass of prevolcanic rocks with alien and perhaps complex structure after deep and widespread denudation has removed all the volcanic cover. Before the destruction of a volcano is complete in such a case, however,

[5a] A. Penck, *Morphologie der Erdoberfläche,* 2, p. 428, 1894.

harder layers and ribs, bedded lavas and interbedded sills of the cone and dykes radiating from the central pipe (Fig. 198A), may stand out as homoclinal and hogback ridges, which will give a definite pattern to the relief features of this stage—the "volcanic skeleton." As a rule the radial dykes live longest as landscape forms, for they may intersect the undermass as well as the volcano itself (Fig. 199).

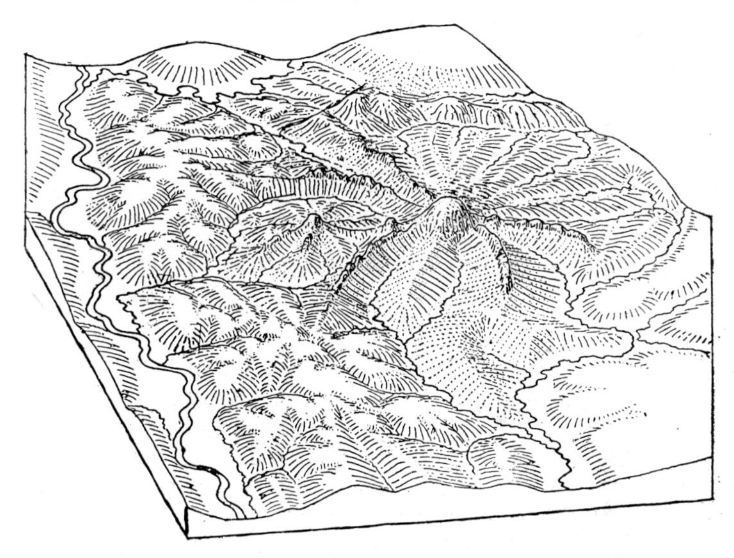

Fig. 199. Volcanic skeletons, consisting of necks flanked by radial dyke-made hogbacks, as exposed after erosion has removed all overlying beds of lava and tuff and long after all surface forms of volcanic construction have been destroyed. (After W. M. Davis.)

Necks

Equally prominent as features of volcanic skeletons are isolated hills—in some cases conspicuous craggy pinnacles—which stand out at those places where circumdenudation leaves standing in relief resistant plugs or fillings of former volcanic pipes. These are "necks" (Figs. 199-203).

It is possible for both necks and dyke-made hogback ridges not only to remain prominent as landforms after the removal of all superficial volcanic rocks of contemporaneous origin but also to persist or to reappear in successive cycles during progressive lowering of the land surface in later ages. Many necks, however, especially some of broad butte-like form, are only slightly denuded plug domes and especially tholoids which have occupied broadly opened funnel-shaped volcanic throats

C. A. Cotton, photo

Fig. 200. The Tokatoka volcanic skeleton, North Auckland, New Zealand, which consists of a prominent neck buttressed by hogbacks on the outcrops of dykes.

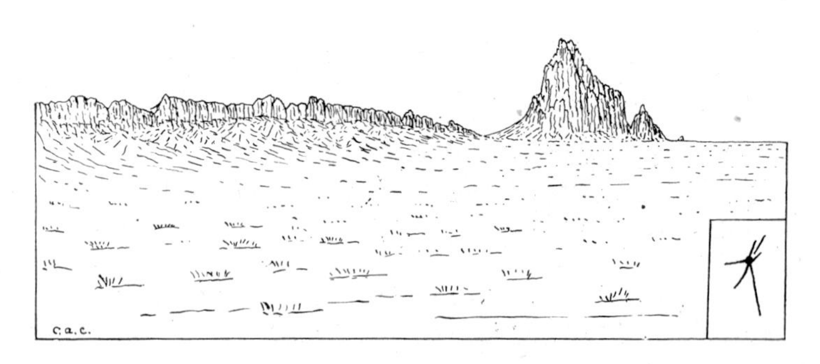

Fig. 201. The Ship Rock volcanic skeleton, New Mexico, showing the central neck and the longest of a group of radiating hogbacks (inset map). (From a photograph.)

Fig. 202. The neck Rocher Saint Michel, at Le Puy, France. (From a photograph by Tempest Anderson.)

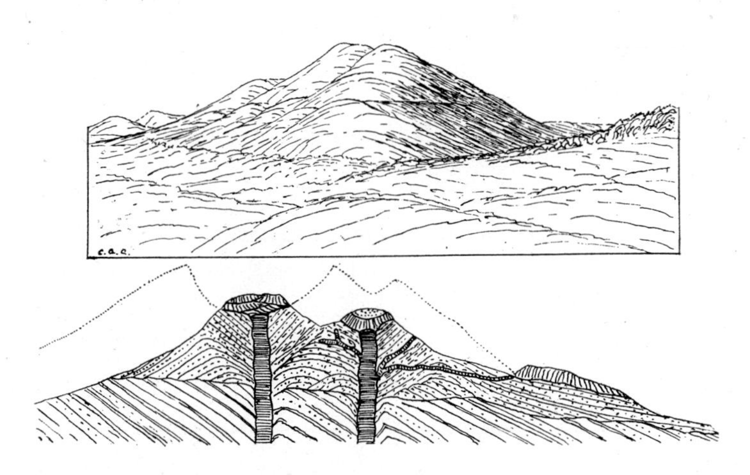

Fig. 203. The neck Largo Law, Scotland, as interpreted by A. Geikie.
Above: Largo Law (outlined from a woodcut).
Below: The structure of Largo Law and a suggested restoration of the profile of the original volcano (after Geikie).

of considerably larger diameter than the deeper root portions of the pipes. This is the "Hopi" type of neck[6] (Fig. 204A).

Some similar necks consist of the breccias and lava benches that have accumulated in the sinks and funnel-shaped enlarged pit craters of basalt volcanoes (Chapter IV), as in the case of some of the ancient (Permian) volcanic remnants in Scotland (Largo Law, for example) which have been described by Geikie.[7] These, though ancient, have been preserved owing to a combination of circumstances in a very slightly denuded condition (Fig. 203). Very similar forms and structures have been described in the Siebengebirge, Germany; Saddle Hill, Dunedin, New Zealand, is a denuded Pliocene volcano in many respects resembling Largo Law;[8] and there are examples of necks of this kind also among the eroded basaltic formations of Oahu, Hawaiian Islands. Here

> several plugs in the Waianae Range taper downward. Some are fills in the craters of cinder cones. . . . In the caldera complex at the head of Lualualei Valley there are massive cup-shaped bodies, identified as crater fills, like the present filling in Halemaumau Crater of Kilauea Volcano, because they rest on talus breccia.[9]

An extensive cluster of necks of various sizes—some very large—associated with the Mount Taylor basalt volcanic field, in New Mexico, which have been described by Douglas Johnson,[9a] are perhaps pit-crater fillings that taper downward, though they have some appearance of being cylindrical pipe-fillings. A number of them occupy enlarged portions of feeding-dyke fissures in the prevolcanic terrain.

Though some are wholly or partly agglomerate, others of the Mount Taylor necks consist almost entirely of thick bodies of solidified lava. In Cabezon Peak, the largest and most

[6] Howel Williams, Pliocene Volcanoes of the Navajo-Hopi Country, *Bull. Geol. Soc. Am.*, 47, pp. 111-171, 1936.

[7] A. Geikie, Geology of East Fife, *Mem. Geol. Surv.*, 1902; *Textbook of Geology*, 4th ed., pp. 749-752, 1903.

[8] W. N. Benson and F. J. Turner, Mugearites in the Dunedin District, *Trans. Roy. Soc. N.Z.*, 70, pp. 188-199, 1940.

[9] H. T. Stearns, Geology and Ground-water Resources of Oahu, *T.H. Div. Hydrog. Bull.*, 1, p. 21, 1935.

[9a] D. W. Johnson, Volcanic Necks of the Mount Taylor Region, New Mexico, *Bull. Geol. Soc. Am.*, 18, pp. 303-324, 1907.

prominent of the group, one such mass is 1400 feet in diameter. It is conspicuously divided into large columns by vertical cracks due to shrinkage, a structure the presence of which has led some investigators to the conclusion that igneous-rock bodies exhibiting it (including the phonolite peak Devil's Tower,

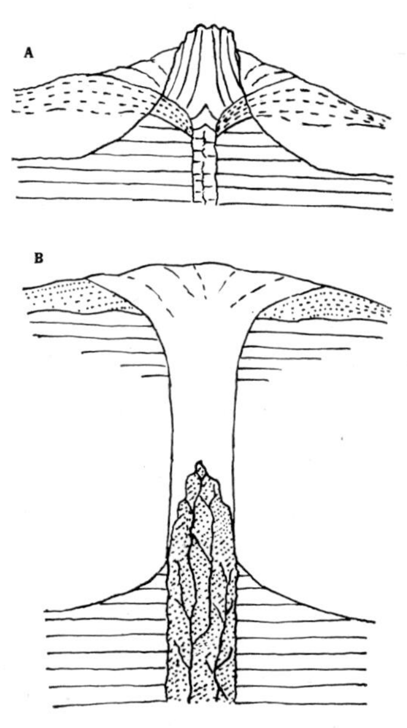

Fig. 204. Two types of neck found in the Navajo-Hopi country, New Mexico (after Howel Williams).

A: "Common type of Hopi neck; diverging columns of lava resting on inward-dipping tuffs and surrounded by . . . sediments" (Williams).

B: "Common type of Navajo neck, Agathla; shaft of tuff-breccia riddled with dykes; . . . conduit probably terminated at surface in an explosion pit or maar" (Williams).

Wyoming, figured in many textbooks) are buttes which are the remnants of thick sills or laccoliths. Such columns develop perpendicular to a cooling surface, and so, with little doubt, these bodies have been lava lakes in pit craters. Below ground such columns are commonly horizontal, where cooling has resulted from conduction through an enclosing wall of perilith. The wall of a lava-feeding pipe or fissure may be at much the same temperature as the lava in it, however, and so the lava is not necessarily chilled by contact with the wall. In the Great Neck, Mount Taylor field, and also in Devil's Tower, though the columns are vertical in the main, they are seen to curve outward towards the horizontal below.

In contrast with necks of the Hopi type the "Navajo" type of Williams (Fig. 204B) is exposed by deeper erosion. It is a spire thus isolated which consists of the filling in the narrower root portion of a volcanic pipe. For such a filling, as Daly remarks, the name "neck" is appropriate, for it has joined "the 'head,' the broad prism of lava in the crater, to the 'body,' the subterranean magma chamber."[10] Such a neck consists of either solidified lava or a well-cemented breccia of volcanic fragments (agglomerate) generally strengthened by dykes. Williams has selected as the type of this form the towering pinnacle Agathla,[10a] in the Navajo Pliocene volcanic field, the relief of which is 1000 feet. Such are the necks that form the central features of some deeply denuded volcanic skeletons. They are never of very large dimensions, and are thus in no danger of confusion with stocks and other like intrusive masses where these have been exposed in relief by differential erosion. Various examples of necks of the Navajo type have been figured erroneously, however, as stocks in textbooks of geomorphology. There is no standard size for volcanic pipes, but Daly[11] describes those necks that are governed as to size by pipe diameters as "always small, even minute, as compared with the sections of the larger intrusive bodies." It is, indeed, the small lateral dimensions of such necks that give them their prominence. Larger intrusive bodies may be steep-sided, and erosion of soft rocks along contacts may leave them standing with walls that resemble fault-line scarps, but circumdenudation commonly fails to reveal the outline of a large body of igneous rock so completely as to make a conspicuous landform at all comparable to a neck or to an exposed dyke.

[10] R. A. Daly, *Igneous Rocks and the Depths of the Earth,* p. 148, 1933.

[10a] See photograph in N. E. A. Hinds, *Geomorphology,* p. 316, 1943.

[11] R. A. Daly, *loc. cit.* (10), p. 149.

Dissection of Basalt Domes

Basalt domes are dissected by valleys, chiefly consequent, that are generally steep-sided and have cirque-like amphitheatre heads.[12] In addition to valleys of the kind normally found entrenching themselves in lava plateaux, especially in cool and comparatively dry climates, there are others of a

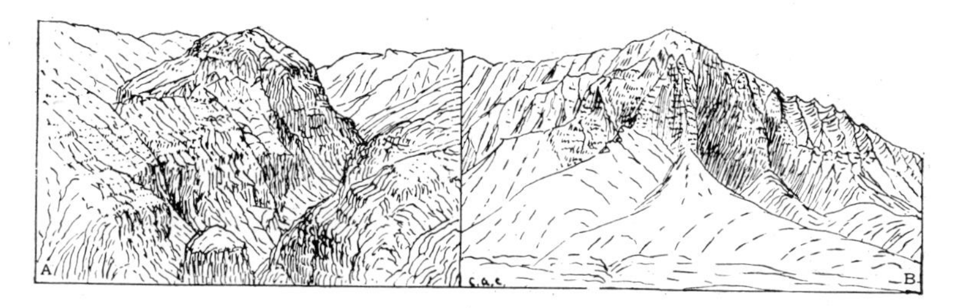

Fig. 205. Contrasting types of dissection in Hawaiian basalt domes.
A: Mature dissection with development of structural escarpments, southern Kauai (Hawaiian Islands).
B: Fluted side wall of U-shaped valley of the Oahu type, Kolalau Valley, north-western Kauai. (From photographs by O. S. Emerson.)

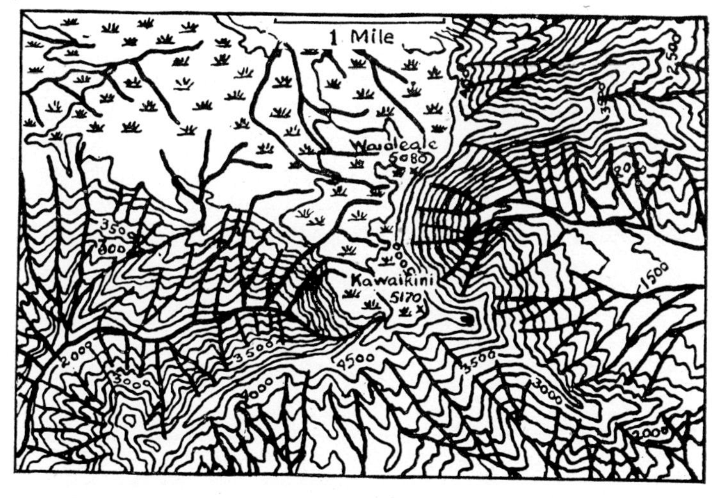

Fig. 206. "Amphitheatre valley heads eroded into an extinct lava dome. Redrawn from U.S. Geol. Survey topographic map of Kauai." (After N. E. A. Hinds.)

[12] N. E. A. Hinds, Amphitheatre Valley Heads, *Jour. Geol.*, 33, pp. 816-818, 1925.

distinct type. The former separate mesas, planezes, and ridges bounded by escarpments which present no unusual features; residual peaks are commonly of stepped-pyramid form; and the valleys are bordered in parts by structural terraces (Figs. 28, 205A, 217). Other valleys, however, are more distinctly U-shaped in transverse profile, with broad floors, steep side walls not stepped but vertically fluted (Fig. 205B), and very conspicuously amphitheatre-shaped heads (Figs. 206, 207A).

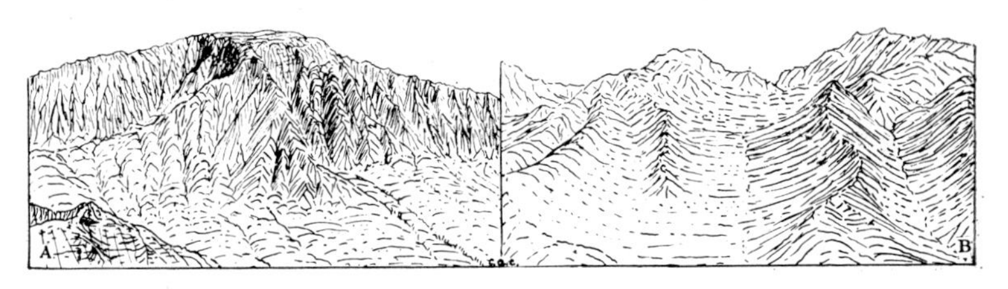

Fig. 207. Dissection of the island of Oahu, Hawaiian Islands.

A: Small remnant of a plateau—part of a constructional dome—the highest point on Oahu (Mount Kaala) surrounded by amphitheatre valley heads (compare Fig. 206).

B: The U-shaped Nuuanu Valley, Honolulu, showing also intercatenary ridges. (From photographs.)

These valleys separate ridges with intercatenary profiles (Fig. 207B) like those of ridges between closely spaced glacial troughs, from which isolated pinnacles like Alpine horns may rise. Such valleys, which are common in the Hawaiian Islands, may be classed as the Oahu type. In the islands of Kauai, Molokai, and Maui they are youthful and still in the process of vigorous development, but in Oahu they have reached full maturity and even old age.

Origin and Development of Oahu Valleys

The radial pattern of Oahu valleys indicates that they are consequents. For this and other reasons they seem to be entirely different in origin from the somewhat similar amphitheatre-headed (but spring-guided and insequent) valleys termed "alcoves" which are tributary to the Snake River in Idaho (Chapter XVII).

It appears that autogenetic falls have developed at an early stage of the dissection of the dome and that retreat of these at the head of a newly opened inner valley has proceeded as a kind of autogenetic rejuvenation which might almost be thought of as fluvial "overdeepening." This has been accompanied by lateral enlargement of the inner valley to a broadly U-shaped or catenary trough form by a wall-retreat in which special processes have been active. The steep side and head walls of the troughs are fluted in a pattern of close-set spurs and ravines, or chutes, which are so precipitous in some cases as to merit the description "vertical valleys" sometimes applied to them; and it is clear that this pattern (Figs. 205B, 207A, 208) is a by-product of the processes that are enlarging the trough both sideward and headward.

As they were explained by Palmer,[13] near-vertical flutings are solution forms, or lapiés on an enormous scale. (Lapiés on the usual small scale, similar to those commonly found on limestone outcrops, occur also on basalt in warm climates and are well developed on some outcrops in Oahu.) Stearns[14] is not in agreement with the solution theory; but, according to Palmer's explanation,

> the small but persistent run-off . . . . supplies a gentle, steady flow of water. . . . The water cannot . . . . fall freely and develop a high velocity, and it is therefore not a strong agent of mechanical erosion. The work of erosion must be done chiefly by solution.

The contrasting view of Stearns is that erosion is here mainly mechanical, taking place when, at rare intervals, "thundering cascades" fill the grooves, scooping out plunge pools in a succession of steps.

Such steep grooves are rather exceptional forms, being limited in their occurrence to near-vertical valley-head walls, which are possible only where rocks are not thoroughly weathered. A somewhat similar pattern of flutings on the upper side-slopes of many Oahu valleys (Fig. 208), which, however, ascend at an inclination rarely much steeper than 45°, is

[13] H. S. Palmer, Lapiés in Hawaiian Basalts, *Geog. Rev.*, 17, pp. 627-631, 1927.
[14] H. T. Stearns, *loc. cit.* (9), p. 62.

attributed by Wentworth[15] to erosion effected by a process of shallow landsliding which from time to time (during very heavy rains) cleans out each gully, or chute, by removing the soil, a foot or two in depth, together with its clothing of forest vegetation, as a "soil avalanche." As the whole of the upper valley-side is here forest-clad, soil is held by the mat of roots on steep slopes, and the fluted slope is maintained at an angle of

C. K. Wentworth, photo

Fig. 208. Typical forest-clad, fluted upper valley-side slope of an Oahu Valley (Kalihi Valley).

about 45° during gradual retreat. (Slopes rather steeper than the average are found at the heads of the chutes, however.) Chemical weathering develops new soil on scars which have been bared by landsliding, and they are colonised by vegetation, which holds the accumulating soil in place until the cycle is repeated by the occurrence of another slide. The sharpened spurs between the deepened chutes are progressively reduced in

[15] C. K. Wentworth, Soil Avalanches on Oahu, Hawaii, *Bull. Geol. Soc. Am.*, 53, pp. 53-64, 1943.

height by the same process, and the whole valley-side slope retreats horizontally without appreciable change of form.

According to an earlier theory proposed by Wentworth[16] to account for the broad floors and also the rather gentle gradients characteristic of Oahu valleys these forms result from control of the erosion level by ground water, above (but only above) which destruction and removal of rock are easy

U.S. 18th Air Base Lab., photo

Fig. 209. Halawa Falls, Molokai Island, Hawaii, "showing that composite plunge pools made by waterfalls form an amphitheatre-headed valley" (Stearns).

because the whole terrain is weakened by chemical decay. This explanation of the valley forms is not entirely satisfactory, however, as the ground-water level in Oahu is in most parts far below the valley floors, being indeed but little above sea-level and ascending inland at a gradient of no more than three feet per mile, whereas the valley floors ascend at 40 or more feet per mile.[17]

Oahu valleys are developed only under an extremely heavy rainfall; and rapid tropical weathering is also essential. These

[16] C. K. Wentworth, Principles of Stream Erosion in Hawaii, *Jour. Geol.*, 36, pp. 385-410, 1928.

[17] H. T. Stearns, *loc. cit.* (9), p. 24.

Fig. 210. Sharp and serrate ridge at the intersection of valleys of the Oahu type at their amphitheatre heads, Nuuanu Pali, near Honolulu, Oahu. The wall-like "Pali" at the right is fluted with "vertical valleys."

are the conditions under which generally concave valley-side slopes and knife-edge divides become the rule on almost every kind of terrain.

According to Stearns,[18] the gathering in of headwater tributaries by abstraction in the course of an early struggle for existence among approximately parallel consequent streams on a rain-soaked surface of considerable slope (the flank of a Hawaiian dome) prepares a master stream thus developed to

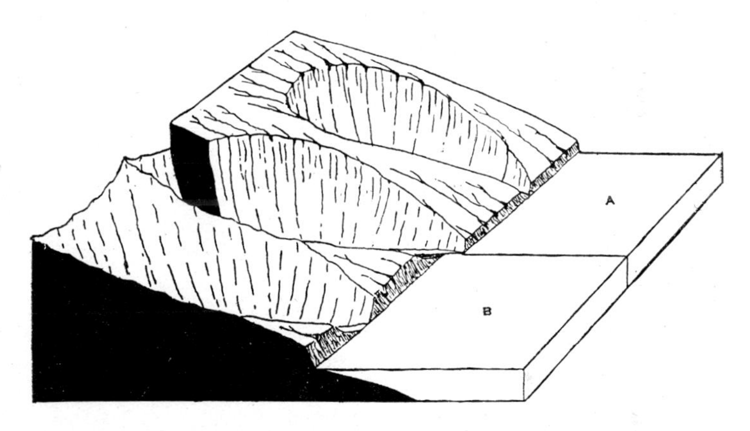

Fig. 211. Two stages in the development of valleys of the Oahu type on the slope of a dome. (Redrawn after diagrams by Stearns.)

take up the task of excavating a steep-headed trough (Figs. 209; 211, A). Small valleys developing somewhat in this manner in the badlands on the ash erupted from Lake Rotomahana, in New Zealand, in 1886 are miniature examples of Oahu valleys at this stage. They have been deepened in a similar way, developing steep heads (Fig. 96). A rough analogy may be traced also between such a valley head and a trough-end in a glaciated valley, which has resulted from overdeepening below the confluence of a group of convergent valley-head cirques.

The next stage is characterised by the processes of slope (or wall) retreat already referred to, which enlarge the "rejuvenated" or "overdeepened" part of the valley (now a

[18] *Loc. cit.* (9), pp. 24-25.

trough) and extend it headward (Fig. 211). It is at this stage that the troughs are widened by the "soil-avalanching" process described by Wentworth, which has already been referred to. According to Stearns, rapid extension of the trough headward takes place by vigorous plunge-pool erosion. This process can continue as long as the tributaries collected by abstraction still pour into the head of the trough; but it must be slowed down when the trough head has been sapped back to a sharpened divide, unless, as is sometimes the case, there is still available a supply of perched or high-level ground water which continues to issue from springs high in the valley head.[19]

Fig. 211A. Coalescing valley heads make a scalloped cliff (the Pali) at the rear of the planed lowland of north-eastern Oahu, Hawaii. The cliff is sculptured in detail by "vertical valleys". (From a photograph.)

While there is still a gathering ground for surface streams which enter the trough at its head these will continue to erode and descend by "vertical valleys," or steep chutes, which flute the head-wall. These flutings (Fig. 209) are not strictly giant lapiés developed as pictured by Palmer, but somewhat resemble them. Such "vertical valleys" are characteristic only of the stage (Fig. 211, B) at which trough heads intersect in very steep-sided knife-edge ridges, or "arêtes."

[19] E.g. along the migrating asymmetrical divide which forms the Pali of Oahu. (H. T. Stearns, *loc. cit.* (9), p. 28.)

The process of headward extension of the inner trough is as follows:

> Captured tributaries entering a master stream form a rim of coalescing plunge pools about the amphitheatre wall at the break in the stream gradient. The narrow ridges between the plunge pools are undercut and fail by their own weight (STEARNS).

It must be remembered that the material which is thus collapsing, like that beside it which is in course of excavation by plunge-pool erosion, is basalt that has long been above the water table and is thus already weakened by chemical decay.

In some valleys of the Oahu type there are perennial streams which are fed in part by ground water locally held up at high levels by dykes, and also by springs issuing from bodies of perched water, as well as by the run-off from rain-soaked uplands. Such streams are vigorous enough to transport much gravel and thus to dispose of the debris of rock falls and landslides, especially that constantly supplied at the head of a rapidly-developing trough. Many valleys on the island of Oahu, however, are heavily encumbered with talus, which has accumulated as a thick apron along each trough side. This accumulation has been attributed by Stearns[20] to a relict condition due to a shrinkage of rainfall which has followed as a consequence of erosional reduction of relief and which has made the streams in the troughs underfit. The talus is very thoroughly decayed by chemical weathering, and it may be that removal of such material in valleys the floors of which are above the level of ground water takes place mainly in solution.

A continuation of the process of wall retreat not only sharpens the intercatenary ridges between Oahu valleys (Fig. 211, B) but eventually destroys them. Peneplanation by some such process has eliminated a former strong relief along the north-eastern seaboard (the windward and therefore the wettest and most rapidly eroded slope) of Oahu. Here the valley-heads, still extremely steep and now coalescing owing

[20] H. T. Stearns, *loc. cit.* [(9)], p. 26.

to the destruction of the ridges between the lower valleys, make a continuous imposing cliff, or "pali" as it is termed (Figs. 210, 211A), which some observers have mistaken for a fault scarp. The crest-line of this "pali" forms a divide which is still migrating and consuming the dissected remnant of a dome.

Erosion Calderas

The central part of a basaltic mountain may be hollowed out by erosion so that the crown of the dome is replaced by either a large amphitheatre or a system of branching valleys. This is an "erosion caldera"; and the hollow-crowned condition has its place as a definite stage in the erosional destruction of some domes (Fig. 212).

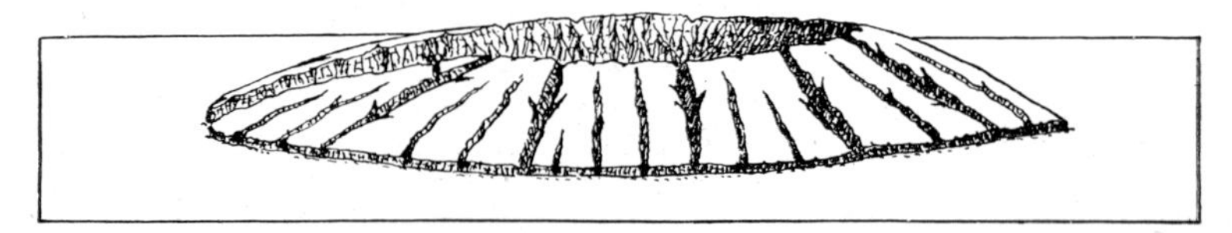

Fig. 212. A basaltic dome with the centre hollowed out by erosion. The ring ridge around the erosion caldera so formed may be broken by a few notches at the heads of beheaded radial valleys as well as by the gap formed by the bottle-necked outlet gorge, or barranco.

As the hollow-crowned stage is not attained, or passed through, by all basaltic domes which undergo dissection, it seems possible that some favourable condition or structure may be necessary for the development of an erosion caldera. A favouring condition may be the presence of an eccentric, or perhaps central, caldera which has been formed by subsidence as a sector graben or a sink (Chapter XVI), the hollow so formed being later much enlarged by erosion. In other cases it may be, however, that no such initial hollow has been present as a guide or control for consequent erosion, but that weaker rocks underlie a dome-building outer shell of resistant lava flows. Such a structure facilitates the erosional retreat of an escarpment at the inner edge of a more or less continuous ring ridge.

Even without such aids it may be possible for the processes of headward erosion which develop escarpment-bounded and amphitheatre-headed valleys in basalt terrains to burrow into the heart of a dome and there expand the head of a master valley.

An erosion caldera may be formed by coalescence of two or more valley heads as the divides between these are sapped away and broken down. In such a case there may be several outlet valleys—as, for example, in the case of the hollow-centred basaltic island of Huaheine (Fig. 217)—but the most perfect examples of erosion calderas have one outlet only—that is to say, the destruction and removal of the crown of the dome have been effected by the streams of a single system. The central lowland which is thus developed may be surrounded by a ring of peaks, the culminating points of planezes, or the planezes may merge into a continuous arcuate homoclinal ridge, the dip slope of which will be the constructional slope of the flank of the dome (Fig. 212).

The Eroded Dome Haleakala

On the eastern and less dissected of the two domes of basalt which make up the island of Maui (Hawaiian Islands) the vast Haleakala "crater" or "caldera," as it has been termed, has been variously explained (Figs. 23, 61, 213, 214). As interpreted after detailed examination by Stearns,[21] who follows a suggestion made earlier by Cross[22], it is

> due chiefly, and perhaps entirely, to stream erosion by the recession of two great amphitheatre-headed valleys . . . . from opposite sides of the mountain. . . . Renewed volcanic eruptions . . . . built large cinder cones and poured out voluminous flows that nearly masked the ancient divide between the two valleys and partly filled both. . . . The lava flows displaced the drainage channels and caused the streams to undercut their valley walls, thereby widening their heads.

This explanation rules out the hypothesis of explosion and that of collapse of the crown of the Haleakala dome to form a

[21] H. T. Stearns, Origin of Haleakala Crater, *Bull. Geol. Soc. Am.*, 53, pp. 1-14, 1942; Geology and Ground-water Resources of the Island of Maui, Hawaii, *T.H. Div. Hydrog. Bull.*, 7, p. 53, 1942.

[22] Whitman Cross, Lavas of Hawaii and their Relations, *U.S. Geol. Surv. Prof. Paper*, 88, p. 92, 1915.

Fig. 213. Deeply eroded valleys meet in the centre of the great dome Haleakala, forming the so-called crater or caldera, Island of Maui, Hawaii. (After H. T. Stearns.)

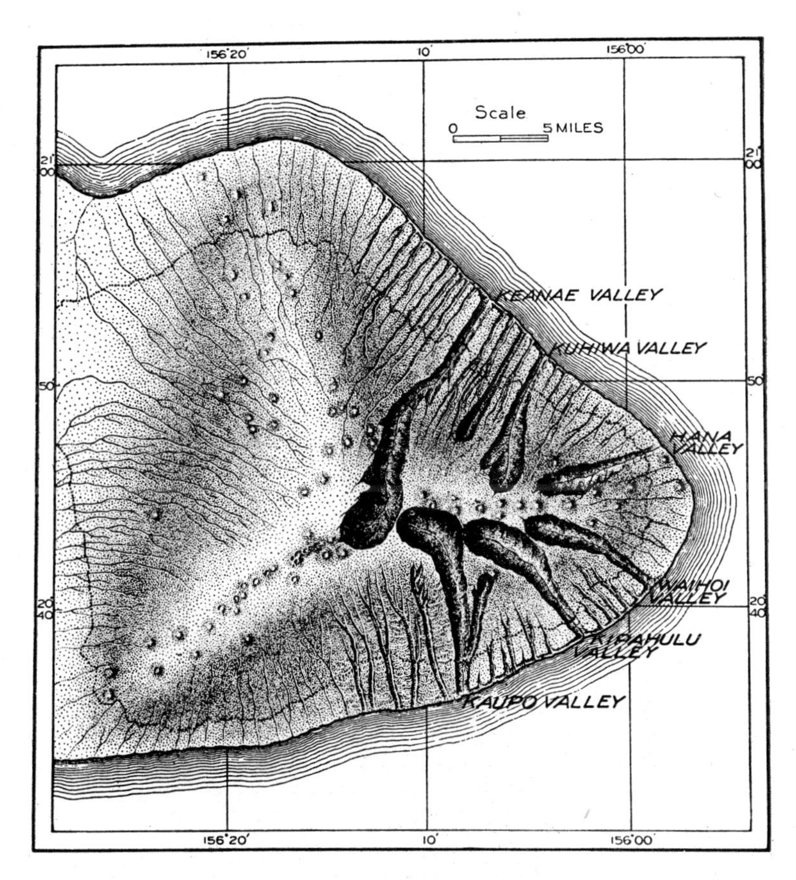

Fig. 214. Restoration of eastern Maui after great valleys have been eroded in the constructional lava dome of Haleakala, but before the occurrence of secondary eruptions which have since obliterated some of these valleys and partly filled others, and prior to a modern submergence the extent of which is indicated by a dotted line at the present-day shoreline. (After H. T. Stearns.)

graben or sink and also that of lateral sliding away of the side of the dome with the formation of what has been termed a volcanic "rent" (a suggestion made by Dana and later adopted by Daly[23]). No evidence has been found of major faulting

[23] J. D. Dana, Geology, *Wilkes U.S. Expl. Exped.*, 10, 1849; *Am. Jour. Sci.*, 37, p. 86, 1889; R. A. Daly, *Igneous Rocks and their Origin*, p. 153, 1914; *Igneous Rocks and the Depths of the Earth*, p. 171, 1933.

with either vertical or horizontal displacement such as would support the graben or the rent hypothesis in explanation of the boundaries of the great depression.

Among various objections which may be raised against the hypothesis that the wide and deep valleys of Haleakala have been opened by erosion, and which have been anticipated and met by Stearns, two in particular may be noted. One of these is based on the question whether valley erosion is competent to produce the observed results in the time available. The origin of the valleys on this dome dates back to the Pleistocene, however, and there is evidence of extensive dissection on the eastern (windward) slopes in that period (Fig. 214), though this is masked now over considerable areas by later lava outpourings. Rainfall is not, and has not been in the past, too scanty for vigorous erosion. There are also many dykes in the central part of the great dome which have held up ground water at a high level until it has been tapped by and has then helped to erode the valleys that now form part of the great depression.

The second noteworthy objection to the hypothesis of erosion is a pronounced bend in the lower course of the northward valley, which might be interpreted as an indication that that valley was in part tectonic instead of being carved out in its entirety by a consequent stream guided by the constructional lava slope of a smooth dome. The initial Haleakala mountain was not a symmetrical dome, however, but rather an assembly of radial ridges built over rift zones, and the whole course seems to be strictly consequent on the slopes of the initial form as Stearns has restored it (Fig. 214).

Young Erosion Calderas

The process of reduction of basalt domes to more or less continuous ring ridges, accompanied by the excavation of extensive interior lowlands, is well advanced in the islands of Réunion and Tahiti. On the great basaltic dome of Réunion, 700 square miles in extent (Fig. 215), several master streams, which isolate broad planezes rising to an elevation of 10,000 feet, are co-operating at their headwaters to excavate an interior

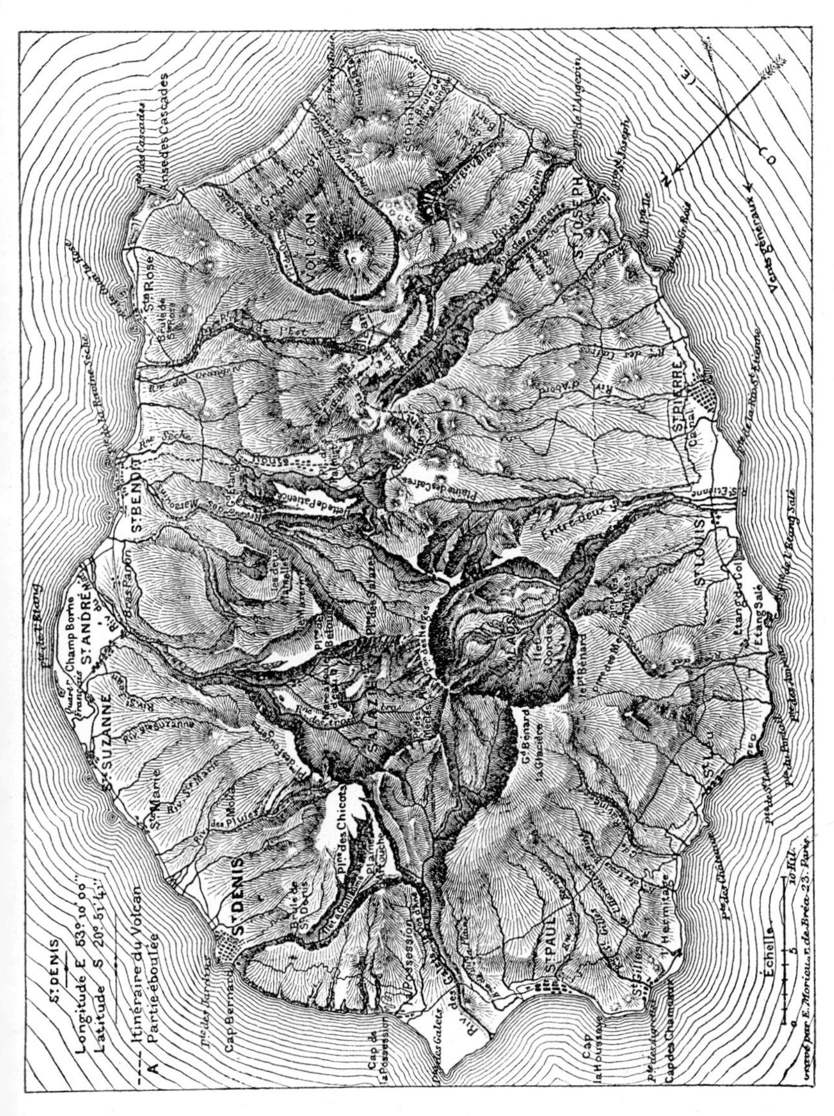

Fig. 215. The island of Réunion, Indian Ocean. (After C. Vélain.)

lowland.[24] A younger and still active basaltic volcano at the south-eastern end of Réunion is contained within a steep-walled depression—the Enclos—which has been termed a sink by Daly.[24a] It may be a sector graben on the main dome, or perhaps an amphitheatre-headed valley, similar to those of Haleakala, excavated by erosion.

In Tahiti, the expanded head of a bottle-necked master valley, that of the Papenoo River, occupies the whole centre of the western and larger of the two domes of which the island is composed. Extension by headward erosion of head-water tributaries of the Papenoo system has hollowed out the core of the great volcano, reducing its interior to a lowland with an area of about ten square miles, within which the hills are little more than 1000 feet high though these are surrounded by an impressive ring of mountains rising to 5000-7000 feet above the sea.[25] The central part of the dome which has been thus eviscerated must have been higher than any of these peaks now are.

The Mature Erosion Caldera of La Palma

The celebrated Caldera of La Palma (Canary Islands) affords an example of a broad and high volcanic dome with a deeply eroded crown. The central amphitheatre, which is between three and four miles in diameter, is walled in for more than half a circle by a high and steep escarpment (Fig. 216). The Caldera, as it is named, supplied Lyell with the term "caldera", to which he gave a general application.[26] The Caldera of La Palma has been described as an eroded valley by various investigators, including Gagel,[27] who has recognised the resemblance of its outlet gorge (Barranco de las Angustias)

[24] W. M. Davis, *The Coral Reef Problem*, p. 244, 1928; A. Penck, *Morphologie der Erdoberfläche*, 2, p. 426, 1894.

[24a] R. A. Daly, *loc. cit.* (23).

[25] P. Marshall, The Geology of Tahiti, *Trans. N.Z. Inst.*, 47, pp. 361-376, 1915.

[26] C. Lyell, *Elements of Geology*, 6th ed., p. 621, 1865. Large crateriform hollows in the Azores, which are regarded as of strictly volcanic (not erosional) origin, are also named *Caldeira*, however, and some prefer to regard these as the type forms of "calderas." (E. de Martonne, *Traité de géographie physique*, p. 521, 1909; R. A. Daly, *Igneous Rocks and their Origin*, p. 145, 1914.)

[27] C. Gagel, Die Caldera von La Palma, *Zeits. f. Erdk. Berlin*, pp. 168-250, 1908; see also A. Penck, *loc. cit.* (24), p. 426.

to the bottle-necked valleys of the Blue Mountains arch in New South Wales.

The question seems incapable of proof whether or not a large crater or caldera was present, as Reck[28] has supposed, on the dome of La Palma before headward-nibbling streams and

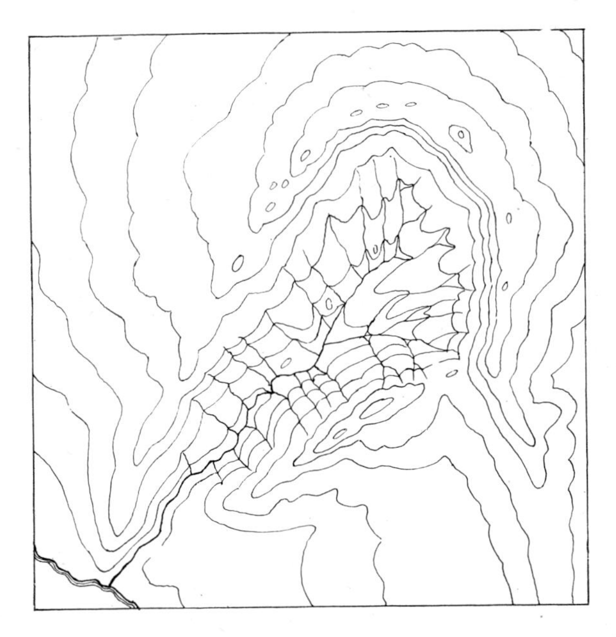

Fig. 216. The Caldera of La Palma, Canary Islands, drained south-westward to the sea by way of the barranco. Cañon de las Angustias. Contour interval, 250 metres. (After C. Gagel.)

escarpment retreat hollowed out its crown. An alternative explanation of the extensive interior erosion is to be found, however, in the presence of older, altered, and distinctly less resistant igneous rocks which underlie the 3000-feet-thick upper lava shell in the escarpment and are eroded to make the spurs descending to the axis of the interior valley.

[28] H. Reck, Zur Deutung der vulkanischen Geschichte und der Calderabildung auf der Insel La Palma, *Zeits. f. Vulk.*, 11, pp. 217-243, 1928.

Drowned Erosion Calderas

On the island of Huaheine (Society Islands) there is an erosion caldera with several outlets drowned to form embayments, two of which are extensive, deep, and landlocked harbours[29] (Fig. 217).

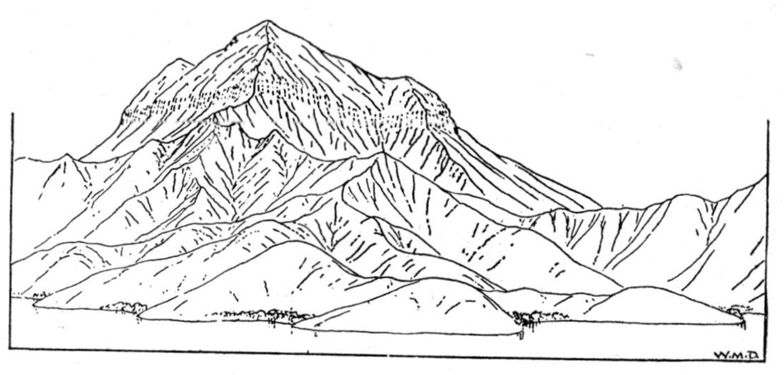

Fig. 217. North wall of the drowned erosion caldera of Huaheine, Society Islands. (After W. M. Davis, *The Coral Reef Problem,* Am. Geog. Soc. N.Y.)

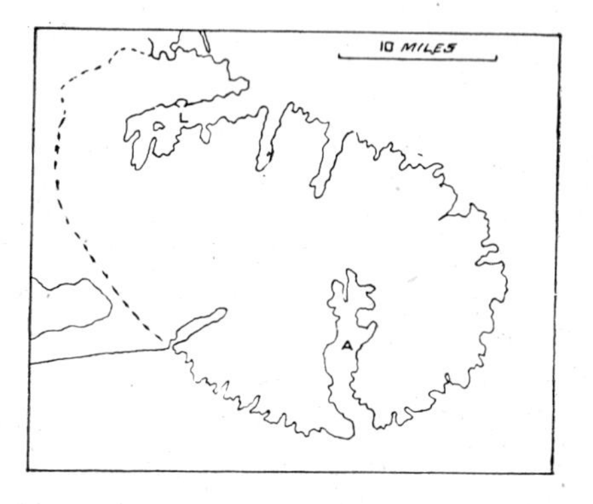

Fig. 218. Banks Peninsula, New Zealand. A, Akaroa Harbour; L, Lyttelton Harbour.

Two submaturely dissected basalt domes comprise the large land-tied island Banks Peninsula, on the east coast of the South Island of New Zealand; and submergence has converted large eroded hollows which have taken the place of their centres into extensive arms of the sea, the twin harbours of Lyttelton and

[29] W. M. Davis, *loc. cit.* (24), p. 291 and Fig. 123.

V. C. Browne, photo

Fig. 219. The Lyttelton erosion caldera. At the left (south side) the symmetry of the amphitheatre is somewhat marred by a lava flow of comparatively recent origin which has poured down the face of the escarpment. In the distance there is a broad gap in the ring ridge.

Akaroa[30] (Figs. 218, A, L; 219; 220). The erosional origin of these escarpment-walled hollows has been demonstrated by Speight.[31]

Escarpment-bounded Erosion Calderas

The process of development of an escarpment-bounded caldera—like that of La Palma, Akaroa, or Lyttelton—in which there is a single, narrowing, drainage outlet ("barranco")

Fig. 220. The Akaroa erosion caldera, Banks Peninsula, New Zealand.

seems simple enough. On the flanks of an infantile dome the usual radial consequent streams follow centrifugal courses, and some of these in time cut deep, escarpment-walled, and necessarily bottle-necked valleys in the gently-dipping lavas. There are now numerous straight radial valleys of this kind in various stages of development on the slopes of the Lyttelton and Akaroa domes, and the more maturely developed of them expand headward into amphitheatres walled in by escarpments.

All around Banks Peninsula these valleys are drowned to form a fringe of bays separated by narrow planezes margined by wave-cut cliffs that are drowned by submergence (Figs. 218, 221). Though infilled with fine silt (abundant rewashed loess) until they contain only a very moderate depth of water, these drowned radial valleys—including the Lyttelton erosion

[30] C. A. Cotton, Some Volcanic Landforms in New Zealand, *Jour. Geomorph.*, 4, pp. 297-306, 1941; Howel Williams, Calderas and their Origin, *Univ. Cal. Publ., Bull. Dep. Geol. Sci.*, 25 (6), pp. 308-309, 1941.

[31] R. Speight, The Geology of Banks Peninsula, *Trans. N.Z. Inst.*, 49, pp. 365-392, 1917; The Geology of Banks Peninsula—a Revision, *Trans. Roy. Soc. N.Z.*, 73, pp. 13-26, 1943.

caldera—have a remarkable box-form transverse profile below sea-level, with floors quite horizontal from side to side between the steep submerged walls of the eroded valleys. This form has resulted apparently from repeated settlings of the silt from re-suspension after it has been stirred up by the ocean swell which enters these straight, radial embayments. In the entrance

V. C. Browne, photo

Fig. 221. The maturely dissected north-eastward slope of the Akaroa basalt dome, New Zealand. Cliffed headlands are separated by bays drowned by partial submergence of the slope dissected by consequent valleys in radial pattern on the flank of the dome.

to the Lyttelton embayment, for example, there is a uniform depth of eight fathoms from cliff-base to cliff-base; a mile within the entrance there is seven fathoms from side to side; and two miles within there is an even six fathoms across the mile-wide barranco (Fig. 222). As a result of this flat-floored form, again, there is practically no refraction of the ocean swell such as takes place in an embayment of more usual form with shallowing water on either side of an axis and causes dissipation of the wave-energy along and against the sides.

Running in towards the land of Banks Peninsula over a perfectly smooth sea-bottom, the waves have undergone no refraction before entering the bays, and the swell rolls on with almost undiminished energy right to the bay-heads. For this reason the smaller bays afford no safe anchorage, and difficulty has been experienced in making a serviceable harbour even in the Lyttelton erosion caldera, notwithstanding the large area of nearly landlocked water it contains (Fig. 219).

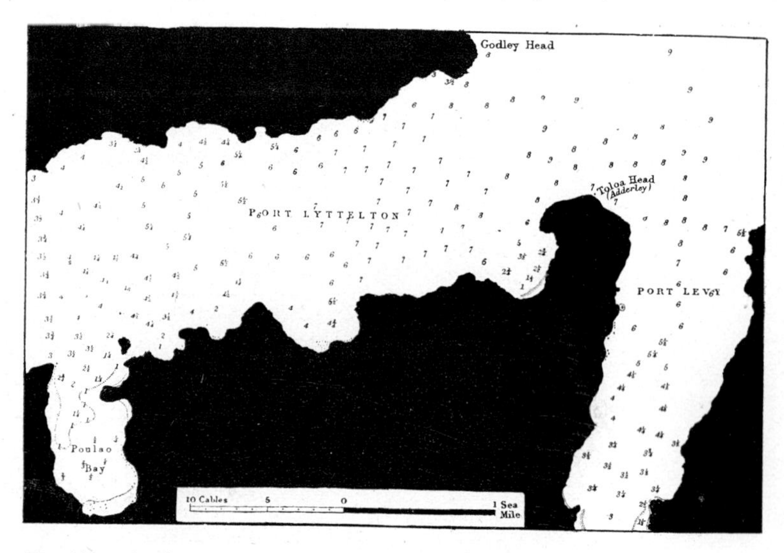

Fig. 222. Flat-floored embayments formed by marine-silt infilling of drowned radial, consequent valleys on the flank of the Banks Peninsula basalt-dome doublet. Depth of water in fathoms, from British Admiralty survey, 1849.

The radial consequent valley which first succeeds in gathering to itself the drainage of the summit of the mountain, or in tapping any central crater which may be there and which may possibly have been already enlarged by explosion or collapse, has a good chance of retaining the mastery and expanding into an erosion caldera with a single outlet. The stream which drains the large central area, though it is cutting down towards the same base-level as its smaller neighbours, will be graded to a gentler declivity and will, therefore, have a deeper valley, becoming the barranco. Headward erosion is stimulated in its headwater tributaries, which have an obsequent relation to the inclined lavas and initial slopes of the dome surface, and these will sap at the base of the wall around the developing central hollow so as to cause it to retreat quaquaversally as an escarpment until the whole broad centre of the volcanic mountain has been quarried out. As in the case of La Palma the domes of Banks Peninsula have other and generally less resistant rocks under the basalt capping.[32]

V. C. Browne, photo

Fig. 223. Details of the escarpment of the Akaroa erosion caldera, showing maturely dissected spurs at the base of the slope, at Akaroa, New Zealand.

The continuity of the three-quarter-circle escarpment around the Lyttelton caldera is broken at one place by a broad gap through the basalt capping (Fig. 219), where a low divide has been developed between the headwater streams of the master system and another consequent valley which opens on the opposite side of the dome. Where the gap now is the basalt was thinnest, for here the pre-basaltic land was highest. As

[32] R. Speight, The Geology of Gebbies Pass, Banks Peninsula, *Trans. Roy. Soc. N.Z.*, 65, pp. 305-328, 1935; The Basal Beds of the Akaroa Volcano, *ibid.*, 70, pp. 60-76, 1940.

Speight[33] has remarked, the Akaroa erosion caldera is a more perfect example of the type form, as the horseshoe-shaped escarpment around is unbroken by any such gap (Figs. 220, 223).

An erosion caldera enclosed by an escarpment has something in common with the eroded hollows popularly termed "pounds" in Australia, which are enclosed by ring ridges in areas of dome-and-basin folding with well-advanced erosional adjustment to structure.[34]

An Erosion Caldera in Old Age

In the island of Mauritius there is a very large erosion caldera in which the original ring ridge is now reduced to a ring of isolated mountains, so that the form may be considered to be so far developed as to be in its old age.

According to the description given by Darwin,[35] who quotes also from Bailly's *Voyages aux terres australes,* Mauritius is a large basaltic dome in a far advanced stage of erosional degradation, though its valleys have been flooded again in more recent times by basaltic lavas. The island is ringed around by precipitous peaks 2000 to 3000 feet high, each of which is the diminished remnant of a planeze or of an isolated sector of a ring ridge. Broad valleys, several miles wide, now separate these remnants, and these are perhaps old-age forms of valleys of the Oahu type. The ring marked out by the surviving peaks is "oval and of vast size; its shorter axis, measured across from the inner sides of the mountains near Port Louis and Grand Port, being no less than thirteen geographical miles in length." The broad lowland in the interior of the island, though its level has been raised by floods of comparatively modern basalt, is still only about 1000 feet

[33] R. Speight, *loc. cit.* (31), p. 370; (1943), p. 25.

[34] D. Mawson remarks: "This term 'pound' may have useful application in descriptive geomorphology; to be distinguished as periclinal or domed pounds and centroclinal or basin pounds" (The Structural Character of the Flinders Range, *Trans. Roy. Soc. S. Aus.,* 66, pp. 262-272, 1942). In volcanic landforms there is an analogy of course, only with "domed pounds."

[35] C. Darwin, *Geological Observations,* Chapter 2, 1844; see also A. Penck, *loc. cit.* (24), p. 428.

above sea-level. Darwin's own observations confirm the statements of Bailly regarding not only the arrangement in ground plan but also the structure of the ring of peaks composing what he terms the skeleton of the island. He quotes from Bailly to the effect that these peaks, all of which are composed of parallel layers dipping seaward, are arranged around the island as a ring of "immense ramparts" sloping at a moderate inclination towards the sea, while, on the other hand, their slopes towards the centre of the island are precipitous escarpments.

APPENDIX

Volcanic Contributions to the Atmosphere and the Ocean

IF WE ASSUME, AS IS NOW AGAIN THE FASHION, THAT THE NASCENT earth passed through a liquid stage, it is obvious that "the molten spheroid . . . retained, occluded within itself, some large part of the water of the present hydrosphere, as well as much of the carbon dioxide represented by the present carbonates and carbonaceous deposits".[1] Most of the carbon dioxide that has become available as a source of carbon is undoubtedly of volcanic origin, being derived from magma.

A useful estimate of the amount of carbon that has been extracted from the atmosphere by various means has been made by Poole,[2] who, however, has unnecessarily revived the theory that this has been all present at one time in gaseous combination in a "primitive" atmosphere. Poole assumes that some of the fixed carbon has been derived from methane so present instead of from carbon dioxide—an assumption he has considered necessary because of an apparent over-supply of the by-product oxygen if carbon dioxide were the sole (or greatly predominant) source of carbon. When, however, oxidation of juvenile hydrogen throughout the long history of volcanic activity is taken into account there is no longer any difficulty in accounting for the disposal of such an apparent surplus of atmospheric oxygen. It is permissible, therefore, to recalculate Poole's estimate of atmospheric methane as carbon dioxide, a more likely atmospheric gas.

When this is done, the total mass of carbon dioxide is $7{\cdot}02 \times 10^{16}$ metric tons; and to this must be added $0{\cdot}4 \times 10^{16}$ metric tons of nitrogen now in the atmosphere and $0{\cdot}18 \times 10^{16}$ metric tons of juvenile hydrogen which has combined with

[1] Chamberlin and Salisbury, *Geology*, Vol. 2, p. 90, 1906.

[2] J. H. J. Poole, The Evolution of the Atmosphere, *Sci. Proc. Roy. Dublin Soc.*, 22, pp. 345-365, 1941.

the surplus of liberated oxygen, giving a total of $7{\cdot}6 \times 10^{16}$ metric tons of gases in what must be regarded as the primitive atmosphere (perhaps very scanty) *together with the gases that have been emitted from volcanoes throughout the history of the earth.* This excludes the juvenile water vapour, which it is impossible to estimate by this method; it undoubtedly has made a substantial contribution to the ocean. For comparison, the mass of the atmosphere at present is given as $0{\cdot}52 \times 10^{16}$ metric tons.

One great advantage of the recognition of the juvenile (magmatic) origin of carbon, as compared with the "primitive atmosphere" theory, is that it is in agreement with the geological doctrine of uniformitarianism, if one is to understand that doctrine in a forward-looking sense. A continuance of the supply from volcanoes will make possible the accumulation of further deposits of coal and limestone in the thousands of millions of years of future geological time, whereas according to the "primitive atmosphere" theory the supply of carbon for this purpose is exhausted, some of the most important of the geological processes have worked themselves to a standstill, and the world has virtually come to an end.

Index

INDEX

PRINTED BY WHITCOMBE & TOMBS LIMITED—G2703